Biometeorology
in Integrated
Pest Management

Academic Press Rapid Manuscript Reproduction

Proceedings of a Conference on Biometeorology
and Integrated Pest Management
Held at the University of California, Davis, July 15-17, 1980

Biometeorology in Integrated Pest Management

Edited by

JERRY L. HATFIELD

Department of Land, Air, and Water Resources
University of California, Davis
Davis, California

IVAN J. THOMASON

Department of Nematology
University of California, Riverside
Riverside, California

1982

ACADEMIC PRESS
A Subsidiary of Harcourt Brace Jovanovich, Publishers
New York London
Paris San Diego San Francisco São Paulo Sydney Tokyo Toronto

ACADEMIC PRESS, INC.
111 Fifth Avenue, New York, New York 10003

United Kingdom Edition published by
ACADEMIC PRESS, INC. (LONDON) LTD.
24/28 Oval Road, London NW1 7DX

Library of Congress Cataloging in Publication Data
Main entry under title:

Biometeorology in integrated pest management.

Based on a conference on biometeorology and integrated pest management which was conceived and sponsored by the University of California Integrated Pest Management Program.
Includes indexes.
1. Pest control, Integrated--Congresses. 2. Meteorology, Agricultural--Congresses. 3. Microclimatology--Congresses. I. Hatfield, Jerry L. II. Thomason, Ivan J. III. University of California Integrated Pest Management Program.
SB950.A2B57 632'.9 81-22780
ISBN 0-12-332850-0 AACR2

PRINTED IN THE UNITED STATES OF AMERICA

82 83 84 85 9 8 7 6 5 4 3 2 1

Contents

Contributors

Numbers in parentheses indicate the pages on which the authors' contributions begin.

Anne J. Anderson (267), Department of Biology, Utah State University, Logan, Utah 84322

J. L. Anderson (449), Departments of Plant Science and Soil Science and Biometeorology, Utah State University, Logan, Utah 84322

C. Barfield (193), Department of Entomology and Nematology, University of Florida, Gainesville, Florida 32601

R. E. Carlson (1), Agronomy Department, Iowa State University, Ames, Iowa 50010

Susan G. Conard (463), Department of Forest Science, Oregon State University, Corvallis, Oregon 97331

J. Davis (193), Department of Geosciences, North Carolina State University, Raleigh, North Carolina 27607

Paul C. Doraiswamy (43), Lockheed Engineering and Management Services Company, Inc., Johnson Space Center, Houston, Texas 77004

J. M. Duniway (307), Department of Plant Pathology, University of California at Davis, Davis, California 95616

Raymond A. Evans (421), USDA/SEA-AR, Reno, Nevada 89505

J. L. Hatfield (147), Department of Land, Air, and Water Resources, University of California at Davis, Davis, California 95616

Carl B. Huffaker (171), Department of Entomological Sciences, Division of Biological Control, University of California at Berkeley, Berkeley, California 94720

C. M. Leach (267), Department of Botany and Plant Pathology, Oregon State University, Corvallis, Oregon 97331

P. D. Lingren (211), Western Cotton Research Laboratory, Agricultural Research, Science, and Education Administration, USDA, Phoenix, Arizona 85040

John M. Norman (65), Department of Agronomy, University of Nebraska, Lincoln, Nebraska 68508

Robert F. Norris (343), Botany Department, University of California at Davis, Davis, California 95616

David T. Patterson (407), U.S. Department of Agriculture, Southern Weed Science Laboratory, Stoneville, Mississippi 38776

S. P. Pennypacker (243), Department of Plant Pathology, Pennsylvania State University, University Park, Pennsylvania 16802

P. J. Pinter, Jr. (101), USDA/SEA-AR, U.S. Water Conservation Laboratory, Phoenix, Arizona 85040

F. L. Poston (229), Department of Entomology, Kansas State University, Manhattan, Kansas 66502

Steven R. Radosevich (463), University of California at Davis, Davis, California 95616

J. Regniere (193), Great Lakes Research Station, Sault Ste. Marie, Ontario, Canada

E. A. Richardson (449), Departments of Plant Science and Soil Science and Biometeorology, Utah State University, Logan, Utah 84322

A. Riordan (193), Department of Geosciences, North Carolina State University, Raleigh, North Carolina 27607

Joseph Rotem (327), Department of Life Sciences, Bar-Ilan University, Ramat-Gan, Israel and Division of Plant Pathology, Agricultural Research Organization, The Volcani Center, Bet Dagon, Israel

Roger H. Shaw (17), Department of Land, Air, and Water Resources, University of California at Davis, Davis, California 95616

R. E. Stevenson (243), Department of Plant Pathology, Pennsylvania State University, University Park, Pennsylvania 16802

R. E. Stinner (193), Department of Entomology, North Carolina State University, Raleigh, North Carolina 27607

S. M. Welch (229), Department of Entomology, Kansas State University, Manhattan, Kansas 66502

K. Wilson (193), Department of Entomology, North Carolina State University, Raleigh, North Carolina 27607

W. W. Wolf (211), Western Cotton Research Laboratory, Agricultural Research, Science, and Education Administration, USDA, Phoenix, Arizona 85040

James A. Young (421), USDA/SEA-AR, Reno, Nevada 89505

Preface

During the past decade, integrated pest management (IPM) has become a well-known phrase; its interpretation, however, depends on the individual. The non-scientist often believes that the goals of IPM are to completely eliminate chemical control of pests, whereas in actuality researchers search for proper methods to control pests more efficiently. IPM has as a major goal the maintenance of pest populations below economic thresholds while utilizing suitable techniques to protect both the environment and nontarget species. This goal can be achieved if an understanding of pest population dynamics can be developed in a quantifiable manner. This could be done for all of our pests—insects, nematodes, pathogens, vertebrates, and weeds—by using the well-known ''disease-triangle,'' which models the initial population, the host, and the environment, the three components required for a pest outbreak.

If we are to understand the population dynamics, we must understand the agroecosystem and its driving forces. It is recognized that the meteorological environment plays a large role in pest population dynamics, yet few biometeorologists or agricultural meteorologists have been involved in IPM programs within the traditional areas of entomology, nematology, plant pathology, and weed science.

No single individual can completely comprehend the pest management system; thus in an effort to open the lines of communication and explore possible common aspects of research, a conference on Biometeorology and Integrated Pest Management was conceived and sponsored by the University of California Integrated Pest Management Program. The goals of this conference were to bring individuals active in the IPM programs together in a forum where both current research and future collaboration could be discussed. This volume is a result of that conference at the University of California at Davis on July 15–17, 1980.

Weather is an important aspect of IPM programs and one of the goals of the conference was to introduce, from a biometeorological viewpoint, the microclimate of agricultural systems in order to describe what is known about the environment in which pests live. Entomologists, plant pathologists, and weed scientists then presented, from their own perspectives, the current state of knowledge in the respective fields in order to outline the current IPM research in each area. This type of forum provided a useful interchange of ideas and will strengthen IPM research if the cooperation continues.

Although much has been written on IPM topics, we have attempted to compile a broad overview of current research areas which will aid in implementation of this research. Norris, in Chapter 15, proposes a pest pentagon which includes all pests and their interactions, with the implicit understanding that the microclimate and the host serve as the foundation for this pentagon. If we proceed with our research along these lines and continue our own interaction, IPM implementation can be realized, with benefit both to agriculture in particular and society in general.

This conference was sponsored by the University of California Integrated Pest Management Program and we are grateful for their support. We are indebted to Drs. Mary Ann Sall, Robert Norris, Nick Toscano, and Ron Hamilton for their help in organizing the conference and suggesting speakers. A special thanks is owed to Joyce Fox for her untiring assistance in helping to organize the conference, assist the conference participants, and help in the preparation of this volume. Carole Krissovich proved to be an excellent typist in preparing the many drafts and the final copy, and DeeDee Kitterman assisted by proofreading each page with diligence and care. Finally, we thank all conference speakers and participants for making the conference a success and for making the goals of IPM seem achievable.

J. L. Hatfield
I. J. Thomason

GENERAL HEAT EXCHANGE IN CROP CANOPIES: A REVIEW*

R. E. Carlson
Agronomy Department
Iowa State University
Ames, Iowa

I. INTRODUCTION

The energy exchanges between crops, the soil, and the atmosphere have been studied intensively, with many of these studies being directed toward the understanding of water use by crop canopies. The pests that infect crop canopies develop and grow in the environment produced by these energy exchanges. It is hoped, therefore, that an understanding of these energy exchanges will provide insight into the management of the multitude of pests that ravage man's food supply.

Tanner (1960) described the energy balance of a crop canopy by

$$R_n + \int_0^z C_p \nabla_H (\rho u T) dZ + \int_0^z \frac{L\varepsilon}{R} \nabla_H \frac{ue}{T} dZ = G + H + LE$$

$$+ \int_0^z C\rho_c \frac{\partial T}{\partial t} dZ + \int_0^z C_p \rho \frac{\partial T}{\partial t} dZ + \int_0^z \frac{L\varepsilon}{RT} \frac{\partial e}{\partial t} dZ \qquad (1)$$

*Journal Paper No. J-10071 of the Iowa Agric. and Home Econ. Exp. Stn., Iowa State University, Ames, Iowa 50011. Project 2397.

ISBN 0-12-332850-0

which is fully illustrated in Fig. 1. The terms on the left (sources of energy) are net radiation and the horizontal divergence of sensible and latent heat, respectively. The terms on the right (utilization of energy) represent heating the soil, heating the air, evaporation of water, the change in heat stored in the crop, the change in heat stored in the air within the canopy, and the change in the latent heat stored in the canopy. Photosynthesis and respiration are assumed small and are not included. If $dz \rightarrow 0$, the model reduces to the classical heat budget equation

$$R_n = G + H + LE \tag{2}$$

In the paragraphs that follow, general principles relating to the components of Fig. 1 and Eq. (1) will be discussed.

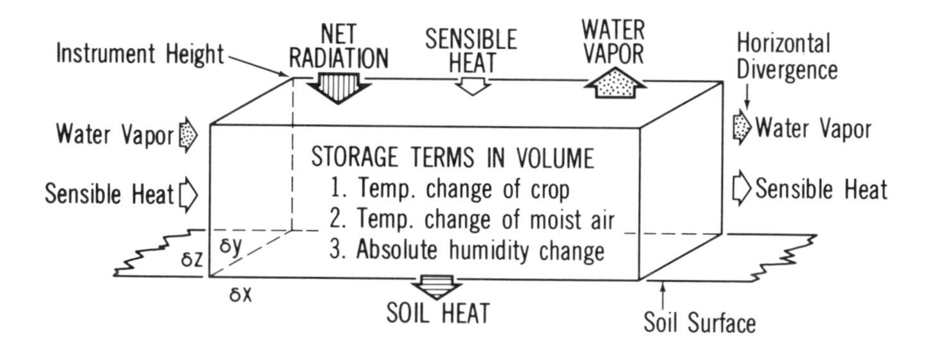

Fig. 1. Complete energy balance of a crop volume; after Tanner (1960).

II. RADIATIVE ENERGY EXCHANGES

A. Basic Laws

All objects possessing temperatures greater than absolute zero (0 K) emit radiant energy. The amount and type of energy emitted is determined by temperature. All energy emission spectrums take the general shape indicated by the 5900 K black body curve indicated in Fig. 2 after Gast et al. (1965). The laws needed to describe these patterns for objects of other temperatures are Planck's Law, Stephan-Boltzmann's Law, and Wein's Law. The amount of energy, E, emitted by an object is given by the Stephan-Boltzmann's Law

$$E = \varepsilon\sigma T^4 \tag{3}$$

where ε is the emissivity, σ is the Stephan-Boltzmann constant (5.67×10^{-8} W m^{-2} K^{-4}), and T is the temperature K. In most

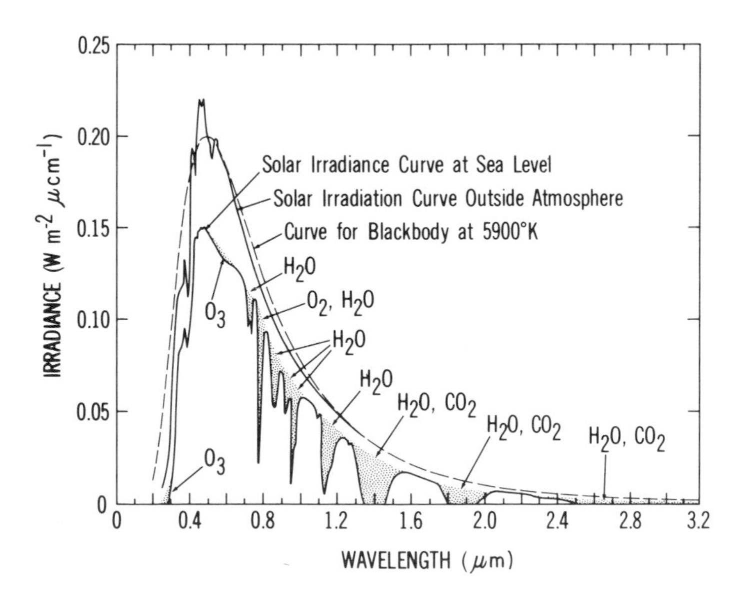

Fig. 2. Spectral distribution curves related to the sun; shaded areas indicate absorption at sea level due to the atmospheric constituents shown; after Gast et al. (1965).

agricultural situations, the emissivity is very near unity, but important variations between individual leaves and vegetated or nonvegetated surfaces can exist, especially when infrared-thermometry techniques are employed. This is discussed by Davies and Idso (1979), and tabulated values are given by Idso et al. (1969). The wavelength of maximum emission (λ_{max}) is determined by Wein's Law

$$\lambda_{max} = 2890/T \tag{4}$$

Because of this inverse relationship, cooler objects emit energy at longer wavelengths as compared with warmer objects.

B. Solar Radiation

Theoretical and actual solar energy spectrum are shown in Fig. 2. The solar energy spectrum shows that the sun acts very much like a blackbody with a surface temperature of approximately 5900 K. Significant absorption is indicated in this figure when comparing extraterrestrial and earth's surface curves, and this will be discussed later. The solar spectrum generally is divided into three major regions: the ultraviolet wavelengths from 0.3 to 0.4 μm, the visible spectrum from 0.4 to 0.7 μm, and the infrared from 0.7 to 100 μm. The approximate percentages compared with the total solar spectrum for each radiation type within each region are 4%, 44%, and 52% for the ultraviolet, visible, and infrared, respectively (Lemon, 1966).

The amount of solar radiation impinging on a horizontal surface is affected by atmospheric and astronomic factors. A general discussion of these factors is given in most general meteorology or climatology texts (Trewartha and Horn, 1980; Byers, 1974, are suggested). Interested readers may find detailed descriptions in List (1966), Monteith (1973), and Rosenberg (1974). For illustrative purposes, the daily receipt of solar radiation on a horizontal surface in central Iowa for three clear days near the vernal equinox (~ March 21) and both the summer (~ June 21) and winter (December 21) solstices is shown in Fig. 3. Of interest is the variation in both daylength and the intensity of radiation over the year as these two factors determine the seasonal distribution of the daily receipt of solar radiation. This pattern varies greatly among locations over the surface of the earth, especially with respect to latitude and altitude.

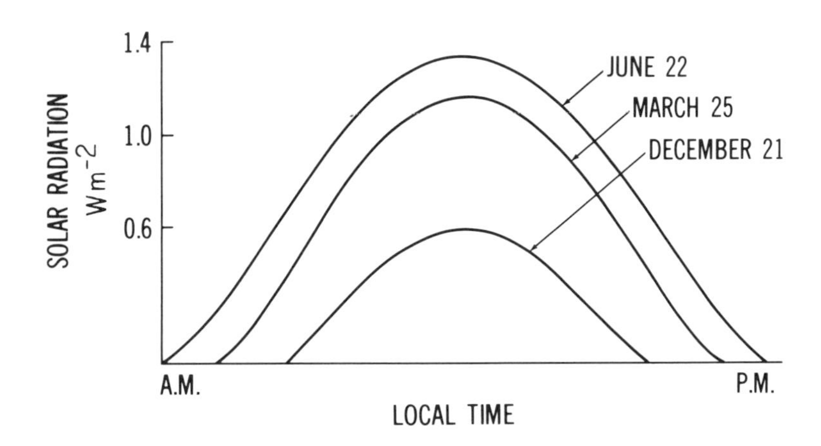

Fig. 3. Daily receipts of solar radiation for three clear days in central Iowa. March 25 is near the vernal equinox, and June 22 and December 21 are near the summer and winter solstices, respectively.

In most meteorological networks, solar radiation is measured on a horizontal surface. Because fields are not always of this orientation, methods for converting solar radiation measured on horizontal surfaces to surfaces of other inclinations are given by Reifsnyder and Lull (1965), Monteith (1973), Rosenberg (1974), Seeman et al. (1979), and Shaw and Decker (1979).

Because energy must be conserved, the solar radiation impinging on a surface must either be reflected, transmitted, or absorbed by the surface. Experience tells us that the magnitude and importance of each component for a surface varies considerably over time. These details are beyond the intent of this chapter; however, surface reflection characteristics are considered by Davies and Idso (1979). The complex interaction of solar radiation with plant communities is dealt with by Anderson and Denmead (1969), Ross (1975), Norman (1975), and others.

C. Terrestrial Radiation

Equations (3) and (4), applied to objects with terrestrial temperatures (approximately 300 K), reveal that the terrestrial radiation spectrum peaks near 10 μm and consists primarily of infrared radiation. This is shown in Fig. 4 by the work of Reifsnyder and Lull (1965), where normalized solar and terrestrial spectrums are compared.

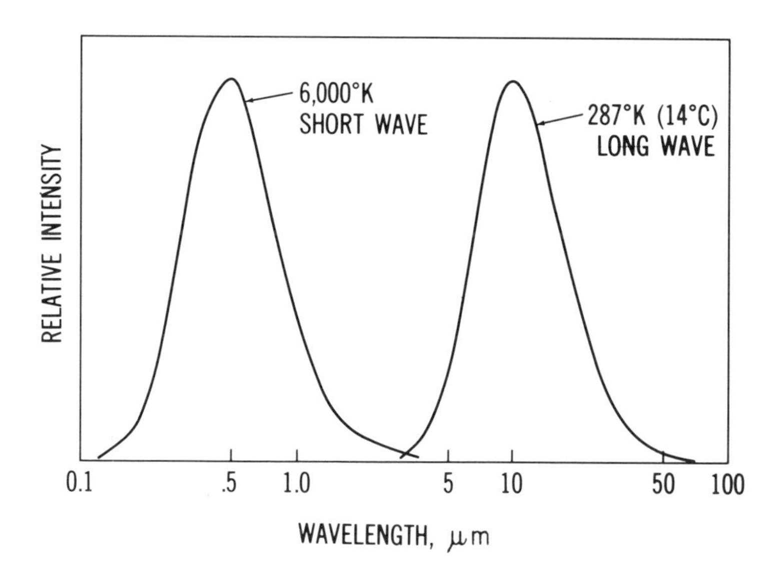

Fig. 4. Comparison of solar and terrestrial radiation, arbitrarily adjusted to the same peak intensity; after Reifsnyder and Lull (1965).

Terrestrial radiation is strongly influenced by cloud cover and atmospheric gases, specifically carbon dioxide and water vapor. Fleagle and Businger (1963) show the detailed absorption bands for atmospheric gases of major importance. When far infrared radiation is emitted by the surface and water vapor or carbon dioxide molecules are present in the atmosphere above the surface, these molecules selectively absorb this infrared radiation and, subsequently, reemit part of this energy back toward the surface. The net result is retarded cooling. Water vapor plays a dominant role in this relationship because its concentration is greater than that of carbon dioxide and varies more from time to time. If the atmospheric water vapor concentration is low, surface cooling at night can be substantial. Desert climates showing large diurnal temperature ranges illustrate this principle. In other climates, nighttime plant temperatures resulting from this loss of terrestrial radiation frequently are sufficient to produce dew. This condition, as is well known, can be very favorable for pathogen development.

D. Net Radiation

The overall radiative energy exchange for a crop surface is, therefore, controlled by the fluxes of solar (shortwave) and terrestrial (longwave) energy fluxes toward or away from the surface. The balance of this energy exchange is termed net radiation (R_n) and is given by

$$R_n = R_S\downarrow - R_S\uparrow + R_T\downarrow - R_T\uparrow \tag{5}$$

where $R_S\downarrow$ and $R_S\uparrow$ are, respectively, downward and upward fluxes of solar energy, and $R_T\downarrow$ and $R_T\uparrow$ are comparable fluxes of terrestrial radiation.

Hourly values of the four components comprising net radiation over a fescue meadow in Missouri are shown in Fig. 5 from Shaw and Decker (1979). If these four components are algebraically summed, net radiation would result. Similar seasonal values are presented by these authors. They point out that other patterns can result, depending upon seasonal-surface characteristics.

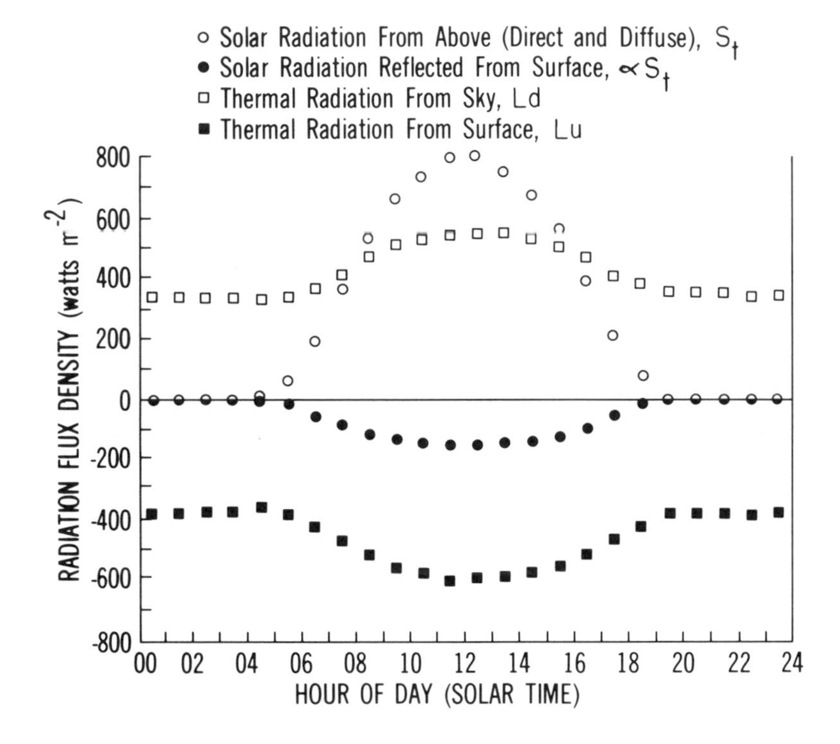

Fig. 5. Radiation balance by hours for a fescue meadow near Columbia, MO. The radiation flux densities are averages for July 1-15 during the years 1965-1974 from measurements horizontal to the surface; after Shaw and Decker (1979).

In Eq. (1), R_n is, in most cases, the dominant source of energy driving the energy balance equation. Typically, on a diurnal basis, net radiation is positive during daylight hours and negative at night. Midday, midlatitude values approach 700 W m^{-2}.

On heavily overcast days, it may be quite small. Examples of the diurnal variation of net radiation will be shown in later sections describing major components of Fig. 1 and Eq. (1).

Because net radiation is not routinely measured in most meteorological data-collection networks, techniques have evolved that can be used to estimate this important variable from other radiation measurements. An excellent review of literature pertinent to this is given by Davies and Idso (1979).

III. NONRADIATIVE ENERGY EXCHANGE

A. General Relationships

Nonradiative energy exchanges between the soil surface, a plant canopy, and the atmosphere are controlled by vertical gradients of humidity, temperature, and air flow. Depending on the direction of these gradients, energy can be transferred toward the surface or away from it. The actual development of these flux-driving gradients is the result of all energy exchanges taking place. Generally, temperature gradients are influenced by radiative energy exchanges, and humidity gradients are determined by atmospheric conditions, soil moisture, and plant factors. The gradients of wind speed are particularly important, and they are influenced by general synoptic weather conditions and surface roughness. Gradients of wind near the surface play an important role in producing the turbulent exchange that carries heat, humidity, and momentum from the surface to the atmosphere. Later chapters in this book deal with this topic. Other excellent sources of information regarding this subject are given by Rosenberg (1974), Gates (1980), and Thom (1975).

B. Individual Components

1. Latent Heat Exchange. The evaporation of water from the soil surface and transpiration from the plant is termed evapotranspiration. Under well-watered conditions, this component is a

large consumer of net radiation. To transform one gram of liquid water to the gaseous state, nearly 600 calories of heat are required. This is illustrated in Fig. 6 from Saugier's (1976) work with sunflowers, where major individual components of the energy budget equation [Eq. (2)] are depicted. Over the course of this day, a major portion of net radiation was apportioned into latent heat exchange. Experiments under drier conditions reveal quite different energy partitioning (Ripley and Redmann, 1976). In their experiments, latent heat exchange was smaller with the residual energy being apportioned into sensible heat and soil heat flux. The seasonal nature of daily latent heat exchanges is presented in Ripley and Saugier (1975).

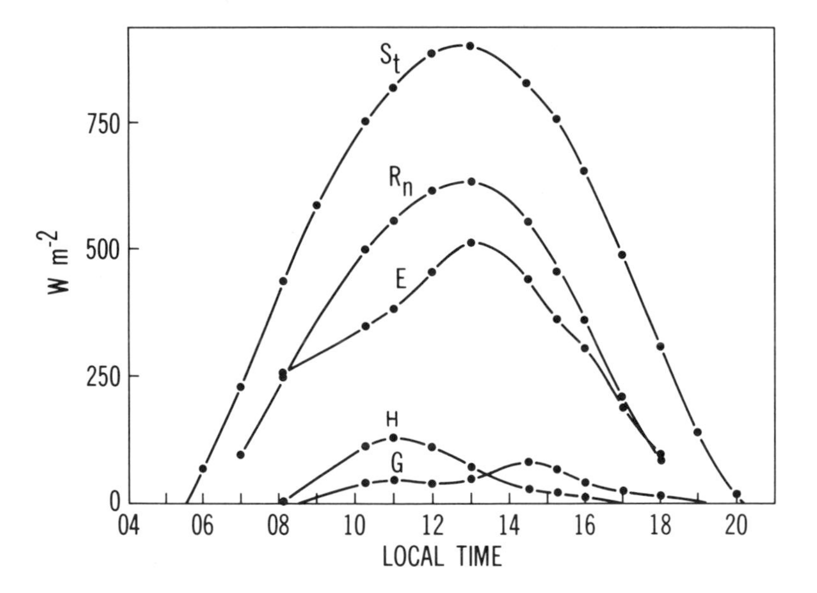

Fig. 6. Energy balance components above sunflower, Montpellier, VT, 12 July 1967; after Saugier (1976). S_T, R_n, E, H, and G are, respectively, solar radiation, net radiation, latent heat flux, sensible heat flux, and soil heat flux.

Although the basic atmospheric and radiative properties that control net radiation, turbulent exchange, and humidity gradients determine the potential for latent heat exchange, plant and soil factors can modify the expected behavior. Leaf area development, stomatal activity, rooting patterns, and soil-moisture levels are examples of these factors that significantly influence transpiration. Limited soil-moisture conditions cause leaf stomates to close, drastically reducing transpiration. Subsequent modifications of the general energy budget equation result, and the canopy or soil is observed to warm. Additional information relating to the influence of plants on the modification of energy transport is given by Verma and Barfield (1979).

2. Soil Heat Exchange. As can be seen in Fig. 6, soil heat flux generally is small when compared with the other components presented, but it can be significant when net radiation is small. On a daily basis, the heat gained during sunlight is mostly lost at night. This is based on the fact that day-to-day soil temperature variations generally are quite small, especially down into the soil away from the surface. Shaw and Decker (1979) show annual soil heat flux patterns collected in Missouri under a fescue meadow. At that location, positive heat flux values commence in March and diminish in September and October.

Kimball and Jackson (1979) review this topic in detail. They show that soil heat flux (G) is given by

$$G = - k_g dT/dz \qquad (6)$$

where k_g is the thermal conductivity, and dT/dz is the vertical temperature gradient. The thermal conductivity is affected by soil density, mineral composition, organic-matter content, water content, and temperature. They also note that Eq. (6) is valid for a porous material like soil as long as there is no large movement of water or air through the pore space. An example of this situation would be with irrigation.

In addition, de Vries (1975) and Rosenberg (1974) present other fundamental relationships and mathematical techniques that apply to this energy balance component. They also describe measurement procedures and discuss the behavior of soil-temperature profiles. These references will be important for those researchers dealing with the soil-borne pathogens.

3. Sensible Heat Exchanges. The transfer of sensible heat to the atmosphere is discussed by Rosenberg (1974), Monteith (1973), and others. Rosenberg describes sensible heat flux taking place in the laminar sublayer via molecular diffusion. Above this layer, he describes the transfer by

$$H = 60\ \rho_a C_p k_h (dT/dz) \tag{7}$$

where ρ_a the air density, C_p the specific heat of air at constant pressure, k_h an exchange coefficient for sensible heat, and dT/dz the vertical temperature gradient. The exchange coefficient, k_h, is a function of turbulence.

Shaw and Decker (1979) show that if only the major components of the energy balance equation are considered, sensible heat transfer is really the residual after the energy required by evaporation is consumed (i.e., $H \cong R_n - LE$). Decker (1964) showed that in irrigated corn in Missouri, on the average, 20% of net radiation was apportioned to sensible heat flux.

4. Advected Heat Exchanges and Time-Related Terms. As noted in Eq. (1), the terms on the left side of the equation, besides net radiation, represent sources of heat. Frequently, heat is brought into a canopy from warmer surroundings. As this occurs, sensible heat advected into the area becomes a source of heat and evapotranspiration can exceed net radiation. The data of van Bavel and Ehrler (1968), and Rosenberg (1969a, b) support this conclusion. The magnitude of this component varies with synoptic conditions, and is most prevalant in drier, warmer climates.

The last three terms in Eq. (1) involve time-related changes of heat and moisture storage in the crop canopy. They are of importance only when these canopy characteristics, temperature and humidity, are changing rapidly. In most studies, they are largely neglected because of the difficulties associated with their measurement. Kanemasu et al. (1979) discuss their measurement and evaluation.

5. Measurements. In any energy balance study, measurement techniques are important considerations necessary to achieve success. The intent of the author in this chapter is only to point out some excellent references relating to these procedures. Besides those already mentioned, Szeicz (1975) and Fritschen and Gay (1979) are excellent sources.

IV. CONCLUSIONS RELEVANT TO INTEGRATED PEST MANAGEMENT

The use of all of the principles discussed in previous sections of this chapter in integrated pest management (IPM) is difficult. The author would suggest, however, that the use of the entire model presented, or portions of it relevant to individual needs, may reap successful rewards. Model building and simulation of systems, whether it be the prediction of crop growth, the prediction of pathogen development or a mix of the two, tells the researcher or user where understanding succeeds or fails. In addition, it forces the individual researcher out of his ivory tower to work with researchers from other disciplines. Looking at the entire system cooperatively with others is a sometimes difficult but needed practice. The challenge is to use and understand those energy balance components that most directly influence the pathogens or systems that you study.

Duncan et al. (1967), Lemon (1969), Murphy et al. (1974), deWit et al. (1978), Norman (1979), and others are making strides in modeling environmental conditions within the canopy, crop growth, and yield relationships. Sinclair et al. (1975) have analyzed errors associated with some of the energy balance techniques. An intermeshing of pest models, crop models, and energy balance models is needed, but much testing and difficult decisions regarding the necessary data input requirements will have to be made. It will not be an easy task but it is a natural mix. Wallin (1967), Waggoner (1975, 1976), and Gutierrez et al. (1976) are making inroads with this difficult task. Cooperation with pure atmospheric scientists and their models also may be very fruitful for those involved with insect and spore transport.

In addition, there is the need to study the accuracy of extrapolation of model results from test sites to general areas and the basic data-collection requirements regarding data. A cautionary note for readers who are users or potential users of weather data. With increasing economic pressures, we must guard against a decline in data availability. This decline has been noted in some areas.

As agricultural production changes, intended or inadvertent modifications to the energy balance may influence pathogen development. Reduced tillage and irrigation practices are well-known examples of this, and an understanding of general energy balance relationships may provide insight and possible prediction of these consequences.

REFERENCES

Anderson, M. C., and Denmead, O. T. (1969). Agron. J. 61: 867-872.

Byers, H. R. (1974). "General Meteorology," McGraw-Hill, New York.

Davies, J. A., and Idso, S. B. (1979). In "Modification of the Aerial Environment of Crops" (B. J. Barfield and J. F. Gerber, eds.), pp. 183-219. Am. Soc. of Agric. Engrs., St. Joseph, Michigan.

Decker, W. L. (1964). Mo. Agric. Exp. Stn. Res. Bull. 854.

de Vries, D. A. (1975). In "Heat and Mass Transfer in the Biosphere—Part I" (D. A. de Vries and N. H. Afgan, eds.), pp. 6-28. John Wiley and Sons, New York.

deWit, C. T. (1978). "Simulation of Assimilation, Respiration and Transpiration of Crops." John Wiley and Sons, New York.

Duncan, W. G., Loomis, R. S., Williams, W. A., and Hanau, R. (1967). Hilgardia 28:181-205.
Fleagle, R. G., and Businger, J. A. (1963). "An Introduction to Atmospheric Physics." International Geophysics Series, Vol. 5, Academic Press, New York.
Fritschen, L. J., and Gay, L. W. (1979). "Environmental Instrumentation." Springer-Verlag, New York.
Gast, P. R., Jursa, A. S., Castelli, J., Basu, S., and Aarons, J. (1965). In "Handbook of Geophysics and Space Environments" (S. L. Valley, ed.), p. 16-2. McGraw-Hill, New York.
Gates, D. M. (1980). "Biophysical Ecology." Springer-Verlag, New York.
Gutierrez, A. P., Falcon, L. A., and van den Bosch, R. (1976). In "Modeling for Pest Management" (R. L. Tummula, D. L. Haynes, and B. A. Croft, eds.), pp. 134-145. Michigan State University, East Lansing.
Idso, S. B., Jackson, R. D., Ehrler, W. L., and Mitchell, S. T. (1969). Ecology 50(5):899-902.
Kanemasu, E. T., Wesely, M. L., Hicks, B. B., and Heilman, J. L. (1979). In "Modifications of the Aerial Environment of Crops" (B. J. Barfield and J. F. Gerber, eds.), pp. 156-182. Am. Soc. of Agric. Engrs., St. Joseph, Michigan.
Kimball, B. A. and Jackson, R. D. (1979). In "Modifications of the Aerial Environment of Crops" (B. J. Barfield and J. F. Gerber, eds.), pp. 211-229. Am. Soc. of Agric. Engrs., St. Joseph, Michigan.
Lemon, E. R. (1966). In "Plant Environment and Efficient Water Use" (W. H. Pierre, D. Kirkham, J. Pesek, and R. H. Shaw, eds.), pp. 28-48. Am. Soc. of Agron. and Soil Sci. Soc. of Am., Madison, Wisconsin.
Lemon, E. R. (1969). In "Physiological Aspects of Crop Yield" (J. O. Eastin, F. A. Haskins, C. Y. Sullivan, and C. H. M. van Bavel, eds.), pp. 117-142. Am. Soc. of Agron. and Crop Sci. Soc of Am., Madison, Wisconsin.
List, R. J. (1966). "Smithsonian Meteorological Tables" 6th Rev. Ed. Smithsonian Misc. Collection, Vol. 114, Smithsonian Institution, Washington, D.C.
Monteith, J. L. (1973). "Principles of Environmental Physics." Edward Arnold, London.
Murphy, C. E., Sinclair, T. R., and Knoerr, K. R. (1974). In "Vegetation and Environment" (B. R. Strain and W. D. Billings, eds.), pp. 125-147. Dr. W. Junk b.v.-Publishers, The Hague.
Norman, J. M. (1975). In "Heat and Mass Transfer in the Biosphere—Part I" (D. A. de Vries and N. H. Afgan, eds.), pp. 187-205. John Wiley and Sons, New York.
Norman, J. M. (1979). In "Modification of the Aerial Environment of Crops" (B. J. Barfield and J. F. Gerber, eds.), pp. 249-277. Am. Soc. of Agric. Engrs., St. Joseph, Michigan.
Reifsnyder, W. D., and Lull, H. W. (1965). "Radiant Energy in Relation to Forests." U.S. Dept. of Agric. Tech. Bull. 1344.

Ripley, E. A., and Redmann, R. E. (1976). In "Vegetation and the Atmosphere, Vol. 2, Case Studies" (J. L. Monteith, ed.), pp. 349-398. Academic Press, New York.
Ripley, E. A., and Saugier, B. (1975). In "Heat and Mass Transfer in the Biosphere—Part I" (D. A. de Vries and N. H. Afgan, eds.), pp. 311-325. John Wiley and Son, New York.
Rosenberg, N. J. (1969a). Agron. J. 61:879-886.
Rosenberg, N. J. (1969b). Agric. Meteor.. 6:179-184.
Rosenberg, N. J. (1974). "Microclimate: The Biological Environment." Wiley-Interscience, New York.
Ross, J. (1975). In "Vegetation and the Atmosphere, Vol. 1, Principles" (J. L. Monteith, ed.), pp. 13-55. Academic Press, New York.
Saugier, B. (1976). In "Vegetation and the Atmosphere, Vol. 2, Case Studies" (J. L. Monteith, ed.), pp. 87-119. Academic Press, New York.
Sceicz, G. (1975). In "Vegetation and the Atmosphere, Vol. 1, Principles" (J. L. Monteith, ed.), pp. 229-273. Academic Press, New York.
Seeman, J., Chirkov, Y. I., Lomas, J., and Primault, B. (1979). "Agro-meteorology." Springler-Verlag, New York.
Shaw, R. H., and Decker, W. L. (1979). In "Modification of the Aerial Environment of Crops" (B. J. Barfield and J. F. Gerber, eds.), pp. 141-155. Am. Soc. of Agric. Engrs., St. Joseph, Michigan.
Sinclair, T. R., Allen Jr., L. H., and Lemon, E. R. (1975). Boundary-Layer Meteor.. 8:129-139.
Tanner, C. B. (1960). Soil Sci. Soc. Am. Proc. 24:1-9.
Thom, A. S. (1975). In "Vegetation and the Atmosphere, Vol. 1, Principles" (J. L. Monteith, ed.), pp. 57-109. Academic Press, New York.
Trewartha, G. T., and Horn, L. H. (1980). "An Introduction to Climate." McGraw-Hill, New York.
van Bavel, C. H. M., and Ehrler, W. L. (1968). Agron. J. 60: 84-86.
Verma, S. B., and Barfield, B. J. (1979). In "Modification of the Aerial Environment of Crops" (B. J. Barfield and J. F. Gerber, eds.), pp. 230-248. Am. Soc. of Agric. Engrs., St. Joseph, Michigan.
Waggoner, P. E. (1975). In "Vegetation and the Atmosphere, Vol. 1, Principles" (J. L. Monteith, ed.), pp. 205-228. Academic Press, New York.
Waggoner, P. E. (1976). In "Modeling for Pest Management" (R. L. Tummala, D. L. Haynes, and B. A. Croft, eds.), pp. 176-186. Michigan State University, East Lansing.
Wallin, J. R. (1967). In "Ground Level Climatology" (R. H. Shaw, ed.), pp. 149-164. Am. Assoc. for the Advmt. of Sci., Washington, D.C.

WIND MOVEMENT WITHIN CANOPIES

Roger H. Shaw
Department of Land, Air, and Water Resources
University of California
Davis, California

I. INTRODUCTION

The change in the intensity, spatial distribution, and quality of solar radiation penetrating vegetation is paralleled to some extent by changes in the movement of air through the canopy (here defined as the region extending from the soil surface to the upper limits of the vegetation). Without taking the analogy too far it is clear that, like the radiation field, air movement is depleted with depth in the vegetation because of the multitude of obstructions to both sunlight and the wind flow. Individual leaves may be shaded from direct sunlight by leaves higher in the canopy and may be sheltered from the direct action of the wind by their immediate neighbors and by the integrated effects of all upwind vegetation.

Air movement within a canopy of vegetation affects, to varying degrees, the growth and development of the plants themselves, insect activity and population growth, and the development and dispersal of fungal and bacterial diseases. The effects of wind can be divided into three categories. Firstly, through the exchanges of heat and mass at the leaf surfaces, and through the diffusion of heat and mass between the canopy air spaces and the atmosphere above, the air flow regulates the microclimate of the

ISBN 0-12-332850-0

vegetation. As an example, water evaporating from the soil and transpiring from leaves acts to increase the humidity of the canopy over and above that of the layers outside the canopy. The degree to which this occurs is dependent on the amount of ventilation, which is determined by the strength of the wind and the extent to which the wind can penetrate the canopy layers.

In the immediate vicinity of components of the canopy and of the soil surface, heat and mass exchanges occur by molecular conduction through a layer that is variable from a fraction of a millimeter to several millimeters. In fact, the major component of the "resistance" to the diffusion of heat and mass between the foliage surfaces and the air above the canopy lies in this thin molecular boundary layer surrounding each element.

In the bulk of the air layer, diffusion is a consequence of the turbulent nature of the wind. This process is called convection to distinguish it from the molecular process of conduction. Turbulence is ubiquitous in the lower part of the atmosphere. It is overcome by the viscosity of the air only in the molecular boundary layer, and is suppressed only, not totally subdued, by the thermally induced stability of the atmosphere at night.

The action of increased wind speed is to increase the efficiency of both the convection and the conduction processes. In the case of the former, convection is increased in a direct manner since increased wind speed results in increased turbulence levels. In the latter case, increased air motion erodes the molecular layer enabling the exchange of heat and mass to proceed more readily. The depth of the molecular layer surrounding a leaf will vary with position on the leaf and will also vary with exposure of the leaf and with time, in response to fluctuating wind speeds.

A second category in which we might class the interaction between the wind field and the canopy concerns the direct mechanical action of the wind on both the plants themselves and objects such as soil or dust particles, small insects, fungal spores, or pollen sitting on, or physically attached to plant parts. Questions that might be asked include: at what wind speeds will trees in a forest or crop plants incur physical damage by leaf or stem breakage, or, at what wind speeds will dust particles be resuspended from a leaf surface? In each case, however, the answer is not straightforward since lodging and resuspension are examples of unique events that occur not because the time-averaged wind speeds exceed a certain limit, but because a single gust of extreme wind speed occurred at the right point in space and time. While useful statistical relationships might be found between, say, lodging and the mean wind speed observed at a nearby weather station, such analysis will contribute little toward understanding the complex relationships between plants and their environment. Closely related problems involve particles already airborne where the primary questions concern suspension times of particles of varying aerodynamic size, and the effect of the wind field on the impaction of particulates on plant surfaces.

The wind acts as a vector for a myriad of gaseous and particulate materials, forming the third and final category. This topic again involves diffusion and transport, but concerns the dispersal of materials from point sources or, at least, from rather confined regions. The rationale for consideration in a separate category stems more from the methodology by which we treat this more complicated three-dimensional problem, than from any differences in the physics of the diffusion processes. Of great interest to plant pathologists is the dispersal of micro-organisms. The rate of spread of fungal or bacterial diseases is wind dependent and, except for long distance transport, the canopy is an important factor in determining the characteristics of the wind field. A second example is insect pheromone release and diffusion. The zone of active space about the source, within which concentrations are large enough to elicit response, is of particular importance.

Review articles on the subject of air movement in plant canopies are limited in number but readers with further interest in the topic are directed toward those by Businger (1975), Thom (1975), Grace (1977), Legg and Monteith (1975), and Raupach and Thom (1981). The first article in this list presents in some detail the scaling laws for the wind flow above the canopy but only a limited account of the within-canopy wind profile. The second treats the canopy problem much more extensively but restricts analysis of momentum distribution to traditional micrometeorological theory and does not go beyond mixing length arguments and K-theory. The text by Grace contains a minimum of micrometeorological theory but is valuable for its treatment of a wide range of responses by plants to air movement. The article by Legg and Monteith is probably the first to question traditional micrometeorological approaches to the air spaces within a structure as complicated as a plant canopy, while a significant step forward is made in the final review paper by Raupach and Thom. These final authors consider contemporary boundary-layer theory in relation to the canopy problem and treat the turbulent component of the wind flow in as deterministic a fashion as current understanding of fluid dynamics allows.

II. AERODYNAMIC PROPERTIES OF VEGETATION

The analogy was drawn between the interception of solar radiation by leaves and their obstruction of air movement but such an analogy has its limitations. While leaves act in a passive manner (at least over reasonably small time scales) in intercepting the rays of the sun, they bend and flex in response to the wind and the cross-section they present to the wind changes continuously. Further, the effect of all other leaves on the radiation impinging on one individual leaf can be treated as a purely geometrical, albeit complex problem, whereas the aerodynamic problem is confused by fluctuating wind directions and meandering wakes in the lee of upwind obstructions.

A. The Drag Coefficient of Isolated Elements and of Leaf Canopies

There are two forces that comprise the drag exerted on an object exposed to a stream of air: viscous drag and form or pressure drag. The former results from the retardation of fluid in the molecular boundary layer due to the finite viscosity of the air, and the latter is caused by the separation of the streamlines from the downwind side of any bluff body, with a resulting difference in static pressure between the upwind and downwind faces. Examination quickly reveals that most plant parts are not aerodynamically streamlined, and form drag is usually taken to be the major component of the drag force. Thom (1968, 1975) has estimated that, on the average, bluff body effects exceed skin friction drag on leaves by a factor of about three. This difference may well be reversed at low wind speeds in the lower portions of some dense canopies, however.

It is normal meteorological practice to express the total drag force F_j on a single element as

$$F_j = \rho C_{dj} a_j U^2 \qquad (1)$$

where ρ is air density and U is wind speed. C_{dj} and a_j are the drag coefficient and the element area, respectively, and although there is some variation in how they are defined, it is most common to take a_j as the area of one side of the leaf and to accommodate changes due to leaf orientation in the drag coefficient, rather than to adjust the exposed area. Published information on single element drag coefficients is extremely limited and of little practical value because individual leaves in a canopy are rarely exposed to a "freestream" velocity and summation of F_j over all leaves would not account for the aerodynamic interference of neighboring leaves.

Equation (1) is extended to the drag force of a collection of leaves or plant parts in a straightforward manner but as a function of height z by writing

$$F(z) = \rho C_d A(z) U(z)^2 \tag{2}$$

where C_d is again a dimensionless drag coefficient now applied to the canopy as a whole to take sheltering into account. A(z) is a leaf area density, the area of one side of all leaves within a unit volume of canopy space and expressed in m^2/m^3, rather than a total leaf area so that F(z) becomes the drag force per unit volume. There is, of course, an implication of spatial averaging over a horizontal plane involved in this concept, which restricts the application of equation (2) to horizontally "uniform" vegetation.

In a nonsteady air stream the average drag force in the direction of the mean wind will be given by

$$F_x = \rho C_d \overline{Au(u^2 + v^2 + w^2)^{\frac{1}{2}}} \tag{3}$$

where u, v and w are velocity components in the direction of the mean wind and the lateral and the vertical directions, respectively. The overbar signifies an average. The wind statistics necessary to evaluate equation (3) are rarely, if ever, available and are usually approximated by either the square of the mean wind speed $\bar{U}$ or the square of the mean longitudinal wind velocity $\bar{u}$. The former is measured with relative ease but theoretical considerations lead to equations for the wind field that are formulated in terms of the latter.

III. MEAN WIND PROFILES

A. Above the Vegetation

The neutrally stratified atmospheric surface layer can be subjected to simple dimensional analysis when the only length scale to enter the problem is the height above the surface. In integrated form, a logarithmic profile results which is usually written as

$$\bar{U} = \frac{u*}{k} \ln (z/z_o) \tag{4}$$

where u* is the friction velocity defined as the square root of the kinematic surface stress, and k is von Karman's constant. The constant of integration is expressed in the form of z_o, the roughness length, which normally turns out to be about an order of magnitude smaller than the height of the vegetation.

As the surface is approached, additional length scales must enter the problem corresponding to the geometrical distribution of the momentum sink, and the wind profile will, in general, depart from a logarithmic shape. Modifying equation (4) to compensate for this by introducing the displacement height d, such that

$$\bar{U} = \frac{u*}{k} \ln \left(\frac{z - d}{z_o}\right) \tag{5}$$

does not totally eliminate the difficulty because the vertical distribution of the drag force within the vegetation is still excluded from the problem.

B. Within the Canopy

Entering the canopy itself, dimensional analysis ceases to be of value because of the multitude of scales that would be necessary to define the structure of the vegetation. Nevertheless, the form of the mean wind profile within the canopy is central to a large number of biophysical problems. Relatively simple single-coefficient expressions exist that are either empirically based or founded on gradient-diffusion theory. The most common expression for the canopy wind profile (Inoue, 1963; Cionco, 1965) states that the wind speed increases as the exponential of the height above the soil surface in a manner such that

$$\bar{U} = \bar{U}(h) \exp[\gamma (z/h - 1)] \tag{6}$$

where γ is an extinction coefficient for the wind speed and h is the height of the vegetation. Figure 1 is a schematic diagram of the vertical profile of mean wind speed assuming a semi-logarithmic shape above the vegetation and an exponential shape within the canopy. Cionco (1972, 1978) has shown that equation (6) provides a good fit to observed wind profiles in a wide variety of plant canopies. Extinction coefficients that he has compiled range over an order of magnitude from 0.44 for a citrus orchard to 4.4 for a gum-maple forest. Agricultural crops are generally intermediate, ranging from about 1.3 to 2.8.

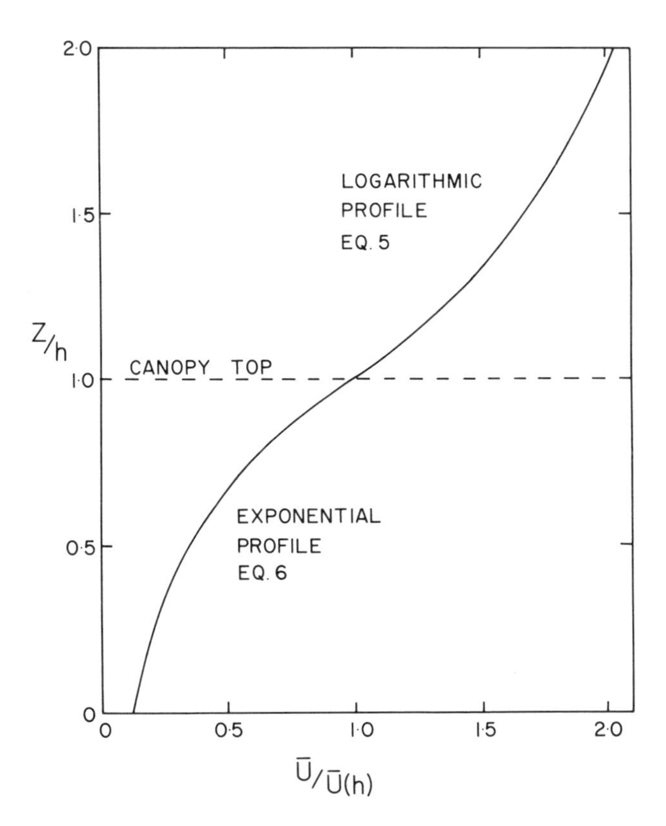

Fig. 1. Schematic diagram illustrating a logarithmic wind speed profile above the vegetation and an exponential profile within the canopy. Profile parameters were chosen as: $d = 0.6$ h, $z_o = 0.12$ h, and $\gamma = 2.08$.

Cionco (1979) has also defined a coupling parameter, R_c, as an index of the degree of transfer of momentum through the canopy. R_c is taken as the ratio of the wind speed at 0.25 h to that at 1.4 h. These heights are selected somewhat arbitrarily but, presumably, are chosen to represent the lower part of the canopy where the wind changes only gradually with height, and the ambient flow at a level not too high, because fetch problems could be introduced, nor too close to the canopy top where the wind speed gradient is large.

The coupling ratio constitutes a useful index of the degree of ventilation of a canopy embodied in a single parameter. The idea could well be extended to complex canopy situations where it might be important to know, for example, the degree of ventilation within a cluster of leaves or in the vicinity of developing fruit. The ratio could be formed of the mean wind speed at the location in question and that at 1.4 h.

Other single coefficient canopy wind profile expressions have been proposed by Cowan (1968), Landsberg and James (1971), and by Thom (1971). All fit observed wind profiles reasonably well in the upper half of the canopy but none is particularly good at matching the wind in the lower reaches of the canopy, where frequently the wind speed varies only slowly with height (Uchijima and Wright, 1964), or contains a noticeable bulge in its vertical profile (Oliver, 1971).

Recently, Pereira and Shaw (1980) have calculated canopy wind profiles using a mathematical model employing second-order closure principles. Results of the study show the rate at which the wind is depleted in canopies of differing density and vertical distribution of leaves. These are reproduced in Figures 2 and 3 for a series of hypothetical, and triangular, leaf distributions in which leaf area density was made to increase linearly with depth from the top of the canopy to some adjustable fraction of the vegetation height and then decrease linearly to the soil surface. As expected, the rate of extinction of the wind speed increased as

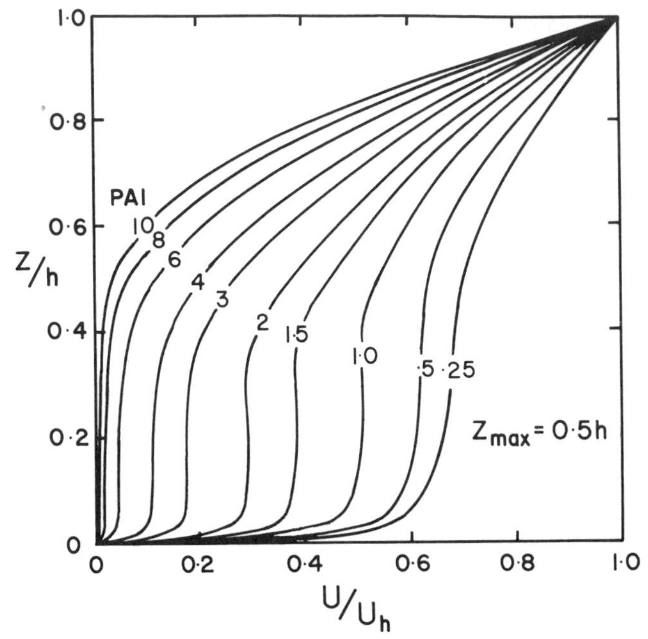

Fig. 2. Relative wind velocities inside canopies of varying foliage density. Density profiles were triangular in shape with maximum density at mid-canopy. Curves are labelled by plant area index (PAI) on the assumption that C_d = 0.2 (from Pereira and Shaw (1980), with permission).

leaf density increased and as more of the total leaf area was distributed toward the top of the canopy. The greatest wind shear occurred at the top of the canopy in all cases, while all exhibited very small or even slightly negative wind shear within the height range 0.1 h to 0.3 h, near the bottom of the canopy.

C. Spatial Variability

The crudest of models used to represent the aerodynamic properties of a plant canopy would probably be one in which the canopy was regarded as a homogeneous medium which was uniform in the horizontal direction but which perhaps was variable in density with height. This medium would resist air motion as a body force proportional to the square of the wind speed at any level.

In reality, the geometrical complexity of any plant stand implies extreme inhomogeneity in the wind field even in the most

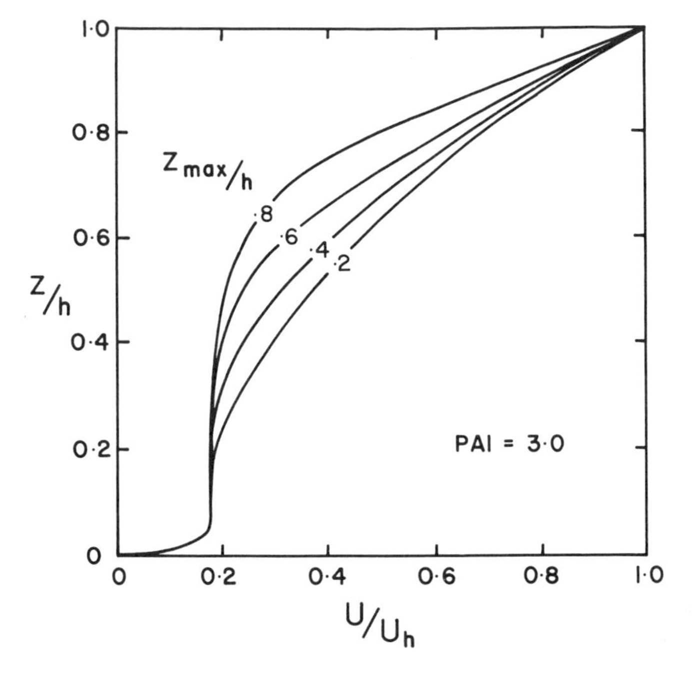

Fig. 3. Relative wind velocities inside canopies of constant PAI but varying structure. Curves are labelled according to the relative height at which plant area density reaches a maximum.

uniform of canopies. The mean wind speed downwind of a leaf, within a leaf cluster, or within bunches of fruit may be an order of magnitude or more, less than that in a location with more open exposure.

Many plant stands, including crops, tree plantations and natural stands with a variety of plant species have obvious spatial variability in plant spacing and distribution. Often, this is in a regular fashion (row crops, for example) but natural stands usually contain considerable randomness. At any level mean winds may well be quite dependent on the specific location, and dependent upon wind direction as well as on the wind speed above. Winds can "channel" or "funnel" between individual plants in more or less the same way air might move through urban canyons. The wind can also adopt a three-dimensional character as, for example, in the vorticular flow that has been observed between vineyard rows (Weiss and Allen, 1976) or between soybean rows (Arkin and Perrier, 1974).

A three-dimensional structure to the time averaged wind field is probably typical even of quite uniform canopies. The mean wind vector will vary in magnitude and direction at different points in the canopy and will not be confined to the horizontal plane. This feature makes it difficult both to take representative measurements of the wind field and to apply mathematical models to canopies with either regular or random variability in plant population or density.

IV. TURBULENCE IN PLANT CANOPIES

Any analysis of the air movement through vegetation that is used to extend our understanding of the formation of the canopy microclimate, or of diffusion within the canopy, must stress the role of turbulence. The turbulent component of the flow is responsible for the diffusion, mixing and exchanges that occur between different regions or layers of the canopy and between the canopy air layers and the atmosphere above.

The theory of turbulence in the atmosphere is featured in a number of texts (Lumley and Panofsky, 1964; Tennekes and Lumley, 1972; Monin and Yaglom, 1971; Haugen, 1973) but will be extended here to an atmospheric layer occupied by an arrangement of obstacles, the elements of the vegetation. Raupach and Thom (1981) have recently provided an in-depth review of this subject.

A. The Origin of Turbulence

Because of dynamical instabilities, the air within a few tens of meters of the surface of the earth is, for all practical purposes, continuously turbulent. Even when air movement is very slight and when a stable temperature gradient imposes a further restriction, diffusion occurs at a rate appreciably faster than it would by molecular activity alone.

Atmospheric turbulence is defined by intensity and scale. The intensity or level of turbulence is a measure of the violence of the motions superimposed on the mean wind flow. Numerically it is expressed as the root mean square of the fluctuations but, since turbulence is three-dimensional, three components of the intensity are often quoted (normally the longitudinal, lateral and vertical components). Often the relative intensity of turbulence is provided in which the intensity is normalized by division by the mean wind speed. The square of the intensity is the variance of the fluctuations and when all three variance components are added together, we obtain twice the kinetic energy of the air flow, per unit mass of air, directly attributable to the fluctuating component. This is called the turbulence (or turbulent) kinetic energy (TKE) and giving it the symbol $\bar{e}$, we can write

$$\bar{e} = \tfrac{1}{2}(\overline{u'^2} + \overline{v'^2} + \overline{w'^2})$$

where u, v, and w are the longitudinal, lateral and vertical components of wind velocity and the primes indicate fluctuations from the time averaged values (Reynolds notation).

As part of the kinetic energy of the atmosphere, TKE is governed by an energy conservaton law. The mathematical expression for this law is the turbulent kinetic energy equation, and expresses the rates at which energy passes to or from the turbulent part of the flow. The equation thus identifies the processes that act to create turbulence, to suppress or eliminate it, and to move it from one place or level in the atmosphere to another. The TKE equation must therefore be considered central to studies of turbulence in the atmosphere, and in the layers of atmosphere within a plant canopy.

In boundary layers free of the complicating influence of the vegetation, and with some simplifying assumptions, the TKE budget may be written (e.g., Lumley and Panofsky, 1964)

$$\frac{\partial \bar{e}}{\partial t} = - \overline{u'w'} \frac{\partial \bar{u}}{\partial z} + \frac{gH}{T\rho C_p} - \frac{\partial}{\partial z} (\overline{e'w'} + \frac{\overline{p'w'}}{\rho}) - \varepsilon \quad (7)$$

in which g is the gravitational acceleration, H is the heat flux density, T is the temperature (°K), ρ is the air density, C_p is its specific heat, p is the static pressure, and ε is the rate of viscous dissipation of turbulent energy.

Any analysis of boundary-layer turbulence must start with the interpretation and evaluation of the terms in this equation. Under steady state conditions the terms on the right hand side will sum to zero. Essentially, production is balanced by dissipation although not always at the same point in space. The first term on the right is the product of the Reynolds stress ($-\overline{u'w'}$, the rate at which momentum diffuses downward toward the surface) and the vertical shear of the mean wind velocity. It is usually positive in sign and is called the rate of shear or mechanical production of turbulence. It is the dominant production term under near-neutral atmospheric conditions. The second term involves the vertical heat flux and expresses the rate at which buoyant forces contribute to turbulence. Unlike the first term, the buoyancy term switches sign, according to the direction of the heat flux. Convective motions in an unstable atmosphere enhance turbulence, but turbulent motions are inhibited when the surface cools below the air temperature and heat passes from the air to the surface. Most stability parameters used in meteorology are closely related to the ratio of the buoyancy term to the shear production rate.

The following term, involving the vertical gradient of two bracketed terms, expresses the rate at which TKE is transported vertically upward or downward by turbulent diffusion and by pressure fluctuations. Normally, these are not considered to be large terms in the overall budget of the atmospheric surface layer. Finally, ε expresses the rate at which the viscosity of the air converts kinetic energy of the turbulent motions into internal energy or heat. It is called viscous dissipation and is the ultimate sink for TKE.

Applying equation (7) to the layers of a canopy we find two very significant differences. First, an additional term enters into the equation, and second, the transport terms attain much greater importance, to the point were they can become the dominant terms in the lower part of the canopy.

The additional term represents the complex motions of the wakes behind the multitude of obstacles to the air flow. The term arises naturally in the equation (Raupach and Thom, 1981) as an extension of the mechanical production of turbulence due to wind shear. This time we are not concerned with the spatially averaged mean wind shear but the smaller scale velocity gradients in the vicinity of each obstacle. The turbulent energy again originates from the mean wind but represents the mechanical work performed by the wind in moving through the canopy against the force of form drag (Wilson and Shaw, 1977). Because turbulence created in this manner is of rather small scale, it tends to dissipate quickly and probably does not contribute greatly to overall turbulence levels even though the rate of production might be comparable with that due to shear of the mean wind (Seginer et al., 1976).

Turbulence levels are generally quite high inside plant canopies despite the fact that low wind speeds and small wind shears result in only small production rates. The reason is that turbulence created at and above the vegetation is "imposed" on the canopy flow. Measurements of the downward turbulent transport of TKE over and within plant canopies (Maitani, 1978; Shaw, 1977) have confirmed that turbulence in the lower part of the canopy, and much of it in the upper levels is not generated locally but originates at higher levels. This has significant consequences on local diffusion models. The pressure component of the transport term has not yet been satisfactorily measured.

B. Observations of Canopy Turbulence

1. Turbulence Intensity. Cionco (1972) has provided a rather extensive summary of turbulence intensities observed in agricultural crops, orchards, deciduous and coniferous forests and in some artificial canopies. In all cases, turbulence levels were large. Some typical values of the relative intensity, i, defined as σ_u/U where σ_u is the standard deviation of the longitudinal velocity component, are listed as follows:

Canopy Type	i
Agronomic crops	0.4
Tree plantations	0.54
Conifer trees	0.6
Deciduous trees	0.65
Jungle forests	0.7 to 1.2

For canopies that Cionco classifies as simple (e.g., rice paddy, wheat, larch tree plantation), the relative intensity tended to remain constant with height and decreased slowly above the vegetation. In other cases, there was considerable variation in i through the canopy. For example, turbulence intensities were highest in the crown layer of a pine forest, wind-tunnel model trees, and immature corn plants, but greatest in the lower levels of jungle canopies and in a deciduous forest.

Like the wind field above the vegetation, the turbulence within the canopy is anisotropic. If a picture could be formed of an eddy, it would appear to be elongated in the downwind horizontal direction. In terms of turbulence intensities, Shaw et al. (1974a) found $i_u > i_v > i_w$ in corn, where the subscripts refer to the longitudinal, lateral and vertical direction, respectively.

2. Turbulence Scale. The physical dimensions of the flow structures of which the turbulence is composed become important when considering the diffusion of material within the vegetation.

Although a small number of examples of two-point correlation studies exist, full spatial correlation studies of the canopy wind field have not yet been completed because of the requirement for a large number of wind instruments. Closely related information is provided much more economically by temporal correlations, which are usually presented in the form of power spectra (Jenkins and Watts, 1968) showing the contribution to the total variance of the velocity components by different frequency intervals. Examples of canopy wind spectra are to be found in articles by Allen (1968), Isobe (1972), Shaw et al. (1974b), Inoue et al. (1975), and Bill et al. (1976). The appearance of the spectra depends on the length of the time period over which the sample is made but, in general, they are more sharply peaked than spectra taken over relatively smooth land surfaces, and do not scale in a simple way with the nondimensional frequency $nz/\bar{U}$, where n is frequency in Hz (Kaimal et al., 1972).

Spectra calculated from records of fairly short duration (e.g., Isobe, 1972) usually exhibit a number of pronounced maxima which probably reflect the input into the flow of turbulence generated by individual leaves, clusters of leaves on a branch, or complete plants. Spectra from longer sampling periods form smoother curves (e.g., Shaw et al., 1974b). This relative smoothness probably results from the wide separation in the scales of motion of the wake generated turbulence and the larger scales imposed by the flow above the vegetation. While the physical dimensions of the wake flow may be reasonably invariant with wind speed, the frequencies of the fluctuations in the wakes will be roughly proportional to wind speed, with the result that wake turbulence will be smeared over a fairly wide range of frequencies.

Translating frequencies of observed power spectra into wave numbers or wavelengths, using Taylor's hypothesis, indicates that, at least in crop canopies, the physical size of the dominant motions is considerably larger than individual leaves, whole plants or the plant spacing.

C. Probability Distribution of Wind Speed

The basic statistical properties of intensity and scale provide an important but limited picture of the flow field. While many phenomena such as the diffusion of heat and water vapor between plants and the air layers above can be adequately described using the time averaged flow properties discussed so far, other important processes depend more on the extreme values of the wind speed. To understand, or perhaps predict, events such as lodging, resuspension of particulates, detachment of spores or insects from plant surfaces, it is necessary to know the probability of occurrence of wind speeds significantly higher than the time averaged value. For wind speeds much greater than one standard deviation above the mean wind speed, a Gaussian distribution grossly underpredicts the probability of occurrence.

The removal of particles, such as the dry spores of some plant fungi, from leaves in a canopy is greater the more turbulent the wind (Eversmeyer et al., 1973). Moreover, Aylor and Lukens (1974) have observed that this removal occurs during winds with average speeds that are seemingly too low. Aylor (1978) concludes that removal occurs during brief but energetic gusts of wind that penetrate the boundary layer surrounding the leaf surface.

A statistical description of a turbulent flow field could be expressed in terms of the statistical moments about the mean. The variance of the wind speed, for example, is the second moment about the mean, while the skewness is the third moment. The higher the moment, the more strongly it is influenced by large excursions from the mean. Large excursions would be relatively rare events and so stable statistics on high moments require long sampling periods, which are usually not available in field situations because of lack of stationarity in environmental conditions. Perrier et al. (1972) have tabulated results of skewness and kurtosis (fourth moment) observed within and above soybeans. They concluded that the wind velocity distribution was largely non-Gaussian but found it to exhibit varying degrees of skewness and kurtosis. Despite the fact that their sampling periods were rather short (4.3 minutes), results show predominately positive skewness with largest values occurring at and near the top of the vegetation. Kurtosis also appeared to be greatest in this region.

Higher moments than the fourth cannot be found for the canopy wind field, but Shaw et al. (1979) show the probability distribution of scalar wind speed at two levels inside maize. An extreme value distribution (Brooks and Carruthers, 1953) fitted the data well in the upper part of the canopy but underestimated the probability of large departures from the mean at about mid-canopy level. The study showed that wind speeds five or six times larger than the mean occurred with reasonable frequency, and that individual gusts penetrating the canopy did not appear to be diminished as rapidly with depth as the mean wind speed itself.

Although no measurements yet seem to have been reported, the structure of the air flow in the immediate vicinity of the leaf surface (within a few millimeters) would be of considerable interest in studies of particle or spore removal. The major experimental problem is leaf flutter but useful information could still be obtained by measuring winds close to a rigidly held leaf. The difficulty in understanding the removal and resuspension processes lies in the fact that small particles on a leaf surface are embedded in a molecular boundary layer that has, in theory, a very rapid adjustment time (Aylor, 1978). It is likely that the time rate of change of speed at the boundary of wind gusts is an important quantity in determining the ability of eddies to penetrate the viscous layer and to remove materials held on the surface.

V. MODELING OF AIR FLOW AND DIFFUSION IN THE CANOPY

Models of the interactions between plants and their aerial environment formalize our understanding of the physical system. They can be either physical models, such as scaled down reproductions of plants in a wind tunnel (or reduced scale field studies), or sets of mathematical equations that may or may not require a computer to provide a solution.

A measure of how well we understand a system is gained from the degree to which we can reproduce a set of physical variables for given external conditions. For our example of the plant canopy, how well can we reproduce the microclimate of the canopy in terms of wind, temperature and humidity, or how well can we reproduce the observed rate of evapotranspiration for given inputs of solar radiation, stomatal conductance, etc.? Comparison between model responses and real situations under a variety of conditions sometimes gives us the confidence to interpolate between, and perhaps even extrapolate from sets of field observations, thereby eliminating the need for further experimentation. The economic advantage of conducting tests on small scale models or of finding a numerical solution to a set of equations can be quite remarkable.

A. Physical Models

Wind tunnel studies of the flow of air through model canopies have made significant contributions to canopy physics (Meroney, 1968; Kawatani and Meroney, 1970; Seginer et al., 1976). Physical simulation of canopy processes is contingent upon meeting certain similarity criteria in relation to geometric, kinematic, dynamic and thermal properties of the system (Sadeh, 1977). While some of these criteria must be satisfied exactly, others may be relaxed to some degree allowing for the general success of wind tunnel modeling.

Wind tunnel models provide the advantage of the ability to control external conditions but, because of their reduced scale, it is not possible to reproduce with any accuracy the source and sink distributions within the canopy of heat, moisture or of other substances exchanged between the atmosphere and the vegetation. Diffusion processes other than that of momentum can generally not be modeled satisfactorily at this time.

B. Mathematical Models

Mathematical models can take the form of a single mathematical equation (Penman's equation is an example) but more unusually, they are presented as a series of interactive equations. Traditionally, mathematical models are rated as being either statistical in character or physically based. However, because of the complexities of the vegetation and of the turbulent air flow through the canopy, no model can be totally deterministic, and must at some point rely on statistically determined relationships between known and unknown properties of the flow. Modelers always start with good intentions from fundamental laws of conservation (momentum, heat, mass, kinetic energy, etc.) which cannot in themselves be criticized. The difficulties arise when expressions of these laws are applied to a geometrically very complex environment. Even without the canopy, turbulence is enormously complicated and can only be treated in terms of average quantities (means, variances, covariances, etc.). How do we take averages within such a heterogeneous medium? If we average spatially over a horizontal plane, are we missing important aspects of the flow such as the downwash behind a crop row or the air pulled into the lee of an individual tree? Raupach and Thom (1981) discuss the averaging process in some detail.

The drag force in the canopy, expressed by equation (2), is balanced primarily by a net downward transport of momentum from above. This can be written as

$$\frac{d\tau}{dz} = -\rho C_d A(z) U(z)^2 \qquad (8)$$

where $\tau(z) = -\rho \overline{u'w'}$. It is quite appropriate to neglect the Coriolis force, considering the scales of motion involved, but Smith et al. (1972) and Kondo and Akashi (1976) have shown that large-scale pressure gradients are significant and can cause a direction change in the wind within the relatively open trunk spaces of some forests.

Conservation equations for heat or moisture, for example, can be formulated in a manner analogous to equation (8), as a balance between the gradient of the vertical flux density and the strength of the appropriate source or sink at each level in the canopy. The budgets so formed cannot be solved directly, however, since each equation contains two unknown quantities. Assuming the leaf area distribution, A(z), and the drag coefficient, C_d, are known for example, equation (8) contains two flow properties, the mean wind speed and the Reynold's stress, which must be treated as unknowns. This is referred to as the closure problem and has arisen because we were forced to average the flow field, thereby losing "information" concerning the nonlinear terms.

The order of the model we formulate to solve the diffusion problem is determined by the level at which we choose to close the equation set. The simplest and traditional way is to express the Reynolds stress in terms of the mean wind speed in a gradient-diffusion approximation which we write as

$$\overline{u'w'} = -K_m \frac{d\bar{u}}{dz} \qquad (9)$$

where K_m is called the eddy viscosity (m^2/s). Similar heat and moisture flux relationships can be formed with the gradients of mean temperature and humidity, respectively.

Micrometeorologists have recently become more critical of this type of approximation but, to see why, we must look into the implications of equation (9). If K_m is a physically realistic quantity, equation (9) states that the turbulent flux of momentum is directly proportional to the rate of change of mean wind velocity at the height in question, and that the velocities and gradients of velocity at other levels are of no consequence. Considering the large scale of the motions involved in the turbulence mechanism and the relatively large vertical distances that momentum can be carried by a single "eddy," the gradient diffusion approximation becomes unrealistic.

Important advances have been made in atmospheric boundary layer studies in the last decade or so. They have stemmed from improvements in instrumentation giving us the benefit of more accurate and stable measurements of turbulence covariance, and higher-order terms, and from the increased speed of modern computers. Relatively simple algebraic manipulation of the basic conservation equations (Donaldson, 1973) can produce a set of equations that represent the rate of change with time of any of the covariance terms, such as the average product of fluctuations of two velocity components, u_i and u_j. Index notation is introduced here for generality. Equation (7) is an example of such a budget equation for the special case when i = j. In the general case, an equation is derived for the rate of change in time of the quantity $\overline{u'_i u'_j}$, which expresses the rate at which a covariance (or variance if i = j) is created, transported and destroyed. Such an equation can be coupled with equation (8), eliminating the need to parameterize the Reynolds stress as in equation (9). Unfortunately, the nonlinearity in the basic relationships ensures that equations for the second moments contain terms of an even higher order, so that the number of unknown quantities expands at a rate greater than the number of equations that contain them. Thus, no solution to the problem can be found simply by extending the set of equations to the next higher order.

Closure can be achieved by parameterizing third moments and equivalent terms, in terms of lower order quantities. While some of the closure assumptions are still controversial, it is generally believed that there is an advantage to including equations for turbulence quantities, rather than modeling the turbulent fluxes directly. The advantage lies not only in improved accuracy in calculated results but also in increasing our ability to interpret observations of turbulent diffusion.

The framework described has been applied to atmospheric boundary layers (Mellor, 1973; Wyngaard et al., 1974) and is now reasonably well established. It appears to be a promising approach to diffusion problems in the plant canopy. Examples of its application to the canopy have been given by Wilson and Shaw (1977) and Lewellen et al. (1979), the latter using a higher-order closure scheme to provide the canopy aerodynamics needed to model dry deposition of aerosols on leaves.

REFERENCES

Allen, L. H., Jr. (1968). J. Appl. Meteor. 7:73-78.
Arkin, G. F., and Perrier, E. R. (1974). Agric. Meteor. 13: 359-374.
Aylor, D. E. (1978). In "Plant Disease: An Advanced Treatise" (J. G. Horsfall and E. B. Cowling, eds.), 2:159-180. Academic Press, New York.
Aylor, D. E., and Lukens, R. J. (1974). Phytopathol. 64: 1136-1138.
Bill, R. G. Jr., Allen, L. H. Jr., Audunson, T., Gebhart, B., and Lemon, E. (1976). Boundary-Layer Meteor. 10:199-220.
Brooks, C. E. P. and Carruthers, N. (1953). Handbook of Statistical Methods in Meteorology. HMSO.
Businger, J. A. (1975). In "Heat and Mass Transfer in the Biosphere" (D. A. deVries and N. H. Afgan, eds.), pp. 139-165. Wiley, New York.
Cionco, R. M. (1965). J. Appl. Meteor. 4:517-522.
Cionco, R. M. (1972). Boundary-Layer Meteor. 3:255-263.
Cionco, R. M. (1978). Boundary-Layer Meteor. 15:81-93.
Cionco, R. M. (1979). Proc. 14th Conf. Agric. Forest Meteor., Am. Meteor. Soc. pp. 105-106.
Cowan, I. R. (1968). Quant. J. Roy. Meteor. Soc. 94:523-544.
Donaldson, C. du P. (1973). In "Workshop on Micrometeorology" (D. A. Haugen, ed.), pp. 313-392. Am. Meteor. Soc., Boston.
Eversmeyer, M. G., Kramer, D. L., and Burleigh, J. R. (1973). Phytopathol 63:211-218.
Grace, J. (1977). "Plant Response to Wind." Academic Press, New York.
Haugen, D. A. (ed.) (1973). "Workshop on Micrometeorology." Am. Meteor. Soc., Boston.
Inoue, E. (1963). J. Meteor. Soc. Japan 41:317-326.
Inoue, E., Uchijima, Z., Horie, T., and Iwakiri, S. (1975). J. Agric. Meteor. Japan 31:71-82.
Isobe, S. (1972). Bull. Nat. Inst. Agric. Sci. Japan, Series A 19:101-112.
Jenkins, G. M. and Watts, D. G. (1968). "Spectral Analysis and its Applications." Holden-Day, San Francisco.
Kaimal, J. C., Wyngaard, J. C., Izumi, Y., and Coté, O. R. (1972). Quart. J. Roy. Meteor. Soc. 98:563-589.
Kawatani, T. and R. N. Meroney (1970). Agric. Meteor. 7: 143-158.
Kondo, J. and Akashi, S. (1976). Boundary-Layer Meteor. 10: 255-272.

Landsberg, J. J., and James, G. B. (1971). J. Appl. Ecol. 8: 729-741.
Legg, B., and Monteith, J. L. (1975). In "Heat and Mass Transfer in the Biosphere" (D. A. deVries and N. H. Afgan, eds.), Wiley, New York. pp. 167-186.
Lewellen, W. S., Sheng, Y. P., and Teske, M. E. (1979). Aeronaut. Res. Assoc. Princeton, Report No. 376. Princeton, New Jersey.
Lumley, J. L., and Panofsky, H. A. (1964). "The Structure of Atmospheric Turbulence." Wiley, New York.
Maitani, T. (1978). Boundary-Layer Meteor. 14:571-584.
Mellor, G. (1973). J. Atmos. Sci. 30:1061-1069.
Meroney, R. N. (1968). J. Appl. Meteor. 7:780-788.
Monin, A. S., and Yaglom, A. M. (1971). "Statistical Fluid Mechanics: Mechanics of Turbulence." MIT Press, Cambridge, Massachussetts.
Oliver, H. R. (1971). Quart. J. Roy. Meteor. Soc. 97:548-553.
Pereira, A. R. and R. H. Shaw (1980). Agric. Meteor. 22:303-318.
Perrier, E. R., Robertson, J. M., Millington, R. J., and Peters, D. B. (1972). Agric. Meteor. 10:421-442.
Raupach, M. R. and Thom, A. S. (1981). Ann. Rev. Fluid Mech. 13: 97-129.
Sadeh, W. Z. (1977). Proc. 13th Agric. and Forest Meteor. Conf. Am. Meteor. Soc. 101-103.
Seginer, I., Mulhean, P. J., Bradley, E. F., and Finnigan J. J. (1976). Boundary-Layer Meteor. 10:423-453.
Shaw, R. H. (1977). J. Appl. Meteor. 16:514-521.
Shaw, R. H., den Hartog, G., King, G. M., and Thurtell, G. W. (1974a). Agric. Meteor. 13:419-425.
Shaw, R. H., Silversides, R. H., and G. W. Thurtell (1974b). Boundary-Layer Meteor. 5:429-449.
Shaw, R. H., Ward, D. P., and Aylor, D. E., (1979). J. Appl. Meteor. 18:167-171.
Smith, F. B., Carson, D. J., and Oliver, H. R. (1972). Boundary-Layer Meteor. 3:178-190.
Tennekes, H., and Lumley, J. L. (1972). "A First Course in Turbulence." MIT Press, Cambridge, Massachussetts.
Thom, A. S. (1968). Quart. J. Roy. Meteor. Soc. 94:44-55.
Thom, A. S. (1971). Quart. J. Roy. Meteor. Soc. 97:414-428.
Thom, A. S. (1975). In "Vegetation and the Atmosphere" (J. L. Monteith, ed.), 1:57-109. Academic Press, New York.
Uchijima, Z., and Wright, J. L. (1964). Bull. Nat. Inst. Agric. Sci. A11:19-65.
Weiss, A., and Allen Jr., L. H. (1976). Agric. Meteor. 16:329-342.
Wilson, N. R., and Shaw, R. H. (1977). J. Appl. Meteor. 16:1197-1205.
Wyngaard, J. C., Coté, O. R., and Rao, K. S. (1974). In "Advances in Geophysics," 18A:193-212. Academic Press, New York.

INSTRUMENTATION AND TECHNIQUES FOR MICROCLIMATE MEASUREMENTS

Paul C. Doraiswamy
Lockheed Engineering and
Management Services Company, Inc.
Johnson Space Center
Houston, Texas

I. INTRODUCTION

The microclimate of a pest may be defined, in general, as the immediate environment in which it survives. The microclimate of a pest associated with an agricultural crop may not only be influenced by the dynamic environment within the crop canopy, but also the climate that prevails a few meters above the vegetation. The pest may inhabit the plant or the soil system for at least part of its life cycle. Under these circumstances, the pest's habitat would be influenced by the physiological characteristics of the crop and soil. An assessment of the pest's microclimate requires an understanding of the soil-plant-atmosphere continuum and the level of measurements to be made. In most studies, an integration at various levels, such as the plant, canopy, and large areas, may be essential to derive a cause-and-effect relationship of insect behavior and development.

The crop microclimate, for the most part, may define the microclimate of the agricultural pests. This would require an assessment of the appropriate factors in the soil-plant-atmosphere continuum system. Each part of the system has a direct or indirect effect on the pest's habitat and its behavior. The pest may, at various stages of development, move from one part of the system to another in search of a microclimate which, in some

ISBN 0-12-332850-0

respects, best meets the needs and considerations of the related energy balance.

Soil factors that may have significant effects on pests, besides the indirect effects on the crop condition, would be the moisture levels that affect the mortality and development of the larvae of certain pests that inhabit the soil during part of their life cycle. Soil texture and compaction, among others, may be secondary factors that would be appropriate to a particular pest infestation.

The physiological condition of the crop may, under certain circumstances, be a significant factor in infestation. Certain crop conditions may increase the susceptibility of the crop to infestation and provide an optimum environment for breeding and spread of infestation. Optimum conditions, however, may not be a requisite for susceptibility under an epidemic. The physiological parameters pertaining to crop condition include crop moisture status, net assimilation, transpiration, and phenological development. These parameters may directly or indirectly influence the pest's habitat.

In the past decade, extensive research and investigations have been conducted that have increased our understanding of crop microclimate and its influence on productivity. Transferring this technology for applications in pest management has been slow and rather difficult because a strong interdisciplinary program is required. Agricultural meteorologists need to understand a pest's behavior in its natural environment and factors that influence its habitat. At the same time, the pest specialists have to appreciate and understand the principles of micrometeorology to properly interpret and assess its influence on the life cycle of the pest.

The instrumentation and techniques used in micrometeorological research that may be suitable for assessment of pest environment are outlined in this paper. The review will not cover technical and theoretical aspects of the instrumentation. Whenever possible, certain meteorological sensors are recommended based on experience. Further information on most of the meteorological sensors are discussed elaborately by Fritschen and Gay (1979). Their work is recommended for detailed reading.

II. SOIL MEASUREMENTS

A. Soil Moisture

Soil moisture is one of the most dynamic properties of the soil. It intensely affects many physical and chemical properties of the soil that influence the growth of plants and other organisms. Soil moisture is also a parameter that has high spatial variability and requires a systematic sampling procedure. This variability is, in part, due to different patterns of root distribution, evaporation rates, and soil physical characteristics. The following two techniques for estimating soil moisture have been used extensively and proven to be reliable when the proper sampling procedure is followed.

The first method is the direct gravimetric or volumetric measurement. Soil samples of known weight or volume are sampled with a sampler tube (Gardner, 1965 and Lutz, 1944) and oven dried (105 C) to a constant weight. The water content is expressed on the bases of g water/g dry soil or g water/cm^3 soil.

The second method is an indirect technique based on the principle of hydrogen atoms in the soil slowing down and scattering fast neutrons from a given source. The slowdown of neutron count in the vicinity of a source of fast neutrons provides a means of estimating the hydrogen content. Water being the most predominent source of hydrogen, the sensor output can be calibrated to determine the water content for a particular soil. Van Bavel, et al. (1963) discuss in detail the principles of the neutron count technique for the measurement of soil moisture.

There are other techniques which provide a measure of the soil matrix potential and may be calibrated to provide soil moisture content. The soil tensiometer and porous conductivity blocks are two sensors that provide direct measurement of soil moisture potential from 0 to 0.75 atm and 0.5 to 15.0 atm, respectively. The sensors may be placed at desired depths, and measurements can be automated for continuous seasonal monitoring.

A tensiometer is a porous cup filled with water and placed in contact with the soil. When equilibrium between water in the cup and the surrounding soil is established, the soil moisture tension is measured (Slatyer, et al., 1961). The porous conductivity blocks were first designed by Bouyoucos and Mick (1940) and operate on the principle of change in electrical conductivity of the soil media with moisture content. The sensor is a gypsum block with two parallell plate-like electrodes imbedded in it. The moisture in the block is always saturated with a solution, such as calcium sulfate, whose concentration varies with the moisture in the soil. The changes in conductivity in the block are in proportion to the amount of solution.

B. Soil Temperature

The biological significance of soil temperature to plants and organisms that grow in the soil is well documented. Certain insects spend a part of their life cycle in the soil, dormant in the winter and at times in the summer to avoid excessive heat. The diurnal amplitude of the soil temperature decreases with depth in both bare and cropped soils. Baker (1965) well illustrates day and night temperature gradients in the soil and the thermal lag of diurnal soil temperatures. Soil temperature profiles may be measured at various depths with thermocouples or thermistor probes. Fritschen and Gay (1979) discuss, in detail, temperature sensors for this purpose and problems associated with placing sensors without disturbing the soil profile.

C. Soil Heat Flux

The soil heat flux is an estimate of the net energy flow into and out of the soil system. Soil reflective characteristics, moisture content, and vegetation density are some of the important factors determining the magnitude of the flux. For a complete crop cover, the soil heat flux is often ignored since it is generally small in magnitude relative to other terms in the energy balance of a crop. The heat flux into the soil is released at night and is of significance in the nocturnal energy balance (Gay and Holbo, 1974), which may be important to the movement and behavior of certain pests.

Soil heat flux may be measured directly by placing heat flux plates in the soil (Fuchs and Tanner, 1968a). The sensor is a differential thermopile connecting the top and bottom portions of the sensor plate. The soil heat flux plate is constructed like a net radiometer without the protective polyethylene domes. Net radiometers will be discussed later in this paper. The sensor should have the same blackbody characteristics as the soil and be calibrated in material with thermal conductivity similar to the soil to be used. Sensor size and placement should be such as to minimize impedance of water movement and vapor trapping (Fritschen and Gay, 1979).

III. TEMPERATURE AND HUMIDITY MEASUREMENT

Temperature and humidity are environmental variables that directly affect an insect's habitat. An insect's ability to seek out its environment may be viewed as a behavioral means of controlling its energy balance. In moving from or to a sunny, shady, moist, or dry environment, the insect may select a combination of heat level and dissipative flux. A combination of the environmental temperature and humidity is essential for maintaining proper internal water balance, which is of great importance to the functioning and survival of organisms. Wellington's work (1949) on the spruce budworm demonstrates this preferendum in larvae.

The microclimate of crops may be assessed by several direct and indirect methods. Direct methods include defining the seasonal and diurnal temperature and humidity of the crop environment. Indirect methods include using meteorological models and gaining an understanding of the relationship of environmental variables. Both methods of assessment may require detailed profile measurements of variables such as radiation, temperature, water vapor pressure, and wind speeds. Micrometeorological models, such as the energy balance, partition the net energy absorbed by the crop into various energy dissipative components. These models are discussed in other papers of this symposium.

Gradients of temperature and vapor pressure are usually greater at the crop canopy surface than above it. In placing sensors above the canopy, some knowledge and considerations about the boundary layer characteristics would be essential. The further one goes above the crop canopy surface, less characteristics of the underlaying crop are measured. For flux measurements of temperature and water vapor, it would be preferable to place sensors close to the surface where gradients are large and thus minimize measurement errors. However, while reducing the size of measurement error, the measurement may not be representative if the surface is nonhomogeneous. Surface roughness greatly determines the size of the boundary layer and the placement of sensors. Rough crop canopy surfaces cause a greater mixing of the air, thus reducing the gradients of temperature and water vapor pressure and hence, requiring sensors with a higher level of sensitivity.

Measurements within the canopy are just as complex, requiring proper placement of highly sensitive instrumentation because gradients are small, especially in densely covered canopies. The sensor size and the horizontal and vertical sampling should not disrupt the natural environment.

A. Ambient Temperature

The daytime temperature above a crop canopy usually decreases with increasing height. The profile below the surface may indicate an increase in temperature up to a distinct zone, and thereafter, decrease toward the soil surface. The most common temperature sensor used in microclimatology is the thermocouple. A copper constantan thermocouple has a suitable range for temperature measurements in biological experiments. Thermocouples have a fairly linear output signal of 0.04 mv C^{-1} and require a signal amplification. Thermocouples have a very stable calibration and require a constant source of reference temperature. The absolute accuracy is between 0.1 and 0.25 C (Rosenberg, 1974). Accuracy may be increased tenfold when measurements are made differentially. The sensor size is very small and very good for measurements in tight locations such as an insect cell.

Two other types of temperature sensors are thermistors and diodes. Both types of sensors require a power source, a proper bridge circuit, and individual calibration for each sensor. A thermistor has a nonlinear output with a large signal from 250 to 500 mV C^{-1} (Fritschen and Gay, 1979). The diode output is linear with a magnitude of 2 mV C^{-1}. Both sensors are expensive, relative to thermocouples and may require calibration checks.

In selecting a particular temperature sensor, one has to study the options such as cost, frequency of measurement, and data recording facilities to match sensor output and calibration requirements. An example of a situation for selecting thermocouples is the need for a small sensor, minimum ventilation, and several sensors to obtain a spatial average. Thermocouples also can excel other sensors because as many as ten junctions may be used in each location for a profile measurement (Black and McNaughton, 1971). There are many precautions that are to be taken in the construction and use of temperature sensors, such as sensor shading and ventilation for adequate spatial sampling. Tanner (1963) discusses some of the sensor placement and sampling problems.

B. Canopy Surface Temperature

The canopy surface temperature is an important parameter that could signal changes in the plant's physiological condition due to the presence of a certain form of temporary or permanent stress. This stress may be induced by disease, insect infestation, or moisture deficit conditions. A plant generally has an energy balance that maintains leaf temperatures below or approximately at air temperature.

The measurement of crop canopy surface temperature is a complex problem. The crop surface is not a two-dimensional plane, but has individual leaves of various orientation with different energy and radiation balances. The crop temperature at various depths in the canopy presents an even greater measuring problem. Measurement of crop surface temperature requires that small-size thermocouples be placed on the dorsal side of the leaf. An average surface temperature may be obtained by measuring a large number of leaf temperatures at the canopy surface. This technique can be tedious, and the result is not necessarily representative of surface temperature.

Fuchs and Tanner (1966 and 1968b) were some of the early researchers who demonstrated the usefulness of a remote sensing technique for measurement of soil and crop surface temperatures. The infrared (IR) thermometer requires no physical contact between sensor and surface and can be mounted on a stationary or a mobile platform above the crop or bare soil to obtain a large measurement sample. This type of sensor has also been used aerially for large area sampling (Bartholic et al., 1972) to detect moisture stress in crops. Recently, Idso et al. (1977) showed that seasonal crop canopy temperatures, which are one of the indicators of crop moisture stress, may be correlated with yield reductions in wheat crop.

The IR thermometer measures terrestrial radiation in the waveband from 8 to 14 μm. Most soils and vegetation have a black body emissivity between 0.90 to 0.99, and temperature measurements can be within ±0.5 C. The sensor is a thermistor bolometer detector which compares the intensity of radiation from a target and an internally controlled blackbody cavity. The sensor integrates all points in its view. The depth of integration is determined by the degree of exposure of lower leaves within the canopy.

C. Ambient Humidity/Water Vapor Pressure

The water vapor pressure profile during the day is similar to temperature and decreases above the canopy. Within the canopy, the vapor pressure increases up to a certain zone of activity, below which it may stay constant or change slightly. The profile within the canopy would be, to some extent, dependent on canopy density, structure, and soil moisture conditions. Ambient water vapor pressure is one of the more difficult variables to measure.

There are several techniques that are used in microclimate analysis. The principles by which water vapor or humidity are measured are different, and each technique may be suitable for a particular type of research and data interpretation. Three of these techniques will be discussed here. Some of the factors that influence the selection of a particular sensor include accuracy, cost of equipment and frequency of sampling.

1. Psychrometer. A psychrometer, one of the most commonly used sensors in micrometeorology, has two temperature sensors that measure the water vapor pressure in the atmosphere. One sensor, called the dry-bulb, measures the air temperature. A second sensor, called the wet-bulb, is covered with a water-saturated wick and measures the depressed temperature. The wet-bulb is aspirated and water from the saturated wick evaporates in proportion to the ambient relative humidity. Evaporative cooling of the wick lowers the wet-bulb temperature. The actual vapor pressure (e_a) of the atmosphere is given by

$$e_a = e_s - AP(T_d - T_w) \tag{1}$$

where

T_d and T_w = the dry- and wet-bulb temperatures (C), respectively
P = the atmospheric pressure (KPa)
e_s = the saturated water vapor pressure at T_w
A = the psychrometric constant dependent on T_w and given as

$$A = 6.6 \times 10^{-4} (1 + 1.15 \times 10^{-3} T_w) \qquad (2)$$

The psychrometer is operational over a range of 5 to 100 percent relative humidity. Any type of temperature sensor may be used, provided it is small and has the necessary resolution.

Rosenberg and Brown (1974) designed a psychrometer with thermocouples as the temperature sensors for use in crop micrometeorology studies. Doraiswamy (1977) used diodes as temperature sensors for psychrometers in micrometeorological studies in forestry. The basic design of the psychrometer requires that the sensors be housed in a solar radiation shield with an aspiration of at least 3 m s^{-1} for sufficient wet-bulb depression. The wet-bulb should be kept clean and saturated with distilled water at all times. Gay and Holbo (1974) discuss the problems in psychrometry instrumentation and its calibration.

2. Dew Point Hygrometer. This instrument operates on the principle of cooling the ambient air to determine the dew point temperature. A mirror-like clear surface in the pathway of a sample airflow is cooled until the dew point temperature is reached. The dew point temperature is determined as the temperature at which a film of moisture is deposited on the mirror surface. An optical arrangement detects the moisture on the surface and the temperature of the cooled mirror is measured. This procedure has been automated by the use of thermoelectric cooling and measurement of mirror temperature as described by Francisco and Beaubien (1963).

An automated dew point hygrometer is expensive and can provide dew point temperature with an accuracy of 0.5° over a range of -45° to 60 C. The response time of the instrument is 2 C s^{-1} when the sample air flow is less than 42 x $10^{-6} m^3 s^{-1}$. The temperature may be recorded in mV for automated measurements. The instrument can be used for measuring water vapor pressure profiles and requires relatively little maintenance. Air samples from various levels above and within the canopy are pumped into the laboratory, and the dew point temperature is measured sequentially by a solenoid switching system. A time constant may desirable for time averaging and can be achieved by mixing the sample air in a chamber before it reaches the hygrometer. One of the major precautions to be taken with this sampling procedure is that there should be no condensation in the sampling tube. Copper tubing is generally used and is placed in a heated jacket to prevent condensation.

3. Electrical Absorption. The effect of moisture content in altering the electrical properties of certain material is exploited in this type of humidity sensor. There are several instruments of this type that would be suitable for one's specific requirements. The Dunmore (1942) hygrometer is a polyethylene rod, wound by a 0.1 mm bifilar of palladium wire. The element is coated with a binder solution of partially hydrolized polyvinyl acetate and a dilute solution of lithium chloride. The resistance of the bifilar winding is a function of the winding geometry and the concentration of the lithium chloride. Several sensors are available for short ranges from 6 to 99 percent relative humidity. The instrument accuracy is about ±1.4 percent relative humidity. The sensing element is inoperable when exposed to 100 percent relative humidity.

IV. RADIATION

The spectral range of solar radiation (shortwave) between 0.35 to 2.5 μm covers most of the waveband that is of significance to plant and animal life. There are numerous sensors that are available for the measurement of shortwave radiation and exclusively visible radiation (0.4 to 0.7 μm), which is known to be in the range of photosynthetically active radiation (PAR). The World Meteorological Organization (1971) and Coulson (1975) present a thorough description of a standard radiation instrument. Some of the desirable characteristics of radiation instruments are: (a) a linear response over a wide range of irradiances; (b) a time constant small enough to follow transient fluctuations in a canopy; (c) a small thermal coefficient of response; (d) an acceptable cosine response; and (e) long-term stability.

The Eppley pyranometer is rated to be one of the better radiation instruments for the measurement of shortwave radiation and is available in several models to suit one's requirements. The sensors have a thermopile with 10 to 15 junctions. The sensitivity of the instrument can be up to 11 μV $W^{-1}m^{2}$ with an excellent response time. The instrument calibration is very stable and is used as a standard for calibration of other radiation instruments. Pyranometers can also be mounted in an inverted position at about 2 m above the crop canopy to measure the reflected shortwave radiation.

Another essential radiation measurement in microclimatology is net radiation. Rosenberg (1974) defines net radiation as the difference between the upward and downward radiation fluxes and also a measure of the energy available at the ground surface. Net radiation includes a balance of both shortwave radiation and longwave radiation. Longwave radiation ranges from 2.5 μm (near infrared) to about 14 μm (thermal infrared) and beyond. Net radiation (R_n) is given by the following equation.

$$R_n = (1-\alpha)\ S_t + L_d + L_u \tag{3}$$

where

α = the shortwave reflectivity

S_t = the total incident, direct, and indirect shortwave radiation

L_d = the incoming longwave radiation

L_u = the terrestrial longwave radiation leaving the earth's surface.

The energy balance analysis of a crop consists of partitioning net radiation into three major energy dissipative processes. These processes are the fluxes of sensible heat, water vapor, and soil heat. The energy used in the photosynthetic and other metabolic processes is considered negligible. The magnitude of net radiation is measured with a net radiometer placed at about 1 to 2 m above the crop with adequate field of view. As with all radiation instruments, leveling of the net radiometer is critical and should be checked periodically.

A net radiometer ideally absorbs radiation of all wavelengths incident to it and activates close to a blackbody. The sensing element is a differential thermocouple located between the top and bottom surfaces of the sensor plate. The measured temperature difference of the top and bottom surfaces of the radiometer are linearly related to the amount of net absorbed energy by the crop canopy. Two models of the Funk designed radiometer (1959 and 1962) have been used as standard radiometers. The miniature version is 14 mm in diameter and has a 45-junction thermopile with a sensitivity of 8 $\mu V\ W^{-1}\ m^2$. The Fritschen net radiometer (1965) is a 22-junction, temperature-compensated thermopile sensor. The instrument is 5 cm in diameter and has a sensitivity of 5 $\mu V\ W^{-1}\ m^2$.

Measurement of radiation within the canopy is a complex problem whether it is at several successive levels or at one level. Gates (1965) presents a classical description of the optical properties of a plant leaf. About 75 percent of the visible radiation is absorbed and 15 percent is reflected. Near IR reflectance and transmission are much greater than in the visible radiation. Attenuation of visible radiation in a plant canopy is complicated by factors such as geometric problems of architecture and natural orientation of leaves. Nonetheless, the approximate amount of visible radiation penetration into a plant canopy has been described by Monsi and Saeki's (1953) adaptation of Beer's law:

$$I/I_o = e^{-kF} \tag{4}$$

where

I and I_o = radiation at the crop surface and at any given level within the canopy, respectively.

k = an extinction coefficient

F = the cumulative leaf-area index from the upper canopy surface.

Lemeur and Blad (1975) review light penetration models that are based on crop geometry and statistical approaches for the arrangement of plant elements in the canopy. For a thorough review of this subject, the reader is referred to their article.

The problem in the measurement of shortwave in the crop canopy lies in the need for adequate sampling of sunflecks from direct solar radiation, superimposed on a background of diffuse and scattered radiation. Norman et al. (1969) and Norman and Tanner (1969) discuss some of the problems in measuring transient light within a crop canopy. To assess the photosynthetic activity and energy balance of the lower leaves of the canopy, shortwave and net radiation measurements must be made within the canopy.

The PAR above and below the crop canopy can be measured with photosensitive cells. Silicon photovoltaic cells have been used by Norman et al. (1969) and Biggs et al. (1971). Federer and Tanner (1966) discuss the use of selenium photovoltaic cells for measurement of PAR. These sensors are small in size (~ 18 mm diameter), relatively inexpensive, and suitable for PAR range. The silicon cell radiometer is recommended because it has several advantages over the selenium cell. Selenium cells are insensitive to low light levels and have a poor spectral response (Norman et al., 1969a).

The transmitted radiation within a crop canopy having a leaf area index of 1 to 3 fluctuates frequently from low and high flux density. Averaging these instantaneous values from stationary points could be misleading and not representative. Spatial and temporal measurements provide a more adequate solution to the problem of representative mean conditions. Although a great deal of detailed information is lost in taking average measurements spatially for certain types of applications, as in a daily analysis of crop energy balance, this level of average measurements would be sufficient.

The horizontal variability inside the crop canopy may be summarized by use of linear radiation probes placed across or along the rows of the crop. Gaastra (1968) considers, among others, the use of phototubes and thermopiles inside the canopy. Several disadvantages in these sensors include specific power requirements, large sensor size, small output signals, and lack of proper spectral response. Muchow and Kerven (1977) used a linear probe incorporating nine selenium cells to measure PAR in crop canopies. Norman and Tanner (1969) used a power-driven telescoping boom to move the sensor and obtained a more representative sample of the PAR. Similar temporal and spatial integration can be made for measurement of net radiation within a crop canopy.

V. WIND SPEED

The intensity of wind speeds in the boundary layer of a crop dictates the turbulent transfer rates of water vapor, sensible heat, and carbon dioxide from the crop surface. The development of a crop canopy boundary layer when wind moves from a short to a tall crop is discussed by Lemon (1960). The mechanical turbulence in the boundary layer is dependent on the wind speed as well as the roughness of the crop surface. Measurements of the wind speed profile are necessary to evaluate the turbulence and boundary layer characteristics for a given crop. Wind speed within the canopy also plays an important role in the canopy microenvironment. Generally, the wind speed is attenuated below the crop surface. Businger (1975) modeled the attenuation of wind speeds within crop and forest canopies.

An additional atmospheric parameter that may be of relevance in microclimate studies is the stability conditions of the atmosphere. The Richardson number (Richardson, 1929) derived from the temperature and wind speed gradients may be used to determine the lapse or inversion condition of the atmosphere. Atmospheric stability can be instrumental in dispersion of pathogenic spores and insect flight behavior on a daily or seasonal scale. Fares et al. (1980) studied the pheromone diffusion rates in a forest ecosystem. The atmospheric stability and turbulent diffusion within the canopy were considered to be two of the major factors influencing pheromone dispersion. Vité et al. (1964) observed that insect behavioral patterns such as aggregation, flight, and infestation were regulated by pheromones. Insect behavior was found to be directly or indirectly related to the existing microclimatic conditions. These studies demonstrated the need for wind speed and turbulence measurements both above and within the crop canopy.

Instruments to measure wind speed under field conditions should be rugged and sensitive over a wide range of wind velocities. Wind sensors used in microclimate research have to be more sensitive with a low starting speed compared to sensors used in standard meteorological stations. The desirable characteristics of wind sensors suggested by the National Weather Service of the USDA, National Oceanic and Atmospheric Administration [(NOAA) Specification No. F460-SP001, October 5, 1970] are that rotating cup anemometers have a starting speed of 270 mm s^{-1} and a distance constant of 1.5 m. The output of the anemometer should be a linear function of the wind speed and operate in wind speeds up to 22 m s^{-1}.

A. Cup Anemometer

The sensor usually has three cups attached to a vertical rotating shaft. The force exerted by the wind on the cups rotates the shaft. The speed of rotation may be determined by a small generator, a magnetic switch pulser, or a photoelectric capacitance chopper mounted at the bottom of the shaft. The wind speed may be calibrated against an output which may be a variation in resistance, a generated voltage, or an integrated pulse count. Commercially available cup anemometers have a starting threshold speed from 9 cm s^{-1} to 22.5 cm s^{-1}. Detailed descriptions of cup anemometers are discussed by Middleton and Spilhaus (1953) and Fritschen (1967). Propeller anemometers described by Gill (1975) are very sensitive to low wind speeds. The sensor has four blades and the capability of being mounted vertically and horizontally to obtain the three-vector components of the wind speed.

B. Sonic Anemometer

The variation in the speed of sound with wind speed is the principle by which this sensor operates. The instrument consists of a transmitter and a receiver, mounted on an assembly maintaining a fixed distance. A phase meter measures the time lapse of sound from the transmitter to the receiver. Wind speed in the x, y, and z directions can be measured with a resolution of 30 mm s^{-1}. Design and operating principles of the sonic anemometer have been presented by Kaimal et al. (1974).

C. Thermal Anemometer

The principle of thermal anemometry is based on the measurement of the rate of convective heat loss from a heated sensor element to the surrounding air. The rate of heat loss from an element is dependent on many factors such as temperature, geometric shape and size of sensor, air movement, pressure, and thermal properties. If all factors affecting the heat loss are maintained constant and only the air movement is altered, then the heat loss may be directly related to the air velocity. The sensing element is a hot-wire (platinum) about 0.013 to 0.13 mm in size, operating in a temperature range of 200 to 500 C. This instrument is very sensitive and is useful for measuring wind speeds from a few mm s^{-1} to supersonic speeds. The sensor requires continuous care as it can be easily contaminated in field conditions. Fritschen and Gay (1979) discuss the theory and operation of the hot-wire anemometer and other types of thermal anemometers.

VI. WIND DIRECTION

Wind direction is a useful parameter in biological studies, such as relating to factors influencing insect movement and spread of infestation. Pheromone diffusion and dispersion of pathogenic diseases by air movement in a given direction may be related to wind speeds as well as direction. Measurement of wind direction may be necessary both within and above the canopy. There are many types of wind direction sensors commercially available. The major differences are in the method of recording and the shape of the wind vane. The shape of the fin determines the stability of the vane. A stable vane should not overshoot or dampen fluctuations in wind direction. Wieringa (1967) discusses the design criteria for a stable wind direction sensor.

VII. CONCLUSION

The instrumentation and techniques for assessment of crop microclimate discussed in this presentation would be applicable in pest management research. Certain references have been made to relate the microclimatic factors influencing the pest habitat conditions and its life cycle. The methodology and sensor requirements would have to be designed to the particular pest of interest. It would be desirable to obtain the measurement of selected parameters that relate to soil and crop conditions in order to establish a cause-and-effect relationship. The measurement and instrumentation requirements in pest management research may be summarized in two categories: (1) measurement of appropriate parameters that describe or define the pest life cycle in its natural habitat; (2) selection of instruments appropriately matched and properly installed to provide the parameter of interest at the desirable level of accuracy.

REFERENCES

Baker, D. G. (1965). Minn. Farm Home Sci. 22:11-13.

Bartholic, J. F., Namken, L. N., and Weigand, C. L. (1972). Agron. J. 64:603-608.

Biggs, W. W., Edison, A. R., Eastin, J. D., Brown, K. W., Maranville, J. W., and Clegg, M. D. (1971). Ecology 52:125-131.

Black, T. H., and McNaughton, K. G. (1971). Bound. Layer Meteor. 2:246-254.

Bouyoucos, G. J., and Mick, A. H. (1940). Mich. Agric. Exp. Sta. Tech. Bull. 172 p.

Businger, J. A. (1975). In "Heat and Mass Transfer in the Biosphere" (D. A. de Vries and N. H. Afgan, eds.), pp. 139-165. Wiley and Sons, New York.

Coulson, K. L. (1975). "Solar and Terresterial Radiation." Academic Press, New York.

Doraiswamy, P. C. (1977). The Radiation Budget and Evapotranspiration of a Douglas-Fir Stand. Ph.D. Thesis, University of Washington, Seattle, Washington.

Dunmore, F. W. (1942). Humidity Resistance Film Hygrometer. U.S. Patent 2,285,421.

Fares, Y., Sharpe, P. J., and Magnuson, C. E. (1980). Proc. of Symp. on "Modeling Southern Pine Beetle Populations" (F. M. Stephen, J. L. Searcy, and G. D. Hertel, eds.) USDA Forest Service Tech. Bull. 1630, February 20-22, Asheville, North Carolina. pp. 75-93.

Federer, C. A., and Tanner, C. D. (1966). Ecology 47:654-657.

Francisco, C., and Beaubien, D. J. (1963). In "Humidity and Moisture," Reinhold, New York. 1:165-173.

Fritschen, L. J. (1965). J. Appl. Meteor. 4:528-532.
Fritschen, L. J. (1967). J. Appl. Meteor. 6:695-698.
Fritschen, L. J., and Gay, L. W. (1979). "Environmental Instrumentation." Springer-Verlag, New York. 216 p.
Fuchs, M., and Tanner, C. B. (1966). Agron. J. 58:597-601.
Fuchs, M., and Tanner, C. B. (1968a). Soil Sci. Soc. Am. Proc. 32:326-328.
Fuchs, M., and Tanner, C. B. (1968b). J. Appl. Meteor. 7: 303-305.
Funk, J. P. (1959). J. Sci. Instrument. 36:267-270.
Funk, J. P. (1962). J. Geophys. Res. 67:2753-2710.
Gaastra, P. (1968). In "Functioning of Terrestrial Ecosystems at the Primary Production Level." UNESCO, Natural Resources Res. 5:467-478.
Gardner, W. H. (1965). In "Methods of Soil Analysis," (C. A. Black, D. D. Evans, J. L. White, L. E. Ensminger, and F. E. Clark, eds.), ASA Monograph No. 9, Madison, Wisconsin.
Gates, D. M. (1965). In "Meteor. Monographs," (P. E. Waggoner, ed.), 6(28):1-26. Am. Meteor. Soc., Boston.
Gay, L. W., and Holbo, H. R. (1974). Studies of the Forest Energy Balance. Water Resources Res. Inst., Oregon State University, Corvallis, Oregon, WRRI-24.
Gill, G. C. (1975). Bound. Layer Meteor. 8:475-495.
Idso, S. B., Jackson, R. D., and Reginato, R.J. (1977). Science 196:19-25.
Kaimal, J. C., Neman, J. R., Bisberg, A., and Cole K. (1974). In "Flow: Its Measurement and Control in Science and Industry." Instrument Soc. Am. 1:349-359.
Lemeur, R., and Blad, B. L. (1975). Agric. Meteor. 12:229-247.
Lemon, E. R. (1960). Agron. J. 52:697-703.
Lutz, H. J. (1944). Soil Sci. 57:475-487.
Middleton, W. E. K., and Sphilhaus, A. F. (1953). "Meteorological Instrument." University Toronto Press, Toronto, Canada. 286 p.
Monsi, M., and Saeki, T. (1953). Japan J. Bot. 14:22-52.
Muchow, R. C., and Kerven, G. L. (1977). Agric. Meteor. 18: 187-195.
Norman, J. M., Tanner, C. B., and Thurtell, G. W. (1969). Agron. J. 61:847-843.
Norman, J. M., and Tanner, C. B. (1969). Agron. J. 61:847-849.
Richardson, L. F. (1929). Proc. Roy. Soc. Am. 97:354-373.
Rosenberg, N. J. (1974). "Microclimate: The Biological Environment." John Wiley & Sons, New York. 315 p.
Rosenberg, N. J. and Brown, K. W. (1974). Agric. Meteor. 13:215-226.
Slatyer, R. O., and McIlroy, I. C. (1961). Practical Micrometeorology. UNESCO, Paris.
Tanner, C. B. (1963). "Basic Instrumentation and Measurements for Plant Environment and Micrometeorology." Dept. Soil Bull. 6., University of Wisconsin, Madison, Wisconsin.

van Bavel, C. H. M., Nixon, P. R., and Hauser, V. L. (1963). "Soil Moisture Measurement with the Neutron Method." USDA-ARS-41-70.

Vité, J. P., Gara, R. I. and Von Scheller, H. D. (1964). "Field Observations on the Response to Attractants of Bark Beetles Infesting Southern Pines." Contrib. Boyce Thompson Inst. 22:461-470.

Wellington, W. G. (1949). Sci. Agric. 29:201.

Wieringa, J. (1967). J. Appl. Meteor. 6:1114-1122.

World Meteorological Organization. (1971). "Guide to Meteorological Instrument and Observing Practice." World Meteor. Organization, Geneva, Switzerland.

SIMULATION OF MICROCLIMATES

John M. Norman
Department of Agronomy
University of Nebraska
Lincoln, Nebraska

I. INTRODUCTION

The dynamic relationship between a single living organism and its environment is so complex that researchers rarely attempt to consider more than a few of the important factors. For example, the apparent dependence of stomatal conductance on leaf environment is exceedingly complex (Rashke, 1979), but we often represent it as dependent on light and possibly some function that is related to leaf-water status or vapor-pressure deficit (Thorpe et al., 1980). These simple descriptions can be very useful if the most important factors are defined clearly; when two living organisms such as a host plant and pest define the system, it becomes difficult to determine the most important factors. The simplest procedure is to study the two organisms separately under isolated conditions, select the most important factors for each independently, and then apply the results when the two organisms occur together. Sometimes this approach works, but if it fails it usually means that the importance of an interaction between factors outweighs the importance of the individual factors. When this occurs it becomes necessary to consider an integrated approach by including as many factors as possible, simultaneously. Using an integrated approach to pest mangement requires a study of the interactions between the physiology and biochemistry of the

ISBN 0-12-332850-0

plant and the pest, as well as the interactions between both organisms and the microenvironment. In this chapter we shall limit our discussion to the interaction between plants and their atmospheric and soil environments. Because the plants can profoundly affect the canopy microclimate, and vice versa, it becomes essential to use an integrated approach that considers most of the important plant and environmental variables simultaneously.

Many organisms only alter their microenvironment indirectly through changes in the plant (of course there are notable exceptions, such as spider mites that alter their immediate environment through the production of webbing). Therefore, with a complete understanding of the interactions between plants and their microenvironments, predictions of pest development may be possible using data directly from isolated, controlled plant-pest experiments. A good example of controlled environment data that can be used in the field with the appropriate microenvironment model is presented by Weiss et al. (1980). The infection of dry edible beans (*Phaseolus vulgaris* L.) by white mold disease (*Sclerotinia sclerotiorum*) is dependent on leaf temperature and leaf wetness duration. Higher leaf temperatures permit white mold to infect a greater percentage of leaves at a given leaf wetness duration (Fig. 1). Clearly, a model that can predict leaf temperatures and leaf wetness duration, as a function of easily monitored ambient conditions above the canopy, can be used to predict the incidence of white mold disease. Unfortunately, leaf wetness is one of the single most difficult microenvironmental factors to predict and to measure. A second interesting example relates humidity and temperature from controlled environment experiments to two-spotted spider mite egg hatch (Fig. 2; Ferro and Chapman, 1979). In this experiment, the percent of eggs hatched and the hours to 50% hatch are both affected by temperature and saturation vapor pressure deficit. Clearly this data could be used directly with a plant-environment model that predicts canopy temperature and vapor pressure reliably.

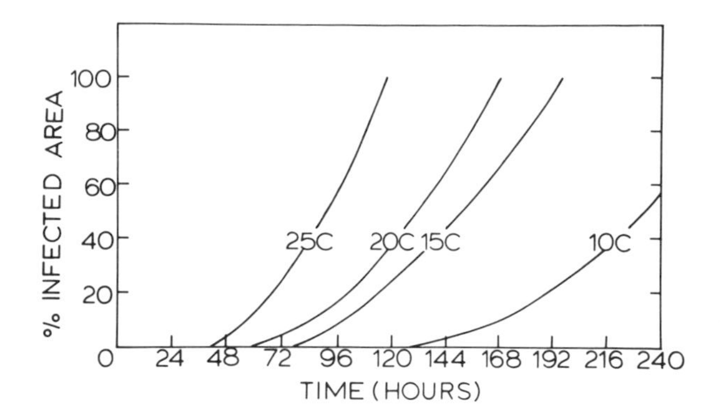

Fig. 1. Percentage of leaf area of dry edible beans affected by white mold as a function of continuous exposure time after inoculation, and temperature at 100% relative humidity (from Weiss et al., 1980).

The microenvironmental variables (from the viewpoint of the pest organism) that may be important to the success or failure of a particular pest can be very specific. For example, it may be important to know the temperature of a large organ such as an apple or a cocoa pod (Monteith and Butler, 1979) or the wetness and temperature within the leaf sheath of corn. The physiological state of the plant may also be important. For example, we know that spider mite populations on corn can reach critical levels more quickly if the corn is under water stress. We also know that sugars and some free amino acids can accumulate in the leaves of stressed plants; whether these coincident events are directly related remains to be proven.

In this chapter, we will consider only some of the more general microenvironmental variables such as air temperature, leaf temperature, air relative humidity, mean wind speed, soil temperature, soil moisture content, radiation in several wavelength bands, and condensation on leaves. The approach is sufficiently basic so that other specific factors can be included when necessary. These factors will be discussed in the context of a comprehensive plant-environment model that is called Cupid.

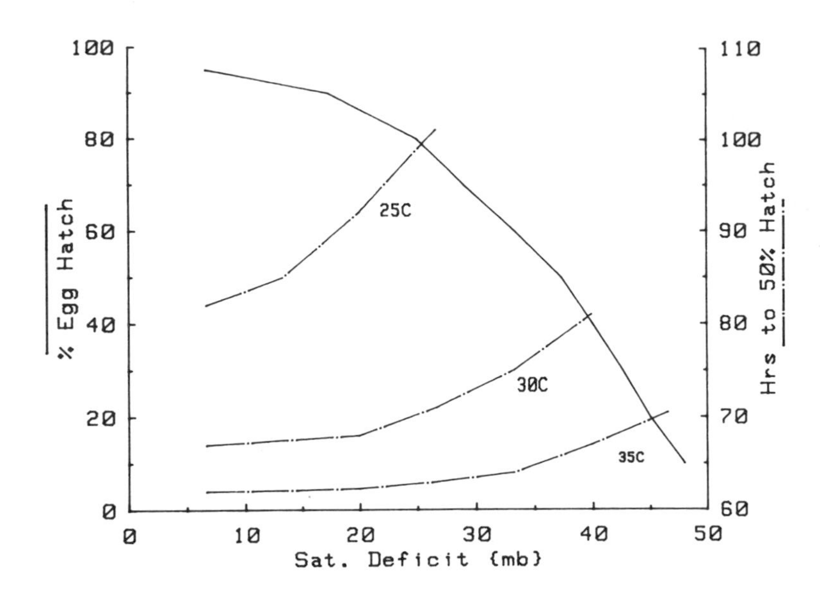

Fig. 2. Percentage of two-spotted spider mite eggs hatched as a function of temperature (average of 25°C to 35°) and vapor pressure deficit for six-hour-old eggs (solid line). Also the time to 50% hatch of all the eggs that hatched is included (broken line). (From Ferro and Chapman, 1979.)

Before we consider this model in detail, it is worth considering how such a comprehensive plant-environment model might fit into the broad, overall view of integrated pest management. The dispersal of many pests on a regional scale can best be described with boundary layer, regional, and synoptic scale meteorological models. The potential success of a particular pest, after it takes up residency in a location, might be described with the comprehensive-environment model. The lower boundary conditions for boundary layer and regional scale meteorological models represent the upper boundary conditions for plant-environment models; therefore, plant-environment models should be compatible with meteorological models to permit their joining in the future. The implementation of a hierachy of meteorological and plant-environment models first requires an objective method for handling the input data required by both kinds of models. Such objective analysis tools for horizontal fields of weather or plant data are well developed in meteorology (Cahir et al., 1981). Figure 3 contains a diagram of how the comprehensive plant-environment model, which we describe in this chapter, might fit into a large scale analysis appropriate for integrated pest management.

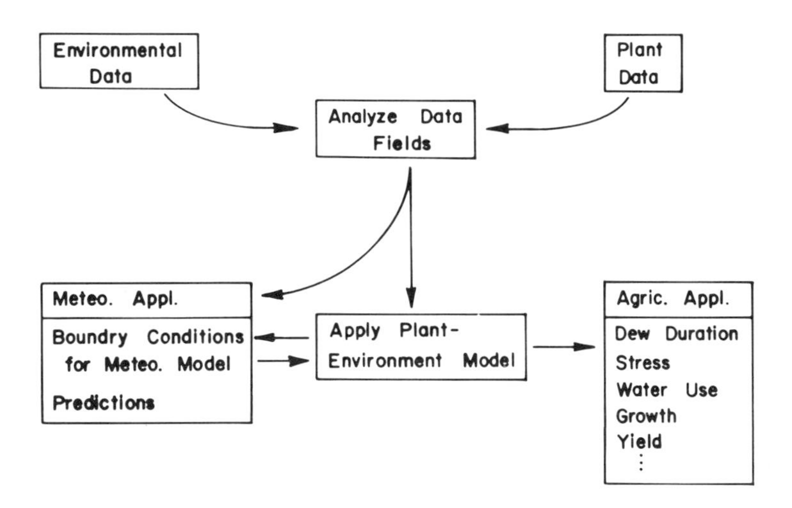

Fig. 3. Schematic diagram illustrating how a comprehensive plant-environment model might fit into a full scale analysis appropriate for integrated pest management.

II. DESCRIPTION OF COMPREHENSIVE PLANT-ENVIRONMENT MODEL

A. Introduction and Model Input Requirements Above the Canopy

The comprehensive plant-environment model, Cupid, is similar to several other comprehensive models described in the literature (Stewart and Lemon, 1969; Goudriaan, 1977). Cupid is designed to include many environmental and plant factors; however, the structure of the model is subject to a number of very rigid constraints: 1) It should be independent of plant type; 2) it should require minimum input information; 3) formulations should be physically based and not use high level empiricisms; and 4) it should be affordable to run in terms of computer time. The first condition of wide applicability is not too difficult to meet; however, the second condition of minimum input information is quite difficult. The minimum environmental input information that Cupid requires is the following: Daily integrated solar radiation, maximum and minimum air temperature, average daily water vapor pressure, total daily wind run, precipitation, a single soil temperature at 50-cm depth, and a soil moisture content at some reference depth. Cupid actually runs in half-hourly or hourly time steps so a series of subroutines have been developed to convert daily inputs into hourly inputs (Fig. 4). With the increasingly wide spread use of low cost remote weather stations, hourly data is becoming routinely available. Even with hourly input data, however, there is a minor problem. Hourly global solar radiation is not sufficient as a direct input to Cupid. There are three main reasons for this: 1) Critical physiological processes in plants are driven by photosynthetically active radiation so use of solar radiation causes unnecessary uncertainties; 2) the visible part of the solar spectrum is largely absorbed while the near-infrared part is largely scattered so that treating the solar spectrum as one wavelength band using averaged leaf spectral properties can lead to sizable errors in radiation penetration estimates; 3) direct beam and sky diffuse radiation are intercepted very differently by canopies so that they should be available separately. Thus, hourly global solar radiation is divided into hourly beam and diffuse, visible and near-infrared components. This is accomplished by first calculating the potential radiation on the horizontal plane at the given time for visible direct

$$R_{D,V} = [600 \exp(-.16\ m)] \cos\theta \tag{1}$$

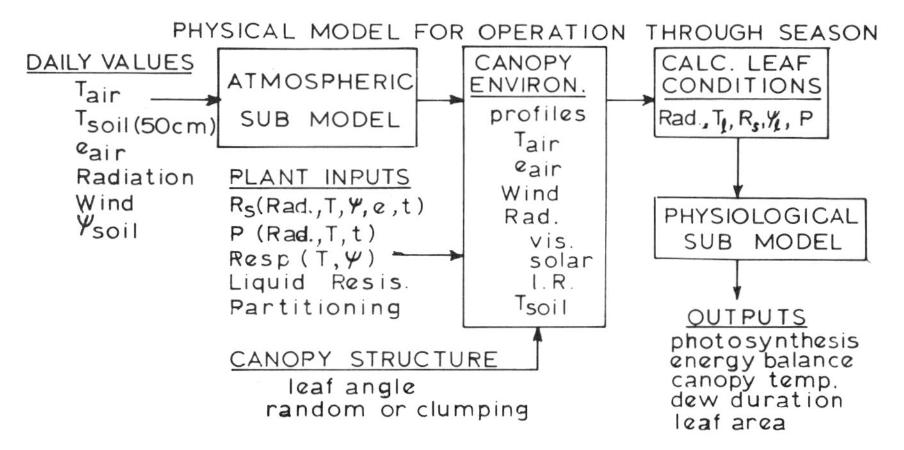

Fig. 4. Block diagram of the model Cupid indicating the flow of information from plant and environmental inputs through canopy environment calculations to specific outputs.

visible diffuse

$$R_{d,V} = 0.4\ (600 - R_{D,V}) \cos\theta \qquad (2)$$

near-infrared direct

$$R_{D,N} = [720 \exp(-0.05\ m) - W] \cos\theta \qquad (3)$$

and near-infrared diffuse

$$R_{d,N}\ \ 0.54\ (720 - W - R_{D,N}) \cos\theta \qquad (4)$$

radiation, where θ is the solar zenith angle calculated from local time, longitude, standard longitude, and solar declination (List, 1966). The water absorption in the near infrared for 1 cm of precipitable water (W) is calculated from Wang (1976)

$$W = 1320\ \mathrm{Antilog}_{10}[-1.1950 + 0.4459 \log_{10} m\ 0.0345\ (\log_{10} m)^2] \qquad (5)$$

where m is the air mass given by

$$m = [(\cos^2 \theta + 0.0025)^{\frac{1}{2}} - \cos \theta]/0.00125 \tag{6}$$

considering curvature of the atmosphere (Kondratyev, 1969) and $m = (\cos \theta)^{-1}$ for a "flat" approximation to the atmosphere. For an actual solar radiation measurement, R*, above the canopy, the estimated total visible radiation on the horizontal is

$$S_V = R^* \left[\frac{R_V}{R_V + R_N} \right] \tag{7}$$

and estimated total near-infrared radiation on the horizontal plane is

$$S_N = R^* \left[\frac{R_N}{R_V + R_N} \right] \tag{8}$$

where $R_V = R_{D,V} + R_{d,V}$ and $R_N = R_{D,N} + R_{d,N}$. The fraction of the visible radiation that is direct beam is

$$f_V = \frac{R_{D,V}}{R_V} - 1.5 \left[0.92 - \frac{R^*}{(R_V + R_N)} \right] \tag{9}$$

and the fraction of near-infrared radiation as direct beam is

$$f_N = \frac{R_{D,N}}{R_N} - 1.2 \left[0.90 - \frac{R^*}{(R_V + R_N)} \right] \tag{10}$$

The fraction of beam radiation defined by Eqs. (9) and (10) has a minimum value of zero, and this is reached when the actual solar radiation is less than about 30% of the potential solar radiation. From data collected in Western Nebraska, 46% of the averaged hourly solar radiation is in the visible and this percentage is essentially independent of zenith angle and cloudiness. Thus estimates of total visible and near-infrared radiation from a measurement of solar radiation were always within ± 15 W m^{-2} of measured visible and near-infrared values and usually within ± 10 W m^{-2}. The beam radiation flux estimates contain larger uncertainties (Fig. 5).

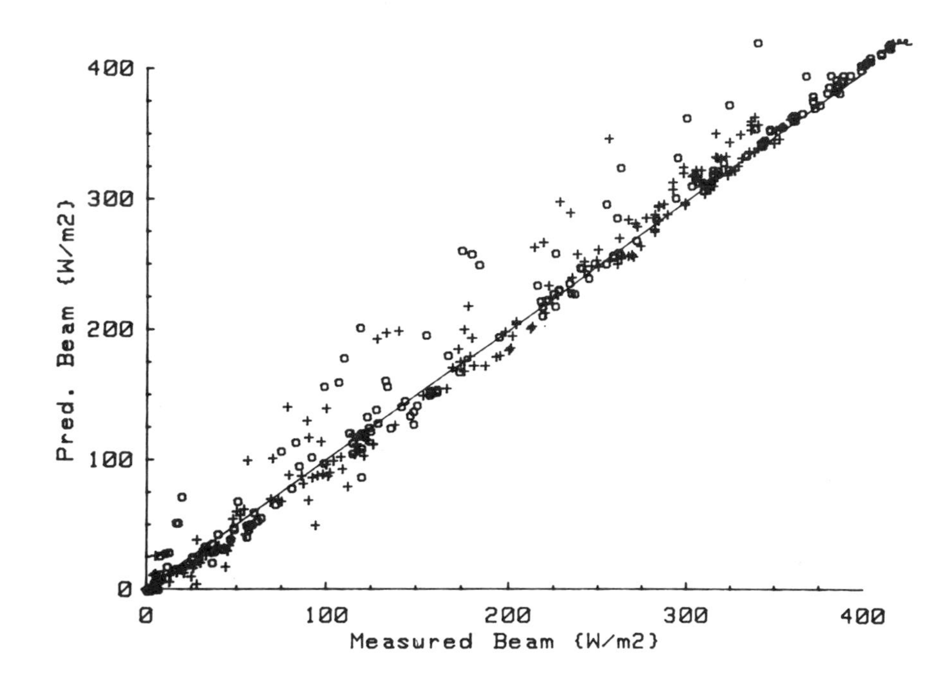

Fig. 5. A comparison of measurements with predictions of hourly-averaged direct beam radiation (on horizontal plane) for visible (+) and near-infrared (o) wavelengths. The predictions are derived from hourly global solar radiation over a wide range of sun angles on clear, partly cloudy and overcast days.

The most desirable radiation input to the model Cupid is derived from direct and diffuse solar radiation measurements. Ideally, direct and diffuse visible and near-infrared radiation measurements would be best, but this is impractical from a measurement standpoint because of data volume and cost. Low cost silicon cell pyranometers and matching shadow bands are available (for example from Li-cor, Inc. Lincoln, Nebraska) for this use and sufficiently accurate if the underestimated clear-sky-diffuse reading is corrected. Diffuse solar radiation estimates accurate within ± 10 W m^{-2} can be obtained by multiplying the following correction factor times the diffuse radiation fraction (f) measured by two silicon cell pyranometers; one under a shadow band and one fully exposed for global radiation:

$$\frac{\text{corrected}}{\text{uncorrected}} = \left[1.17 - \frac{1}{1.2 + 11.8\ f}\right]^{-1} \qquad (11)$$

Of course the correction in Eq. (11) must be applied after correcting for the shading effect of the shadow band.

Given the appropriate input information, profiles of radiation are computed from radiative transfer equations; profiles of temperature, humidity and wind speed are derived from convection formulations, and soil temperature profiles are evaluated from conduction theories (Fig. 4).

B. Canopy Structure

The structure of the canopy details the distribution of foliage and stem area with respect to spatial position and angular orientation. For canopies of full cover, the assumptions of random leaf positioning and symmetry about the azimuth are excellent so that only the leaf inclination angle distribution and leaf area distribution with height are necessary. Leaf area index profiles can be measured or approximated (Norman, 1979). In the absence of detailed information, the spherical leaf angle distribution (deWit, 1965) is usually the best choice. In fact, there is some doubt about whether direct measurements of leaf angle distributions are sufficiently accurate to be worth considering seriously. The problems of leaf movement due to water stress, heliotropism, wind, and measurement inaccuracies could be most serious in such angle distribution measurements. Since leaf-angle distributions are most relevant to radiation interception, it seems most reasonable to try to estimate such distributions from radiation information. Integral inversion techniques have been used successfully to obtain <u>both</u> leaf area index and several classes in a leaf angle distribution (Fig. 6) from measurements as simple as the fraction of the length of a meter stick that is in sunflecks (Norman et al., 1979). In fact, there is good evidence that the mean angle between sunlit leaves and the direction of the sun is sufficient to provide good integrated canopy estimates of even the most nonlinear light dependent physiological processes (Sinclair et al., 1976; Norman, 1980), provided that sunlit and shaded leaves are treated separately.

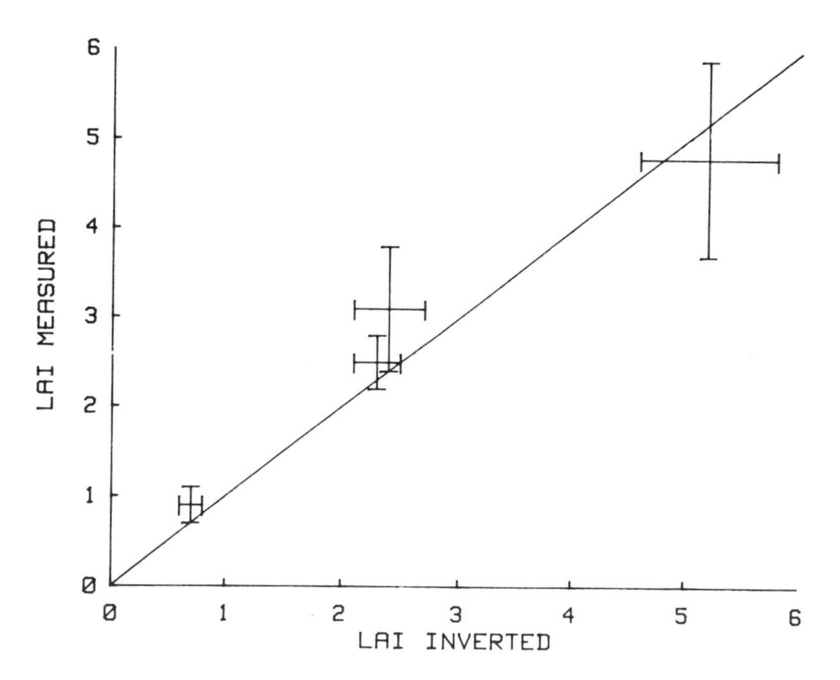

Fig. 6. A comparison of leaf-area-index measurements from leaf sampling with leaf area index estimated from the mathematical inversion of sunfleck-area-fraction measurements. The means and standard deviations are indicated for two heights in two different corn canopies. The input data for the inversion estimates was the fraction of the horizontal area in sunflecks for a range of sun angles.

If a plant canopy approximates full cover but the individual elements obviously are not randomly distributed, then modified one dimensional treatments are useful (Nilson, 1971; Norman and Jarvis, 1975).

If the plant canopy obviously does not approximate a full cover, such as widely spaced rows or individual trees, then horizontal position information must be included in the canopy description. General treatments of clumped vegetation are available for rows (Allen, 1974), for a single isolated plant (Mann et al., 1979), and for any array of individual plants (Welles, 1978; Norman and Welles, 1982).

C. Radiation Penetration in Canopies

The penetration of radiation into vegetation is well understood (Lemeur and Blad, 1974; Norman, 1975) and uncertainties rarely account for any serious limitation in the application of a comprehensive plant-environment model. The simplified radiation treatment by Norman (1979) is adequate for most Cupid applications except, perhaps, in the case of isolated plants when equations from Mann et al. (1979) or Welles (1978) are more appropriate. Because of serious limitations in our understanding of turbulent convection in heterogeneous canopies, the difference between the simpler one-dimensional equations of Norman (1979) and the complex three-dimensional equations of Welles (1978) are not nearly as serious as might be suggested from comparisons based only on the radiation models. In this section, we will consider several refinements and simplifications and refer the reader to the literature for detailed radiation penetration equations.

The spectral properties of the foliage can play a significant role in the calculation of radiation penetration, particularly in dense canopies. In general, errors in spectral properties of leaves translate directly into comparable errors in radiation absorbed by the leaves or the soil. Although most vegetation has surprisingly similar spectral properties (about 85% of the visible radiation is absorbed and 15% of the near-infrared is absorbed), the spectral properties can vary by a factor of two or three on occasion so that some care must be taken to insure a reasonable estimate of leaf reflectance and transmittance (Gausman and Allen, 1973). Most of the radiation equations for scattering that appear in the literature (deWit, 1965; Idso and deWit, 1970; Norman et al. 1971; Norman, 1979) assume that individual canopy layers are so thin that the probability of leaf overlap is negligible. This permits an estimate of reflected (or transmitted) radiation from the product of the intercepted radiation and the leaf reflectance (or transmittance). If layers are not chosen very thin (leaf area index increments less than 0.2) scattering will be overestimated. Unfortunately in canopies of large leaf area indices, thin layers mean a large number of layers with commensurate increases in computer execution time. Two methods are available for increasing the thickness of individual canopy layers without significantly degrading the accuracy of radiation penetration estimates. The layer thicknesses can be increased from a leaf area index increment of 0.1 to 1 with a factor of ten improvement in execution. The first method is to derive thick-layer reflection and transmission properties by dividing that thick layer into very thin sublayers and solving the same set of equations that were used for the whole canopy. Since this need be done only once for all layers (each of the same leaf area index increment), this saves much computation time. A second method, which is simpler, is to derive the layer transmission and reflection coefficients using higher order terms in the Poisson distribution, thus considering leaf overlap. Using this method, the layer transmission (T_L) becomes

$$T_L = \sum_{j=o}^{3} \tau^j \quad I_d \frac{(-\ln I_d)^j}{j!} \tag{12}$$

and layer reflection (R_L) becomes

$$R_L = \sum_{j=1}^{3} \rho^j \quad I_d \frac{(-\ln I_d)^j}{j!} \tag{13}$$

Where I_d is the noninterception factor for layer j for diffuse radiation (Norman, 1979; Norman and Jarvis, 1975), τ and ρ are leaf transmittance and reflectance. The sum is only carried to three because the contribution of more than three leaf overlaps is negligible even for leaf reflectances of 50% and layer thicknesses of one unit of leaf area index. Obviously this correction is most important in the near-infrared where τ and ρ are large.

Simplified theories often are sufficient for many purposes. One special purpose that has wide-spread application is the calculation of integrated canopy photosynthesis or stomatal conductance as a function of light. Since both stomatal conductance and photosynthesis depend on light in a nonlinear fashion, some care must be taken in integrating over leaves receiving differing amounts of radiation. It has been shown that good estimates of integrated canopy photosynthesis (or stomatal conductance) can be obtained by dividing the canopy into two classes of leaves; those that are sunlit and those shaded (Sinclair et al., 1976; Norman, 1980). For a canopy with a spherical leaf angle distribution, the sunlit leaf area index is given by

$$F^* = [1 - \exp(-0.5\ F/\cos\theta)]\ 2\cos\theta \tag{14}$$

and the shaded leaf area index by

$$F - F^* \tag{15}$$

where F is the total leaf area index and θ the solar zenith angle. The illumination of shaded leaves is difficult to estimate simply. However, from comparisons with the radiation model presented by Norman (1979), the average <u>visible</u> irradiation received by all shaded leaves can be approximated by

$$S_{shade} = R_{d,V} \exp(-0.5\ F^{0.7}) + C \tag{16}$$

where C arises from multiple scattering of direct beam radiation and is given by

$$C = 0.07\ R_{D,V}\ (1.1 - 0.1\ F) \exp(-\cos\theta)$$

The radiation above the canopy in the horizontal plane is given by $R_{d,V}$ for diffuse and $R_{D,V}$ for the direct beam. Conceptually, the 0.07 represents a scattering coefficient, the term in brackets accounts for the decrease in multiple scattering with depth and the last exponential term accounts for the increased scattering at high zenith angles. Equation (16) agrees to within a few watts/m^2 of the model of Norman (1979) for any sky conditions and leaf reflectances (assumed equal to leaf transmittances) from 0.03 to 0.10. Although shaded leaves near the top of the canopy receive more light than shaded leaves at the bottom, the average calculated from Eq. (16) works very well because most light-dependent physiological processes are linearly dependent on light over this irradiance range. The irradiation received by sunlit leaves can be derived from the mean leaf-sun angle (α) which is 60° (and independent of sun angle) for a canopy with a spherical leaf angle distribution and may vary from 45° to 75° for a wide range of sun angles and over many canopies (Ross, 1975). The sunlit-leaf irradiance is

$$S_{sun} = R_{D,V} \cos\alpha/\cos\theta + S_{shade} \tag{17}$$

Where $R_{D,V}/\cos\theta$ is the visible irradiance on a plane perpendicular to the direction of the sun. This simple method can produce good estimates of canopy photosynthesis or canopy stomatal conductance from leaf characteristics as a function of light. An example calculation is carried out in section II.D.2.

A second simplified relationship can be derived from the radiation model presented by Norman (1979). The average daytime net radiation below the canopy, which can be a primary factor in determining soil evaporation (Tanner and Jury, 1976), can be calculated as a function of daily solar radiation above the canopy. It is somewhat surprising how large the net radiation can be below a canopy of large leaf area index (Fig. 7).

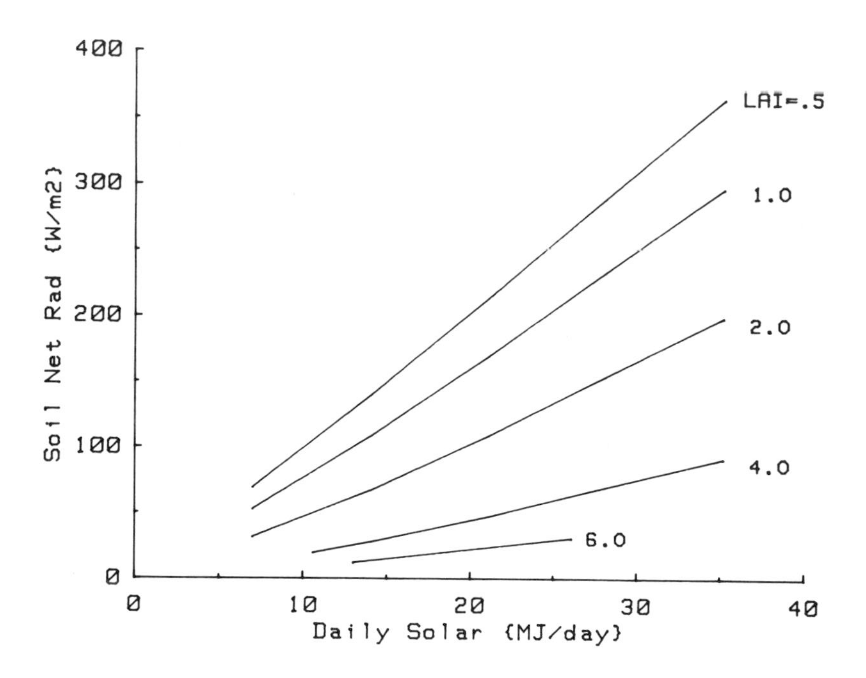

Fig. 7. Predicted net radiation immediately above the soil surface averaged over the daylight hours as a function of daily integrated global radiation above the canopy and leaf area index.

These two simplified results serve to demonstrate how a comprehensive plant-environment model can be used to derive simple relationships. Because the comprehensive plant-environment model considers nearly all of the important factors, a real fringe benefit of this approach is a clear understanding of when the derived simplifications might fail. Thus, a comprehensive plant-environment model not only contributes to our basic understanding of the plant-environment relationship, but it also facilitates the development of simple, but reliable, relations for limited applications.

D. Convection and Conduction

The present state of our understanding of turbulent transfer in plant canopies is wholly inadequate, particularly under conditions of low wind speed when free convection becomes significant. Although present formulations are very useful for canopies of full cover, they remain one dimensional approximations based on eddy transfer coefficients that are only marginally understood in homogeneous canopies and totally baffling in heterogeneous canopies (Garratt and Hicks, 1973).

The description of any canopy profile begins with a knowledge of the vertical source-sink distribution based on leaf considerations. Of course the leaf energy and mass exchanges depend on the canopy profile so that complete solutions involve either iteration (Norman, 1979), or matrices (Waggoner, 1975), or a combination of both.

1. Leaf Energy Balance. The leaf energy balance has been discussed extensively in the literature (Stewart and Lemon, 1969; Gourdriaan, 1977; Norman, 1979) and Fig. 8 contains a schematic representation. We will not deal with a complete development here, but rather deal with minor departures from the description of Norman (1979). The key part of solving the leaf energy budget is solving for the temperature difference, ΔT, between the leaf and the air that satisfies the energy exchanges occurring on that leaf. Using the linearization procedure described by Norman (1979) and evaluating constants at 25°C, yields ΔT for a particular leaf

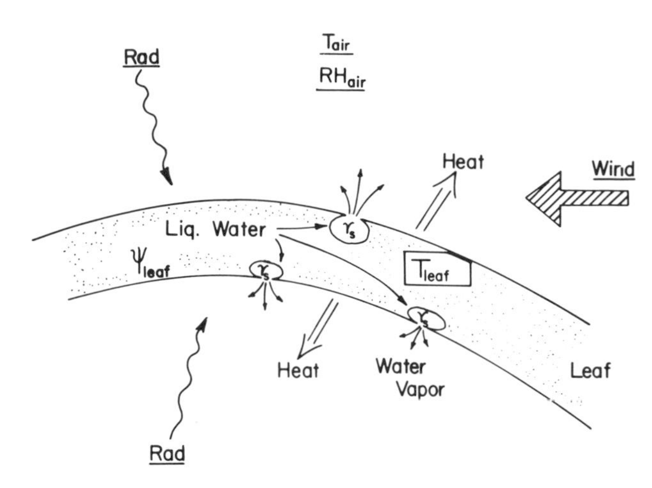

Fig. 8. Illustration of the leaf energy budget.

$$\Delta T = \frac{R'_n - \frac{1825}{r_\ell}[e^*_s - e_a] - 1.077 \times 10^{-7} [T_a + 273]^4}{\frac{1825}{r_\ell} S_a + \frac{2400}{r_h} + 4.308 \times 10^{-7} [T_a + 273]^3} \quad (18)$$

Where R'_n is the total incoming radiation in all wavelength bands absorbed by the leaf (W m^{-2}), e^*_s the saturation vapor pressure of the air (mb) and e_a the actual vapor pressure of the air near the leaf (mb), T_a the temperature of the air near the leaf (°C), S_a is the slope of the saturation vapor pressure versus temperature relation (mb $°C^{-1}$) and r_ℓ the total leaf diffusion resistance given by the parallel combination of stomatal (r_s) plus boundary layer (r_h) resistance for each side of the leaf (s m^{-1}). This leaf diffusion resistance will be different for amphistomatus and hypostomatus leaves. The boundary layer resistance is calculated from the wind speed near the leaf and leaf size for forced convection (Monteith, 1973) and from ΔT in free convection. Three complications arise in using Eq. (18): 1) the R'_n depends on the temperature of other leaves in the canopy because of their thermal emission so R_n' must be recalculated after all leaf ΔT's are known and a new ΔT computed; 2) the calculation of r_h requires the velocity profile in the canopy and e_a and T_a represent the profile of vapor pressure and temperature in the canopy; and 3) the leaf stomatal resistance must be known.

Recent studies have shown that under a wide range of conditions the stomatal resistance may be slaved to photosynthesis. Thus, if photosynthesis is known, the stomatal resistance can be calculated from

$$r_s' = \frac{C_a - C_i}{P} - r_h' \tag{19}$$

where $r_s' = 1.6\ r_s$ and $r_h' = 1.3\ r_h$ are the resistances to CO_2 diffusion for both sides of a leaf, C_a is the ambient air CO_2 concentration (mg m^{-3}), C_i is the CO_2 concentration in the intercellular spaces of the leaf (mg m^{-3}) and P is the photosynthetic rate (mg $m^{-2}s^{-1}$). In this case, r_h' is the boundary layer resistance in the leaf chamber used to get the photosynthetic rate (P) and not the boundary layer resistance in the canopy. Wong et al. (1980) have shown that the internal CO_2 concentration C_i can be approximated by

$$\begin{aligned} C_i &= 0.80\ C_a \qquad \text{for } C_3 \text{ plants} \\ C_i &= 0.37\ C_a \qquad \text{for } C_4 \text{ plants} \end{aligned} \tag{20}$$

Equations (19) and (20) amount to assuming that stomata sense C_i and attempt to hold it constant at the values given by Eq. (20). The leaf photosynthetic rates for many plants are in the literature as functions of light, CO_2 concentration, leaf temperature, and leaf water potential; equations are available for describing these relations (Reed et al., 1976). Under some conditions, stomata can function independently of photosynthesis; 1) when leaf turgor is very high the stomata may remain open day and night, 2) when leaf temperature exceeds 35 to 40°C, stomata may remain open when photosynthesis stops, 3) if leaves are chilled to near-freezing temperatures, stomata may remain closed when photosynthesis might occur, and 4) leaves with a structure that favors high peristomatal transpiration rates may exhibit stomatal closure when high vapor pressure deficits cause sufficient water loss from guard cells (Tyree and Yianoulis, 1980). To account for the direct stomatal closure effects, the appropriate direct stomatal response functions (f [direct effect]) must be obtained and the stomatal resistance calculated from Eq. (19) can be altered as:

$$r'_{s_2} = r'_{s_1} \cdot f\,[\text{direct effect}] \tag{21}$$

where r_s, is obtained from Eq. (19). When stomata are closed by direct effects, the new photosynthetic rate (P_D) can be calculated from

$$P_D = P\,\frac{r'_h + r'_{s_1} + r_m}{r'_h + r'_{s_2} + r_m} \tag{22}$$

where r_m is defined by

$$r_m = \frac{C_a - \Gamma}{P} - r'_h - r'_s \tag{23}$$

and r'_s is calculated from the appropriate equation (Eq. 19 or 21). In Eq. (23), Γ is the CO_2 compensation concentration, approximately 10 mg m^{-3} (5 $\mu\ell/\ell$) in C_4 plants, about 60 mgm^{-3} in C_3 plants at 25°C and dependent on temperature. It is interesting to note that when stomata are slaved to photosynthesis [Eqs. (19) and (20)] mesophyll resistance and stomatal resistance change together as Dubé et al. (1974) observed. However, direct stomatal effects result in higher stomatal resistances with approximately constant mesophyll resistance (Troughton and Slatyer, 1969).

Equations (19) and (20) are applicable to a wide range of moderate conditions for plant growth including nutrient stress conditions and slowly imposed water stress. Since many response functions are available for photosynthesis, Eqs. (19) and (20) provide ready access to reasonable stomatal responses as a function of light, temperature, CO_2 concentration, and slowly imposed water stresses for both C_3 and C_4 plants. Figure 9 contains data from a comparison of stomatal conductance (k_s) calculated from Eq. (19) with measurements of stomatal conductance using a leaf chamber. Both photosynthesis and stomatal conductance were measured simultaneously on the same leaf with the same chamber. The soybean conductances and photosynthesis values are unusually high (all conductances include both sides of the leaf) but these values were verified independently with a steady state porometer. The two soybean varieties and both of the corn varieties had experienced moderate to severe stress during their development and some of the corn measurements were made on stressed leaves.

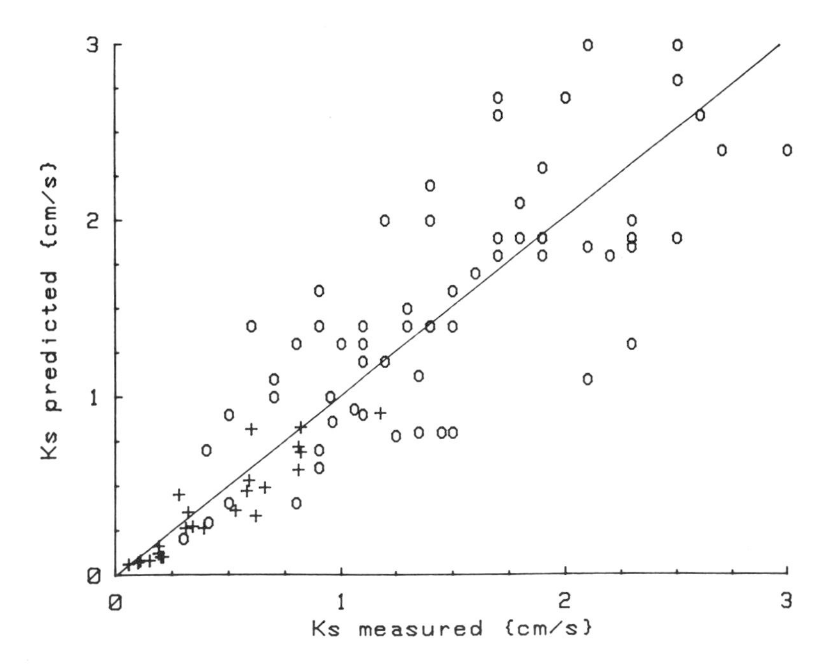

Fig. 9. Stomatal conductance predicted from Eq. (19) using measured photosynthesis rates compared to direct stomatal conductance measurements for corn (+) and soybeans (o). Stomatal conductance values are for water vapor.

2. Canopy and Soil Exchanges of Heat and Water

The solution of the leaf energy balance requires knowledge of the air temperature and humidity (vapor pressure) immediately surrounding the leaf. These air temperature and humidity profiles, in turn, depend on the wind speed in and above the canopy, the heat and water source-sink distribution defined by the leaf energy balance, and the soil as a source or sink of heat or moisture. The simplified analysis of Norman (1979) is reasonable if soil surface boundary conditions are available. However, for a comprehensive plant-environment model to be most useful, soil surface boundary conditions should not be required as inputs. Furthermore, the equations describing heat and mass exchange should include storage terms and not be limited to steady state. Ideally a single set of equations should be used to describe heat and water movement from below the root zone to above the canopy so that the soil surface boundary conditions are predicted by the model. The transient heat conduction equation can be used to accomplish this

$$\rho C_p \frac{dT}{dt} = \frac{d}{dz}\left(k \frac{dT}{dz}\right) + Q \qquad (24)$$

where ρC_p is the volumetric heat capacity (J $m^{-3}K^{-1}$), T is temperature (K), z is height (m), k is the thermal conductivity near the height z (J $m^{-1}s^{-1}K^{-1}$), and Q is the source or sink distribution of heat (W m^{-3}). Using a finite element approximation to Eq. (24), we can solve for the temperature as a function of height by inverting a tridiagonal matrix (Chakranopakhun Tongyai, 1976). In the soil, the thermal conductivity (k) is a function of soil type and water content (Baver et al., 1972). The soil heat capacity is mainly a function of water content. In and above the canopy the conductivity (k) is the turbulent eddy diffusivity (m^2s^{-1}) multiplied by the volumetric heat capacity for air (about 1200 J $m^{-3}K^{-1}$). The volumetric heat capacity of the canopy can be adjusted for the presence of foliage, but this only becomes significant with tall forest canopies. In the soil, the heat sources are zero except at the soil surface where the heat source is the net radiation above the soil surface minus the soil surface latent heat flux from evaporation. In the canopy, the source (or sink) terms, Q, are obtained from the solution of the leaf energy budget equations in each layer. Equation (24) is most useful because only a single tridiagonal matrix need be inverted to obtain the temperature profile from below the root zone to above the canopy. The most difficult task in applying Eq. (24) is obtaining the profile of turbulent eddy diffusivities. Methods for obtaining eddy diffusivities are described by Goudriaan (1977) and Norman (1979); note there is a subscript error in Eq. (3.6.43), both using stability corrections from Businger (1973).

Water movement can be described by equations analogous to Eq. (24); however, it now is desirable to use slightly different formulations for the soil and for the canopy. The equations for soil-water movement and root-water extraction, which are formulated in terms of the soil-water potential, have been described in detail by Nimah and Hanks (1973) and Chakranopakhun Tongyai (1976) so they will not be described here again. The equation analogous to Eq. (24) for water vapor movement in the canopy is

$$\frac{\rho\varepsilon}{P}\frac{de}{dt} = \frac{d}{dz}\left(\frac{\rho\varepsilon k}{P}\right)\frac{de}{dz} + E \qquad (25)$$

where ρ is the density of air (kg m^{-3}), $\varepsilon = 0.622$, P atmospheric pressure (mb), e water vapor pressure (mb), z height (m), k the eddy diffusivity ($m^2 s^{-1}$) and E the vertical mass-flow source-sink distribution of evaporation or condensation derived from the leaf energy balance equations (kg $m^{-3} s^{-1}$). The solution of Eq. (25) for the vapor pressure profile requires inversion of another tridiagonal matrix. The eddy diffusivity for water vapor is assumed equal to that for heat in and above the canopy.

The use of Eqs. (24) and (25) to solve for temperature and water vapor profiles (and thus heat and water fluxes) has an added benefit. Since the canopy displacement height and roughness are only required for momentum, we escape some of the uncertainties associated with different displacement heights and roughnesses for heat and water (Campbell, 1977) and we also avoid the peculiarities associated with extrapolated surface temperatures and vapor pressures (Businger, 1973).

Clearly the simultaneous solution of all of these heat and water flow equations in the canopy and soil is formidable. In the model Cupid, a special iteration technique is used to obtain rapid convergence of all the relevant equations including wind profiles, heat and mass flow, radiation profiles, and the leaf energy balance. A 24-hour simulation with a 38-layer model, having six leaf angle classes, can be run for less than two dollars on an IBM 370 computer.

This complete model may be too complex for some applications, but it can be used to help develop simple models. For example, we can derive a simple method for estimating transpiration from a canopy that is transpiring at less than potential because of partial stomatal closure. Often measurements of stomatal conductance are available, or at least the dependence of stomatal conductance on light is known. If leaf photosynthesis values are known, or at least their dependence on light known, Eq. (19) may be used to estimate stomatal conductance as a function of light. If the stomatal conductance versus light relation is known, or if stomatal conductance measurements are available for sunlit and shaded leaves separately, then an effective, bulk canopy resistance to water vapor can be determined with the help of Eqs. (14-17). This idea is not new (Monteith, 1965), but in the past a major obstacle has been lack of a reasonable estimate of the illumination level for shaded leaves. The average illumination of shaded leaves is given by Eq. (16) where the beam and diffuse-visible radiation can be derived from global solar radiation using Eqs. (1-10). The illumination of sunlit leaves can be calculated from Eq. (17) using 60° for α, the mean leaf-sun angle. Knowing these illuminations, one can estimate stomatal conductance for sunlit (K_s^{sun}) and shaded (K_s^{shade}) leaves. Thus, the canopy stomatal conductance, K_s^{cpy}, (based on a unit of ground area) is given by

$$K_s^{cpy} = F^* K_s^{sun} + (F-F^*) K_s^{shade} \tag{26}$$

where F is the total leaf area index and F* the sunlit leaf area index given by Eq. (14). If the potential transpiration can be estimated from Penman (1948), Priestly and Taylor (1972) or some other method (Jensen, 1973), then the actual transpiration for a crop of full cover can be approximated by (Monteith, 1965)

$$\frac{T_r\ (\text{Actual})}{T_r\ (\text{Potential})} = \frac{S + \gamma}{S + \gamma\left[1 + \dfrac{K_a}{K_s^{cpy}}\right]} \tag{27}$$

where S is the slope of the saturation vapor pressure versus temperature relation (mb K^{-1}), γ the psychrometer constant (0.66 mb K^{-1}), and K_a the aerodynamic boundary layer conductance approximated by

$$K_a = \frac{0.16\ U}{\left[\ln \dfrac{(z-d)}{z_o}\right]^2} \tag{28}$$

for near-neutral conditions. In Eq. (28), U is the mean wind speed (m s^{-1}) at height z (m), d and z_o are the canopy displacement height and roughness (Monteith, 1973). Equation (28) can be adjusted with the diabatic profile corrections if the sensible heat flux above the canopy is known (Businger, 1973). This sensible heat flux can be estimated from the energy balance if the net radiation above the canopy is available.

3. Condensation. In principle, condensation can be modeled by setting the stomatal resistance to zero when the vapor pressure of the air exceeds the saturation vapor pressure at the temperature of the leaf (Goudriaan, 1977). Thus only boundary layer resistance inhibits the flow of water from the atmosphere to the leaf. In practice, the onset of condensation can present numerical difficulties if large time steps of an hour or more are used. This occurs because the temperature depression of the leaf below the air provides the driving force for vapor movement to the leaf, but condensation heats the leaf and tends to diminish this driving force. Because condensation depends on the leaf boundary layer resistance, dew deposition might not be expected to be large under very calm conditions. Alternatively, the leaf temperature depression below air temperature would be small under high wind speed conditions. Thus, intermediate wind speeds would be expected to produce the greatest amount of dew.

Condensation is not simply "evaporation in reverse." The dicotomous nature of its occurrence or nonoccurrence is evidence for that because evaporation, by contrast, is a continuous process. A second difference between evaporation and condensation arises from the surface area involved and the definition of boundary layer resistance. Normally leaf transpiration occurs from the stomata and epidermis so the epidermal surface area is appropriate. When condensation occurs, the appropriate surface area may include leaf hairs as well as the epidermis so pubescence may affect dew deposition by increasing surface area and effectively reducing the leaf boundary layer resistance for condensation. More research is required for this to be understood fully.

III. SOME RESULTS AND NEEDS FOR FUTURE RESEARCH

The model Cupid has been applied to a typical 2.3 meter corn canopy in Eastern Nebraska in July. The stomatal response functions are similar to those described by Norman (1979) except that the minimum stomatal resistance in full sunlight is about 80 $s\ m^{-1}$

for both sides of a leaf. The corn is not under water stress. The canopy has a leaf area index of 4.0, and there are 25 layers in the canopy and atmosphere above the soil with 13 soil layers to make a total of 38 layers in the model. The radiation, wind, temperature, and vapor pressure boundary conditions over the 24 hours are depicted in Fig. 10. The top 10 cm of loam soil was at 40% water content by volume and below that at 30%. Some runs were made with the top 10 cm very dry and the results are much different from those presented here.

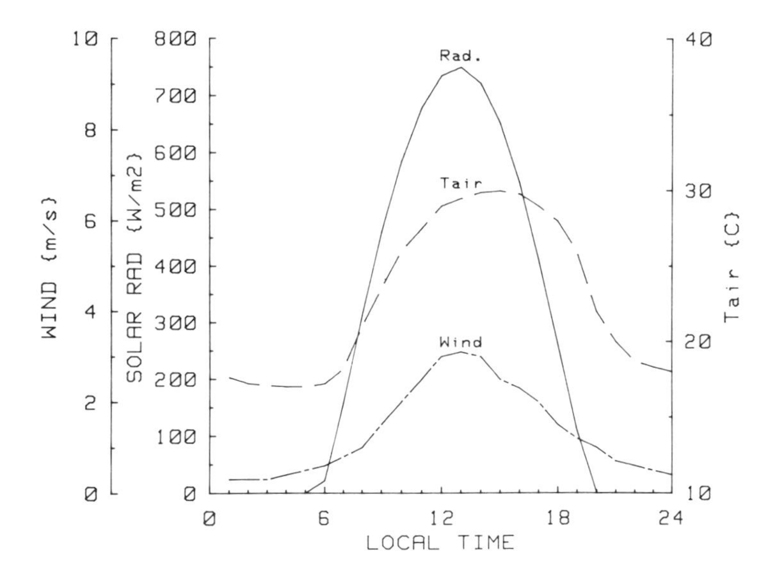

Fig. 10. Input data at a height of four meters used in applications of the model Cupid. Vapor pressure of the air was constant at 12.4 mb and the 50 cm soil temperature was constant at 20°C.

The soil surface energy budget below the canopy is presented in Fig. 11. It is somewhat surprising that the net radiation below a canopy of leaf area index equal to four is about 25% of the net radiation above the canopy (Fig. 12) because of the cool wet surface and the large near-infrared multiple scattering. With the top 10 cm of soil dry, the soil net radiation is about 18% of

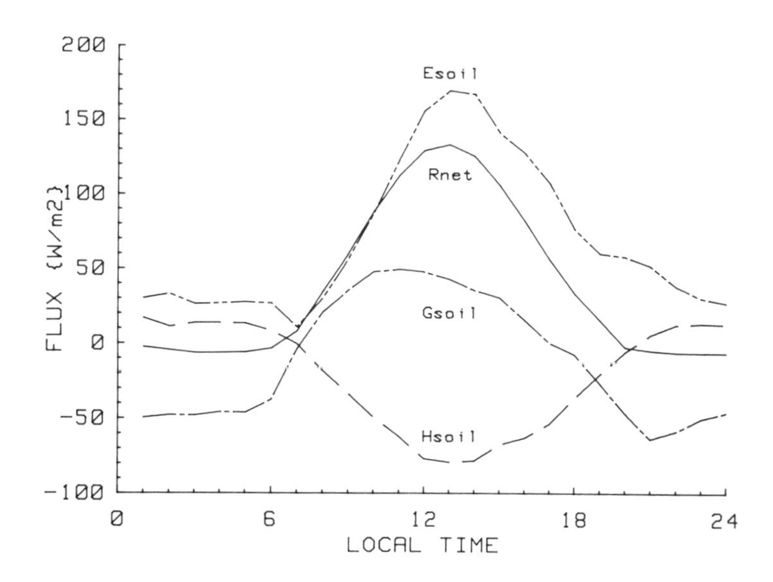

Fig. 11. Diurnal soil surface energy budget components below a corn canopy of leaf area index 4 using input from Fig. 10 with a wet soil and soil surface. R_{net} is net incoming radiation immediately above the soil surface, E_{soil} is soil evaporation, H_{soil} is sensible heat flux from the surface evaporation, H_{soil} is sensible heat flux from the surface to the air and G_{soil} is the heat conducted downward and into the soil.

the net radiation above the canopy. The evaporation at the soil surface can be larger than the net radiation there just as above the canopy. Unfortunately, the soil surface energy budget depends on the eddy transfer coefficients just above the soil surface and these are not well understood. This is an area in need of much research. Even though condensation occurs at night, soil evaporation remains positive day and night. At night, soil evaporation, radiation loss, and heating of the canopy air are all sustained by the conduction heat loss from the soil. Thus, during these hours

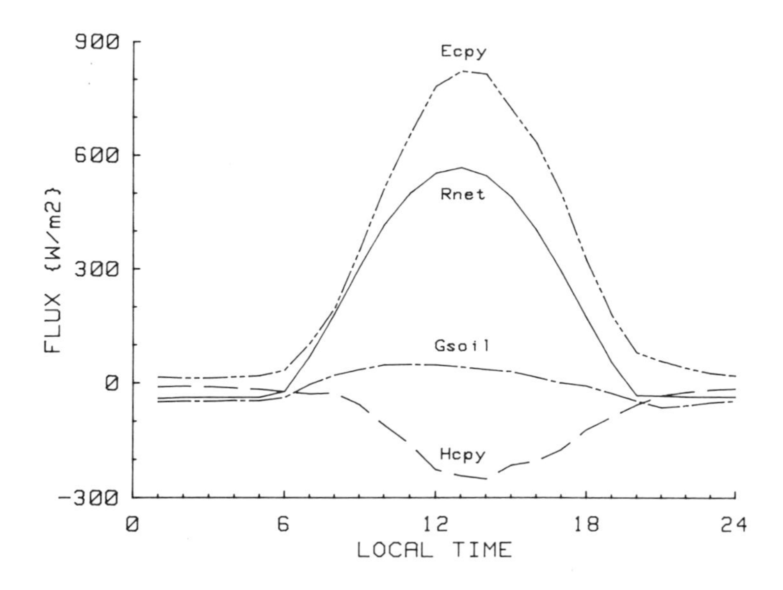

Fig. 12. Diurnal energy budget components above a corn canopy of leaf area index 4 experiencing no water stress and having a wet soil and soil surface. R_{net} is the incoming net radiation above the canopy, E_{cpy} is the total evaporation flux above the canopy, H_{cpy} is the total upward sensible heat flux above the canopy and G_{soil} is as in Fig. 11.

the soil heat conduction component is of primary importance. Soil heat conduction below dense canopies often is neglected and this seems an undesirable simplification for diurnal energy budgets. The magnitude of the soil heat conduction component at night also depends on soil-water content because the soil thermal properties depend on water content. When the top 10 cm of the soil surface is dry, the sensible heat flux from the soil remains positive throughout the day, ranging from 20 W m^{-2} at night to 50 W m^{-2} during midday.

Figure 12 contains the diurnal distribution of energy fluxes above the canopy. This pattern is quite common for well-watered

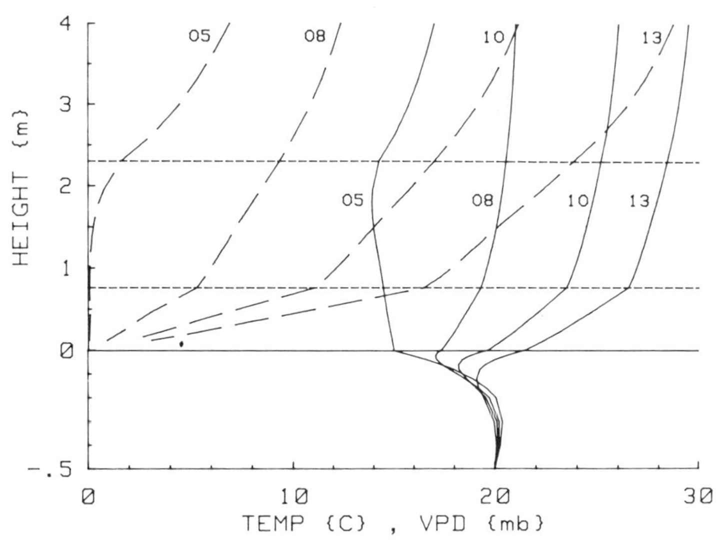

Fig. 13. Vertical profiles of soil and air temperature (—) and air vapor pressure deficit (- - -) in corn having a wet soil surface. The four profiles are from before sunrise (05) to about solar noon (13). The soil vertical scale is expanded and the dotted lines indicate the region of the corn canopy containing live foliage.

crops in hot, dry atmospheric environments with the sensible heat contributing to evaporation because of the large vapor pressure difference between the crop and the atmosphere (30 mb in this case at midday). A comparison of Figs. 11 and 12 indicates that about 20% of the evapotranspiration originates from soil evaporation. If the top 10 cm of the soil surface is dry, the total evapotranspiration is only 10% less than with a wet soil. Thus, the extra sensible heat from the dry soil surface (and the drier canopy environment) causes greater canopy transpiration with a dry soil surface. In this example, the savings in evapotranspiration that result from a dry soil surface are only about half of what might be expected from a simple argument that ignores enhanced transpiration because of the dry soil.

Profiles of temperature and water vapor in the canopy are most important to integrated pest management studies (Fig. 13). Even though the atmospheric humidity before sunrise is only 65%, condensation occurs throughout most of the canopy because of water evaporating from the wet soil surface (Fig. 14). With a wet soil surface, large gradients of vapor pressure and temperature can occur throughout a canopy, and the mean soil temperature can be considerably below the mean air temperature (Fig. 13). If egg hatch (Fig. 2) were estimated from above-canopy conditions, one might expect 75% egg hatch with about 85 hours to 50% hatch; whereas, based on in-canopy conditions near 100% egg hatch with 50% hatch in under 75 hours would be more realistic.

Figure 14 contains the amount and duration of leaf condensation in four canopy layers on a relatively humid day (100% relative humidity just before sunrise) and a relatively dry day (65% relative humidity just before sunrise). Although the amounts of dew may be small under the drier atmospheric conditions, the presence of very small amounts of liquid water can be significant to undesirable pathogens. The information in Fig. 1 could be applied in the field using results similar to those shown in Figs. 13 and 14. However, before this could be done some controlled environment experiments would have to be carried out to determine how the intermittancy typical of day-night conditions affect infection times. This points out another area in need of more research; cooperative efforts of research teams to attack a single pest-environment problem comprehensively. Since cooperative efforts depend heavily on the scientists' personalities, this kind of work can only be encouraged, not forced. Considering this fact, along with the great strength of traditional disciplines, we cannot expect great things to happen in team research in the university environment.

The model Cupid also can be used to determine the source (atmosphere versus soil) of leaf condensate. With the drier atmosphere, all of the condensate comes from the soil. In fact, in the dry-atmosphere example, only about 40% of the soil evaporation is condensed and the remaining water has been lost to the atmosphere by evaporation. Thus, great care must be exercised in

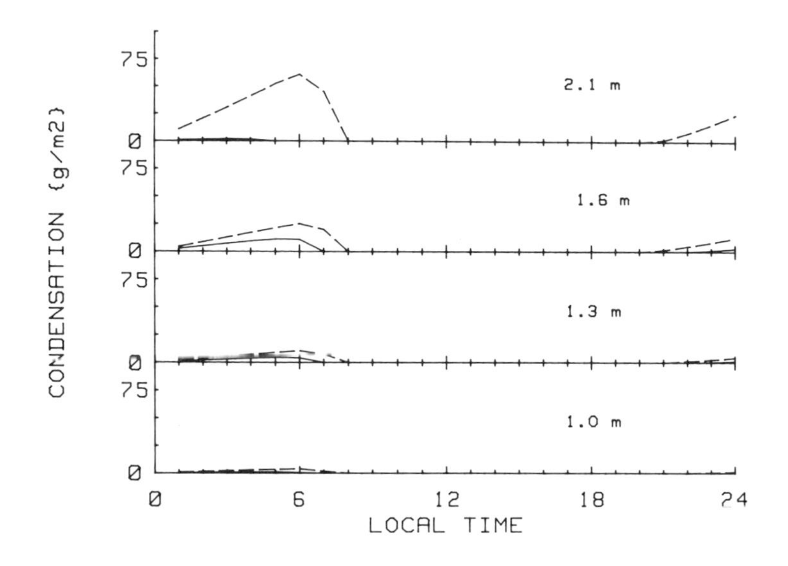

Fig. 14. Amount and duration of condensation on leaves in 4 layers of a corn canopy with a wet soil surface and relatively dry (vapor pressure = 12.4 mb ———) and moist (vapor pressure = 19.4 mb - - - -) atmospheric environments above the crop. The condensation is expressed as g m^{-2} of ground area. At 0600, 0.3 mm of condensate accumulated in the moist case and 0.1 μm accumulated throughout the whole canopy in the dry case.

the use of weighing lysimeters to measure canopy condensation with an unsaturated atmospheric environment. In the example with the more humid atmosphere, about 80% of the condensate is distilled from the soil. The amounts of dew predicted by this model are in good agreement with the measurements reviewed by Sharma (1976).

The condensation process itself deserves much more attention, particularly the influence of pubescence and leaf exudates. However, progress in this area probably will be slow until we improve our understanding of turbulent transfer in canopies under moderate to low wind speed conditions.

ACKNOWLEDGMENTS

I would like to thank Dr. Al Weiss for sharing his radiation measurements and Jon Welles, Larry Toole, and Dr. Charles Sullivan for their help in the gas exchange experiments. Some of the research reported in this chapter was supported by USDA Competative Grant No. 7900231 (Neb17-035).

REFERENCES

Allen, L. H. (1974). Agron. J. 66:41-47.
Baver, L. D., Gardner, W. H., and Gardner, W. R. (1972). "Soil Physics." John Wiley and Sons, N.Y.
Businger, J. A. (1973). In "Workshop on Micrometeorology" (D. A. Haugen, ed.), pp. 67-100. Am. Meteor. Soc., Boston, Massachusetts.
Cahir, J. J., Norman, J. M., and Lowry, D. A. (1981). Monthly Weather Rev. 109:485-500.
Campbell, G. S. (1977). "An Introduction to Environmental Biophysics." Springer-Verlag, N.Y.
Chakranopakhun Tongyai, M. L. (1976). "An Evaluation of Finite Difference Numerical Methods for Soil-Plant Water Relations Studies." M. S. Thesis, Washington State University, Pullman, Washington.
deWit, C. T. (1965). Agric. Res. Rep. No. 662, Center Agr. Publ. Doc., Wageningen, The Netherlands.
Dubé, P. A., Stevenson, K. R., and Thurtell, G. W. (1974). Can. J. Plant Sci. 54:765-770.
Ferro, D. N., and Chapman, R. B. (1979). Environ. Entomol. 8:701-705.
Garratt, J. R., and Hicks, B. B. (1973). Quart. J. Roy. Meteor. Soc. 99:680-687.
Gausman, H. W., and Allen, W. A. (1973). Plant Physiol. 52: 57-62.
Goudriaan, J. (1977). "Crop Micrometeorology: A Simulation Study" Centre for Agric. Publ. and Doc., Wageningen, The Netherlands.
Idso, S. B., and deWit, C. T. (1970). Appl. Optics 9:177-184.
Jensen, M. E. ed. (1973). "Consumptive Use of Water and Irrigation Water Requirements." Am. Soc. Civil. Engr., New York. N.Y.
Kondratyev, K. Ya. (1969). "Radiation in the Atmosphere." Academic Press, N.Y.
Lemeur, R., and B. Blad. (1974). Agric. Meteor. 14:255-286.
List, R. J. (1966). "Smithsonian Meteorological Tables" Smithsonian Misc. Collections, Vol. 11, Smithsonian Institution, Washington, D.C.
Mann, J. E., Curry, G. L., and Sharpe, P. J. H. (1979). Agric. Meteor. 20:205-214.

Monteith, J. L. (1965). In "The State and Movement of Water in Living Organisms, XIX Symposium," pp. 205-234. The Soc. for Expt. Biol., Cambridge University Press.
Monteith, J. L. (1973). "Principles of Environmental Physics." Amer. Elsevier Publ. Co., N.Y.
Monteith, J. L., and Butler, D. R. (1979). Quart J. Roy. Meteor. Soc. 105:207-215.
Nilson, T. (1971). Agric. Meteor. 8:25-38.
Nimah, N. M., and Hanks, R. J. (1973). Soil Sci. Soc. Amer. Proc. 37:522-527.
Norman, J. M., Miller, E. E., and Tanner, C. B. (1971). Agron. J. 63:743-748.
Norman, J. M. (1975). In "Heat and Mass Transfer in the Biosphere" (D. A. de Vries and N. H. Afgan, eds.), pp. 187-206. Scripta Book Co., Washington, D.C.
Norman, J. M., and Jarvis, P. G. (1975). J. Appl. Ecol. 12: 839-878.
Norman, J. M. (1979). In "Modification of the Aerial Environment of Crops" (B. Barfield and J. Gerber, eds.), pp. 249-277. Am. Soc. Agric. Engr. Monograph No. 2, ASAE, St. Joseph, Michigan.
Norman, J. M., Perry, S. G., Fraser, A. B., and Mach, W. (1979). In "Fourteenth Conference on Agriculture and Forest Meteorology," pp. 184-185. Am. Meteor. Soc., Boston, Massachusetts.
Norman, J. M. (1980). In "Predicting Photosynthesis for Ecosystem Models," pp. 49-67. CRC Press, Inc., Boca Raton, Florida. Volume 2.
Norman, J., and Welles, J. (1982). Agron. J. In press.
Penman, H. L. (1948). Roy. Soc. (London) Proc. A. 193:120-146.
Priestly, C. H. B., and Taylor, R. J. (1972). Monthly Weather Rev. 100:81-92.
Raschke, K. (1979). In "Physiology of Movements" (W. Haupt and M. E. Feinleib, eds.), pp. 383-441. Springer-Verlag, N.Y.
Reed, K. L., Hamerly, E. R., Dinger, B. E., and Jarvis, P. G. (1976). J. Appl. Ecol. 13:925-942.
Ross, J. (1975). In "Vegetation and the Atmosphere" (J. L. Monteith, ed.), pp. 13-55. Academic Press, N.Y.
Sharma, M. L. (1976). Agric. Meteor. 17:321-331.
Sinclair, T. R., Murphy, Jr., C. E., and Knoerr, K. R. (1976). J. Appl. Ecol. 13:813-839.
Stewart, D. W. and Lemon, E. R. (1969). "The Energy Budget at the Earth's Surface: A Simulation of Net Photosynthesis." Interim Report 69-3. Tech. Report ECOM 2-68, I-6, Cornell University, Ithaca, N.Y.
Tanner, C. B., and Jury, W. A. (1976). Agron. J. 68:239-243.
Thorpe, M. R., Warrit, B., and Landsberg, J. J. (1980). Plant, Cell and Environ. 3:23-27.
Troughton, J., and Slatyer, R. (1969). Aust. J. Biol. Sci. 22:815-827.
Tyre, M. T., and Yianoulis, P. (1980). Ann. Bot. 46:175-193.

Waggoner, P. E. (1975). In "Vegetation and the Atmosphere" (J. L. Monteith, ed.), pp. 205-228. Academic Press, N.Y.
Wang, W-C. (1976). J. Appl. Meteor. 15:21-27.
Weiss, A., Kerr, E. D., and Steadman, J. R. (1980). Plant Disease 64:757-759.
Welles, J. (1978). "A Model of Foliage Temperatures for a Heated Orchard," M.S. Thesis. Dept. of Meteor., Pennsylvania State University, University Park, Pennsylvania.
Wong, S. C., Cowan, I. R., and Farquhar, G. D. (1979). Nature 282:424-426.

REMOTE SENSING OF MICROCLIMATIC STRESS

P. J. Pinter, Jr.
USDA-SEA-AR
U.S. Water Conservation Laboratory
4331 East Broadway
Phoenix, Arizona

I. INTRODUCTION

Over the past decade, integrated pest management (IPM) has become the environmentally acceptable method to solve many problems in agricultural resource management. The cultural, chemical and biological methods formerly used with success to manage pests, weeds, and disease must now be merged in concert to effect control where alone they have become either inappropriate or ineffectual. More than ever, resource managers are finding this approach mandates more timely and accurate predictions of pest infestations, disease outbreaks, nutrient deficiencies and microclimatic-related stress phenomena over larger expanses of land. It is likely that this requirement for early detection will be met in coming years by remote sensing technology.

In a similar fashion remote sensing can provide the basis for improving the predictive capabilities of pest populations, plant disease, and plant growth models used in IPM. In as much as temperatures and vapor pressures from standard meteorological shelters bear no simple relationship to the microclimatic conditions within a plant canopy, predictions based on these parameters at times contradict field observations. Site specific microclimatic data are better suited as model inputs. They can be

ISBN 0-12-332850-0

obtained via aircraft and satellite-mounted sensors for remote areas far from the nearest meteorological station and then used to drive, validate, and provide intermediate updates of various types of IPM models.

The need for reconnaissance of potential military targets fostered the early development of remote sensing and has been responsible for many of its recent technological advances. Over the past several decades, however, there has been an increasing awareness of its utility for nonmilitary objectives, especially for monitoring and managing natural resources. For example, geographers immediately found it useful in cartography. It soon became indispensable to meteorologists in mesoscale weather analysis and short range forecasting, and geologists utilized it for mineral exploration. Among others, agriculturalists are now recognizing its potential for surveying extensive acreages and revealing agronomic and physiological information not readily available by other means.

Operationally, we may define remote sensing as any noncontact means by which information about a distant object is conveyed to an observer. Although the human body is equipped with a rather limited noncontact sensory repertoire, it is capable of perceiving incredible detail about its surroundings. A farmer, for instance, can qualitatively predict yield by mentally integrating a visual assessment of crop vigor with past and expected future weather conditions. However, there is much information about that crop and the microclimatic conditions governing its growth which goes largely unnoticed. Bound by a sensitivity to visible light in a relatively narrow band (0.4 to 0.7 μm), the farmer is unable to detect reflected near-infrared light where subtle changes in a crop's physiological status are often mirrored. Transparent to him also, are variations in crop canopy temperatures which provide an early warning of water stress and loss of vigor. Armed with those two additional inputs, he could irrigate more efficiently and better control pests and disease, important considerations if he wishes to maximize profits. An important goal of agricultural remote sensing is to provide him with this type of extravisual information.

The innovative exploitation of all regions of the electromagnetic spectrum has enabled us to infer more from agricultural targets than we thought possible just a few years ago. And, given the rapid pace of technological advances in this field, the utility of remote sensing for agricultural resource management should expand dramatically in the coming decade. The purpose of the following discussion is to provide a general review of the utility of remote sensing for agricultural purposes and to highlight specific instances where certain techniques are applicable to IPM.

A. Hierarchy of Approaches

Within the context of our original definition, the word "remote" specifies a noncontact approach rather than implying any constraints on the distance required between observer and target. Accordingly, remote sensing instruments are used at various levels ranging from ground-based, hand-held infrared thermometers where the target may be only centimeters away to sophisticated multispectral scanning devices on board statellites at a height of 36,000 km above the equator. These levels of deployment offer one method to categorize and compare different remote sensing approaches (Table 1). Each higher level of deployment has associated with it the benefits of greater geographical coverage. Not surprisingly, each level also requires progressively more expense and technological competence to collect, process and interpret the data.

Developing the basic relationships that exist between a remotely-sensed parameter and a vegetation feature are probably most efficiently carried out on the ground using either hand-held or truck-mounted instruments. Under these conditions, researchers can be confident of the identity of their target, manipulate

Table I HIERARCHY OF REMOTE SENSING APPROACHES

LEVEL	DEPLOYMENT	COVERAGE	EXAMPLE	UTILITY	ADVANTAGES	DISADVANTAGES
GROUND	Hand held	Plot Field	Camera Multi-band radiometer IR thermometer	Agronomic research Model validation Small scale resource management	Low Cost One Man Operation Unrestricted Mobility Positive Target Identification Rapid Delivery of Information to user	Generally low data rate Limited spatial coverage Limited applicability for tall crops or widely spaced row crops
	Boom or truck mounted	Plot Field	Above plus Scanning spectrometer Microwave Radar FLD	Identification of useful spectral regions Model validation	Many wavebands available Preliminary data reduction moderately rapid Appropriate for tall & row crops Positive target identification	Reduced mobility Some targets less accessible Moderate to high costs Several man operation
AIR	Low altitude	Farm Irrigation district	Same as above	Large scale resource monitoring & management Examination of spatial variability	Permits simultaneous comparison of targets over a relatively broad region	Team operation High Costs Delivery of data to potential

	High altitude	Watershed		Location of meteorologically representative areas for continuous monitoring by ground stations	Resolution sufficiently small that target identification is rarely a problem	consumer 1 to 6 months Data reduction often requires large computer facility
SPACE	Geosynchronous satellite (fixed position)	Fixed regional Hemisphere scale	GOES Sensors in visible, near IR thermal	Regional stress assessment land use patterns Regional weather Solar Radiation	Network exists for near-real time processing and delivery to potential consumer Daily or even hourly coverage permits more flexibility and frequent coverage for identification of phenological events Repetitive imagery from fixed target	Poor spatial resolution Atmospheric interference Volumes of data create storage problems Target identification difficult
	Sunsynchronous satellite (orbital)	Repetitive regional Global scale	LANDSAT MSS thematic mapper	Mapping of cultural and natural features Regional stress assessment Land use patterns Crop identification	Coverage of inaccessible areas Acceptable spatial resolution Target identification possible High data rate	Infrequent coverage precludes full exploitation for agricultural resource management Slow delivery to user Illumination angle and intensity change from image to image Atmospheric interference

experimental treatments more effectively and, perhaps most importantly, make measurements ad lib. Such flexibility is not present with more sophisticated aircraft- and satellite-mounted sensors. However, those platforms offer the most logical and cost effective means for implementing an applied program over areas larger in size than single fields or farms. Development of a successful applied program will probably require a multistage approach at first, with data from each level serving to calibrate the next higher level. Operationally, it may be possible to by-pass certain intermediate levels, but some ground truthing information will almost always be required.

It is curious that much of the early impetus in remote sensing for agricultural purposes came following the launch of LANDSAT 1 in 1972. Starting with a rather complicated multispectral scanning device at an altitude of 905 km, researchers were faced with the task of interpreting data from 185 x 185 km segments of the earth's surface. That they were able to catalog land use, discriminate crop types, detect diseases, and much more is a testament to the ingenuity of remote sensing specialists. All at once they were confronted with the problems associated with 0.45 hectare ground pixel size, once every 18 day target coverage, varying sun angles and illumination intensity and changing atmospheric conditions. At times it was not possible to get more than one or two cloud-free images per growing season. Such drawbacks made LANDSAT data somewhat less than optimum in a research mode but did not eclipse the demonstration that orbital platforms have great monitoring potential.

Many of the problems which plague satellite and aircraft systems can be circumvented in ground-based operations and this is one reason much of the basic research is being conducted with them today. Moreover, ground-based techniques are becoming increasingly more important in supportive roles within various agricultural disciplines such as physiology, genetics, pathology and entomology. Imaginative application of noncontact techniques can offer a short cut to the more tedious conventional methods of monitoring plant vigor and characterizing growth patterns.

B. The Electromagnetic Spectrum

Remote sensing technology has exploited almost every region of the electromagnetic spectrum, from x-rays and gamma rays and neutron particles in soil moisture studies, to the range finding radars of aircraft and sonars of submarines. Although some sensors with potential for stress assessment rely on reflected ultraviolet light (Reeves, 1974), many involve measurements of reflected radiation in the visible and near IR regions. These, along with the thermal infrared, will comprise much of the discussion in later sections. Microwaves will not be explored in detail in this paper; both active (radar) and passive systems hold much promise in soil moisture and crop inventory studies. This is mainly because of their all weather capability, ability to penetrate beyond surface features, and good sensitivity to soil moisture (Jackson et al., 1981a; Schmugge, 1980; and Ulaby et al, 1980). The unique geometric characteristics of an active microwave approach (i.e., shadows, layover, foreshortening) (Reeves, 1974) at times emphasize surface textural features which are not otherwise readily apparent; however, they also make the interpretation of radar data exceedingly complex. Rainfall rates and windspeed are among certain unique meteorological events which can be monitored by microwave (Beer, 1980). Readers are advised to consult Reeves (1974) for a comprehensive review of other remote sensing techniques.

II. REMOTE SENSING IN THE REFLECTED PORTION OF THE SPECTRUM

A. Photographic Techniques

For many years, both ground- and aerial-based photography have been the mainstay of remote sensing. Today they remain perhaps the most cost-effective means to record the "lay of the land," assist in the selection of field sites, and provide baseline data against which subsequent photos can be compared. Although mainly qualitative in nature, the information present in a photograph can be digitized using a random dot grid overlay or sophisticated densitometric analysis thus providing quantitative data on scene parameters such a percent ground cover or mean angle of leaf inclination.

Various manufacturers offer black and white and color films with emulsions that vary in their sensitivity to reflected light (Heller, 1970). Conventional color films cover the visible region of the spectrum (0.4 to 0.7 μm). Color infrared (CIR) films are sensitized to record not only visible light but also a portion of the near-IR wavelengths (0.7 to 0.9 μm). Contrary to popular belief, CIR film is not capable of monitoring target temperatures at normally encountered field temperatures (Fritz, 1967). Instead, the color of the final image results from differential exposure of three emulsion layers to green, red and near-IR reflected light. Because each emulsion is overly sensitive to blue light, a deep yellow filter is usually employed to exclude it and provide suitable rendition of biological targets (Eastman Kodak, 1972). With this filter, the image of vigorous green vegetation typically appears magenta due to a low reflection of light in the red portion of the spectrum and high reflectance in the near-IR. Stress interferes with the way plants partition solar radiation, decreasing the efficiency with which the photosynthetic mechanism captures visible light. The net result being that the reflectance of red light increases while near-IR decreases. This changes the color of the final film image from magenta to blue green (Fritz, 1967).

This shift forms the basis for CIR use in the remote detection of stress. Some researchers, however, have concluded that CIR and conventional color films possess equal powers of discrimination between healthy and stressed vegetation. This may be due to changes in visible reflectance which occur just as rapidly as those in the near-IR (Bauer, 1975). Fox (1978) contends that the emulsion of CIR film accentuates slight changes in reflectance which might not be noticeable to an observer of color photographs. The unnatural, striking colors of CIR undoubtedly contribute to its utility and popularity. Furthermore, CIR is better suited for aerial photography because of the superior haze penetration of reflected near-IR light.

The number of reports on CIR in the literature attests to its widespread use for agricultural purposes. These range from the monitoring of soil moisture (Curran, 1979) and drought stress (Lillesand et al., 1979) to detection of plant diseases (Colwell, 1956; Wallen and Philpotts, 1971). Although insects themselves are not normally detectable using aerial photography because of their small size, often their metabolic by-products or defoliating habits are, and this provides the basis for their detection (Hart and Meyers, 1968; Hart et al., 1971; Green et al., 1977; and Talerico et al., 1978). Chiang et al. (1976) were able to document armyworm defoliation in corn provided it occurred on plants that were already under some moisture stress and thus had reduced biomass.

Henneberry et al. (1979) demonstrated qualitative CIR image differences that were attributed to cultivar, disease, fertilizer and growth regulator treatments in cotton. Based on the presence of salt tolerant vegetation or bare salt crust, Dalsted et al. (1979) reported CIR could differentiate between intermediate and mature saline seeps in North Dakota. Other stress related uses of CIR include monitoring dune encroachment in the Sahara using orbital photographs (El-Baz, 1978) and prediction of potential soil erosion from CIR-derived information on plowing operations and crop residue (Morgan et al., 1979).

B. Multispectral, Nonimaging Techniques

The first in a series of satellites designed to inventory earth resources, LANDSAT 1 (launched in 1972) ushered in a new era in remote sensing. With worldwide, 18-day coverage, the 4-band multispectral visible and near-IR data from this and subsequent satellites, gave new impetus to the discipline by promising regional application of previous smaller scale stress studies such as the Corn Blight Watch (Bauer et al., 1971; MacDonald et al., 1972). Many of the early agricultural applications, however, were for crop inventory rather than stress assessment. The Large Area Crop Inventory Experiment (LACIE) for example, used LANDSAT crop acreage data in a historical, meteorologically driven wheat yield model (MacDonald and Hall, 1978).

Thompson and Wehmanen (1979, 1980) demonstrated that multispectral satellite data could be used to detect moisture stress in semi-arid wheat growing regions and, when combined with ancillary ground-based meteorological and crop growth calendar data, in more humid areas where corn and soybeans are the predominant crop. The use of multispectral data in stress identification has been limited though because its detection is a far more formidable task than simple crop identification where dates of planting, maturity rates, geographic location, etc. narrow the possibilities considerably (Bauer, 1975). It is not difficult to recognize why stress detection techniques have been slow to develop. Canopy geometry, row orientation, phenology, percent cover and background soil color all contribute to canopy spectral characteristics. Small changes in any one of these alone can obscure subtle spectral changes induced by an incipient or perhaps even advanced plant stress. Of course, such factors also limit the universal application of many of the photographic techniques mentioned earlier and contribute substantially to a need for extensive ground verification.

1. Meteorological events

The direct detection of certain environmental conditions is possible with multiband spectral data in the reflected portion of the spectrum. For example, papers presented by Rango (1975) illustrate the role of remote sensing in the evaluation of snowcover to predict runoff for watershed management programs. Patterns of nighttime surface winds and radiative cooling are revealed in the spatial variability of frost. Since the spectral reflectance of most soils decreases significantly when wet (Idso et al., 1975), areas of recent rainfall can be identified from repetitive multispectral coverage. Such information would be useful in the prediction of optimum planting dates and the development of native annual vegetation on rangelands for grazing purposes. On the other hand rainfall events are often harbingers of incipient desert locust outbreaks where advance warning techniques could improve control efficiency (Heilkema and Howard, 1976).

Solar radiation is an input required for many crop growth simulation models, yet due to the sparse network of solar meteorological stations, it is not readily available for many regions. Gautier et al., (1980) conducted a regional solar radiation intensity experiment in southern Canada. Their results demonstrated that frequent satellite observations of cloud distribution and density patterns adequately described solar flux at the surface. Cloud movements also convey information on vertical and horizontal wind patterns above the condensation level (Lindqvist and Mattsson, 1975; Novak and Young, 1977). These, of course, have a bearing on meso- and synoptic-scale weather forecasting. Sprigg (1978) commented on the potential use of multispectral data for long-term global climate analysis.

2. Plant vigor assessments

The use of multispectral reflected light in the monitoring of various green agronomic parameters such as biomass, percent ground cover, canopy height, etc. is currently the focus of much research effort. The basis for its utility in the detection of stress is much the same as discussed for CIR film with the important exception that reflectance in several wavebands is monitored instead of being lumped together in a single wide waveband. A dense green canopy reflects very little of the red light (i.e., MSS5 = 0.6 to 0.7 μm) incident upon it but reflects very highly in the near-IR (MSS7 = 0.8 to 1.1 μm). Then as a canopy becomes stressed, senesces or experiences a reduction in biomass the reverse becomes true, visible reflectance increases while near-IR decreases (Gates, 1970).

Numerous indices that take advantage of these spectral relationships have been proposed to characterize the amount of green vegetation present (Deering et al., 1975; Kauth and Thomas, 1976; Richardson and Wiegand, 1977; Tucker, 1979; Jackson et al., 1980) and predict crop yields (Colwell et al., 1977; Nalepka et al., 1978; Tucker et al., 1980b; Aase and Siddoway, 1981). Employing different approaches to predict yield in drought-stressed small grains, Idso et al. (1980) found a positive correlation between final yield and the spectrally-measured rate of canopy senescence and Pinter et al. (1981) showed a strong relationship between yield and the integral of a spectrally derived vegetation index over the grain filling period.

An observer can determine whether growth is progressing as expected by tracking biomass levels with a vegetation index. Any departure from expected trends would flag potential physical or biological stresses. Although their precise identification might remain obscure, clues would exist in the temporal and spatial structure of the data. As an example, Tucker et al. (1979) showed that the presence of weeds between rows of soybeans caused the vegetation index to peak much earlier in the season than was observed in treatments where weed control was practiced. Aase and Siddoway (1980) suggested spectral data be used to monitor winterkill of wheat in Montana. They were able to detect when simulated stand survival fell below 40%, the threshold level used to establish whether reseeding is required.

There are obvious advantages in going directly to a vegetation target and monitoring its growth response to a complex of environmental conditions. Relatively short-term adaptation to an unfavorable situation (e.g., drought hardening) enables plants to optimize growth and development during conditions which otherwise might be deemed stressful. Monitoring the growth responses of the plants via a spectral approach would reveal this process whereas measuring the environmental stress by itself might lead to erroneous conclusions. Then, too, there is always the question of whether a measured meteorological parameter is relevant to conditions to which the plant is actually exposed in its microenvironment.

The spectral signature of an entire canopy is very much different than reflectances exhibited by single leaves measured *in vitro*. The three-dimensional spatial distribution of individual components within a canopy, ensures that patterns of sunlit and shaded vegetation and soil will be dependent upon such properties as plant density and row width, orientation, and spacing. As a result, the literal transfer of *in vitro* acquired reflectance data to the field is at best, very risky. Slight changes in the amount of sunlit soil present in a scene can alter overall canopy reflectance dramatically. Nevertheless studies on individual leaves are valuable in understanding the basic reflectance characteristics of leaf aggregates and in the identification of particular wavebands which may be useful in the field detection of stress. Such studies have shown that increased reflectance in visible, near-IR and middle-IR light accompanies a reduction in leaf water content (Knipling, 1970). Water absorption bands (1.55-1.75 μm) in the middle-IR may hold exceptional promise for the sensing of canopy water status (Tucker, 1980).

In vitro studies on pollution and nutrient stress show changes in leaf reflectance also. Gausman et al. (1978) showed that cantaloupe seedlings exposed to ozone could be distinguished from controls by higher reflectance in the 1.35-2.50 μm waveband. Likewise, Younes et al. (1974) reported that phosphate-deficient corn plants had different visible and near-IR characteristics than healthy plants. It appears that these laboratory and greenhouse studies can be extended to the field. Vegetation growing in heavily mineralized soils can often be identified by characteristic reflectance properties (Horler et al., 1980). Drewett (1975) used remotely sensed spectral data to monitor phosphate deficiency in spruce trees and Vickery et al. (1980) concluded that LANDSAT type data could be used to assess fertilizer levels in Australian pastures. Jackson et al. (1981c) found that nutrient deficiency in a sugarcane canopy was accompanied by increased visible and decreased near-IR reflectance.

In field studies it is difficult to determine whether the observed reflectance anomaly is due to stress-induced spectral changes in scene components or simply different levels of green biomass. The latter may prove true in many cases. If so, then the appropriate question becomes "Are the spectral properties resulting from a second-order effect of stress on biomass or from nonstress related differences in canopy density?" Certainly, this is a more difficult question, yet it must be addressed if remote sensing is to assume a major role in the detection of plant and/or microclimatic stress. Two specific examples from multispectral research at the U.S. Water Conservation Laboratory in Phoenix serve to illustrate these concepts. The first, a direct method, explores changes in reflectance, for wheat canopies of similar structure and density, that are due to different meteorological stress events. In the second, it is shown that similar spectral information can be a result of either normal low biomass levels (as during early canopy development) or those induced by drought stress. To deduce the cause in this indirect method, a multitemporal sequence of data collection is required.

3. Dew on wheat canopies

Qualitative observations on the effect of dew on spectrally derived vegetation indices from wheat canopies (*Triticum durum* Desf. var. Produra) were made during the past several years.

Measurements were taken at the same time each morning using a hand-held, 15° fov radiometer with a bandpass configuration similar to that on present LANDSAT platforms. Inspection of these data revealed that the ratio of radiances in MSS7 to MSS5 (i.e., 0.8 to 1.1 μm/0.6 to 0.7 μm) was significantly lower on days when dew was present. It returned to original levels when dew accumulation was prevented by wind movement or nighttime cloud cover.

Accordingly, an experiment was designed to quantify the relationship between the amount of dew present and the change in spectral data. On 19 March 1980 removable 2 x 2 m, opaque "dew-out" shelters were positioned about 50 cm above a section within each of five experimental plots, which due to staggered planting dates had wheat canopies at different stages of growth and biomass levels. The shelters eliminated dew formation by altering the down-welling thermal radiation regime in a fashion analagous to nighttime cloud cover. They were removed shortly after dawn on 20 March and spectral measurements over dewless sheltered and unsheltered targets with natural dew were begun at 0700h and repeated at 13 intervals throughout the day. Beginning at 0630h and continuing at approximately hourly intervals until 1100h, samples of the above ground vegetation in each plot were harvested for determination of the amount of dew present. This was accomplished by gently transferring the clipped vegetation to a tared plastic bag, which was then sealed and weighed in the laboratory. Regardless of the amount of dew present, each sample was blotted dry for the same amount of time (seven minutes) and reweighed. The amount of water present in each sample was converted to dew density (g m^{-3}) by dividing the weight of water by the ratio of the sample's green leaf area to the leaf area per m^2 of the plot, and then dividing that quotient by the height of the leaf portion of the canopy. This value was crudely adjusted to account for tissue water loss during the blotting period by subtracting the 1100h "dew density" (all visible dew had dissipated by that time) from each value.

The natural dew density is shown for each of the plots in Fig. 1. Curves through the data points were drawn by eye. The data show that dew accumulation was greatest on the plot with the highest green leaf area index (4B), and tapered off gradually with

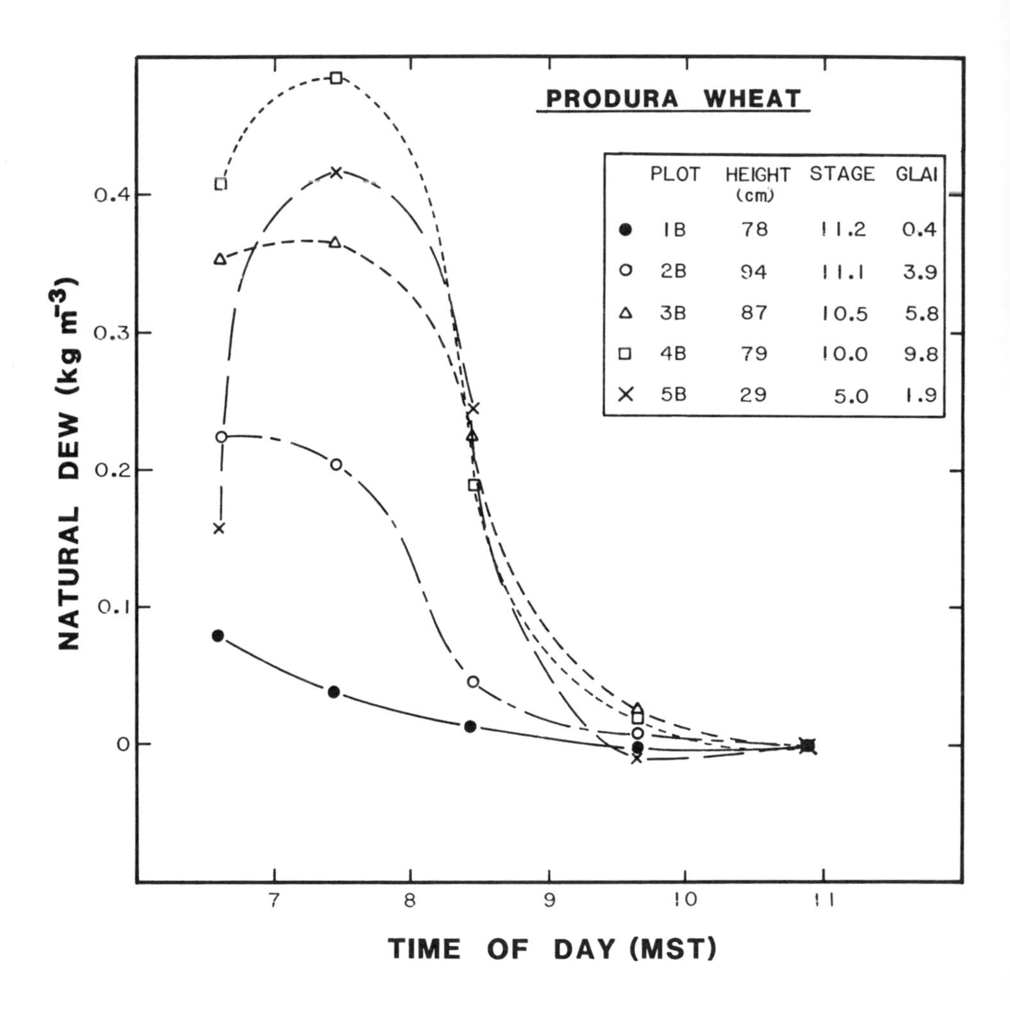

Fig. 1. The natural dew density on experimental wheat canopies at Phoenix, AZ, on 20 March 1980. Data on plant height, stage of growth (Feekes scale) and green leaf area index are also shown.

advancing phenological age and decreasing green leaf area. Plot 5B is an apparent exception. However, its canopy height was about 1/3 that of the other plots and increased dew density may be a result of substantially different micrometeorological conditions. The increase in dew density at 0730h in plots 4B and 5B was confirmed visually and on photographs. Dew dissipation followed time trends typical of late winter in Arizona with most noticeable dew disappearing shortly after 0900h.

Spectral data for the control (3B east with dew out shelter) and treatment areas (3B west, natural dew in the morning), are shown in Fig. 2. Note that the scale for the west subplot has been shifted upwards approximately 1.50 units, an alignment required to normalize the subplots for slight differences in green biomass. The lines represent 2 point moving averages through the data and the trends shown for 3B are representative of those observed in other plots. The area having natural dew formation in the morning (3B west) exhibits a lower IR/red ratio than the control. In the afternoon, an artificial dew condition was created in 3B east by misting the canopy with water 5-10 minutes prior to each set of readings. This caused a similar depression in the ratio. In both the natural and artificial dew cases, the change induced in the IR/red ratio is caused primarily by an increase in red radiance caused by the specular reflection from tiny dew droplets.

When the data for each subplot were aligned and the graphic separation [i.e., the difference between the upper (dew absent) and lower (dew present) lines of Fig. 2] calculated for each time period, a pattern emerged that was related to natural dew density. Figure 3 shows this relationship is linear and appears independent of green leaf area and phenological stage of development. A heavy dew density (0.4 kg m^{-3}) depressed the IR/red ratio by three regardless of its original value. By approximately 0930h the visible dew had almost disappeared from the canopies and the IR/red ratio had essentially returned to control levels.

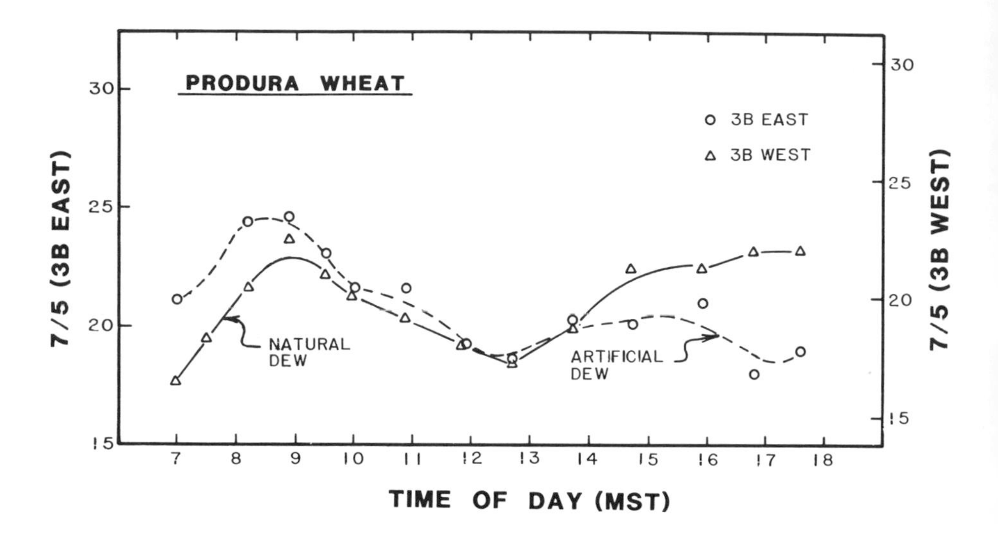

Fig. 2. The diurnal trend in a spectrally derived vegetation index, the ratio of MSS7 to MSS5, showing the influence of natural dew in the morning and simulated dew conditions in the afternoon. The wheat in plot 3B comprised a dense canopy at the time of heading.

Thus, it appears feasible to use remote spectral measurements to detect and quantify the formation and dissipation of dew. Dew assessment may also play a role in the prediction of dew-related plant diseases. In addition, the effects of dew on spectral vegetation indices merit serious consideration when timing aircraft or satellite data acquisitions over agricultural areas where dew phenomena occur. Failure to take dew into account could lead to serious errors in the estimation of agronomic parameters from spectral data.

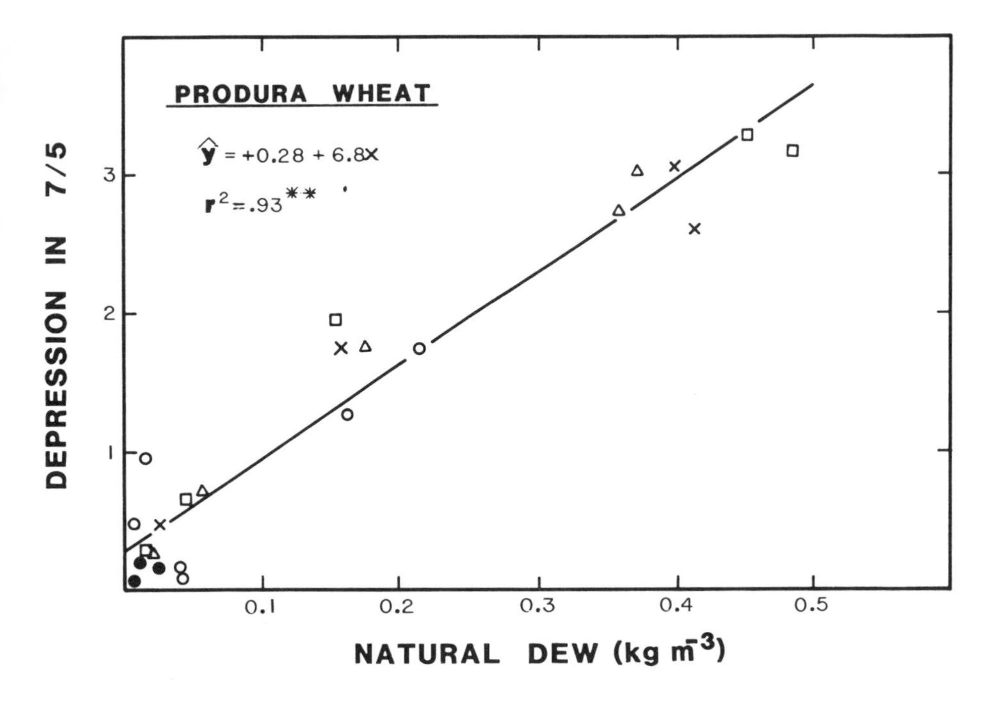

Fig. 3. The relationship between the depression in the vegetation index and the amount of dew present in the five wheat plots during the morning. The symbols are the same as those used in Fig. 1.

4. Moisture stress in alfalfa

Experiments were conducted at the University of Arizona Experiment Farm in Phoenix, to examine the spectral reflectance of alfalfa (*Medicago sativa* L.) grown under two different irrigation treatments. Beginning in late May 1980, observations were made in five borders, each with a uniform, vigorous nine-month-old stand of alfalfa. Then midway into the June harvest cycle, three borders were irrigated. The two remaining borders were not irrigated until the middle of the following harvest cycle. This provided an opportunity to examine the effects of limited soil moisture on spectral data obtained at low and high biomass levels.

Canopy development was characterized by clipping four random 0.25 m^2 above-ground vegetation samples from each border at twice weekly intervals, drying the samples at 70°C for 48-72h and then weighing for total dry biomass. Samples from borders with the same irrigation treatment were pooled (thus n = 12 for wet borders; n = 8 for the dry borders) and presented in Fig. 4. Biomass levels were very similar for both treatments prior to day 162 when borders 2, 3, and 5 were irrigated. Differences in biomass accumulation appeared soon after this time and persisted until harvest on day 176 when both canopies were reduced to about 100 g m^{-2}. (The number of unharvested samples on day 176 was reduced to four for both the wet and dry plots because of early arrival of the custom hay operators. Thus, the apparent downward trend of the irrigated biomass is suspected of being a sampling artifact.) Six days following the June harvest, the biomass in the irrigated treatment showed signs of rebound, while the biomass in the dry borders had little regrowth until after the mid-July irrigation.

Spectral data were collected at frequent intervals during this period using the hand-held, four-band radiometer described in the dew study. Measurements were taken along a continuous south to north transect which transversed all borders. Approximately 75-125 four-waveband scans per border were collected at evenly spaced intervals using a microprocessor-based data acquisition system (Allen et al., 1980). Data were collected at a morning solar altitude of 33° to minimize the effects changing sun angles might have on measurements. Results (Fig. 5) show the trajectory of a vegetation index, the IR/red ratio, with time for the two treatments. These data support visual and biomass sampling observations that the wet and dry treatments were very similar prior

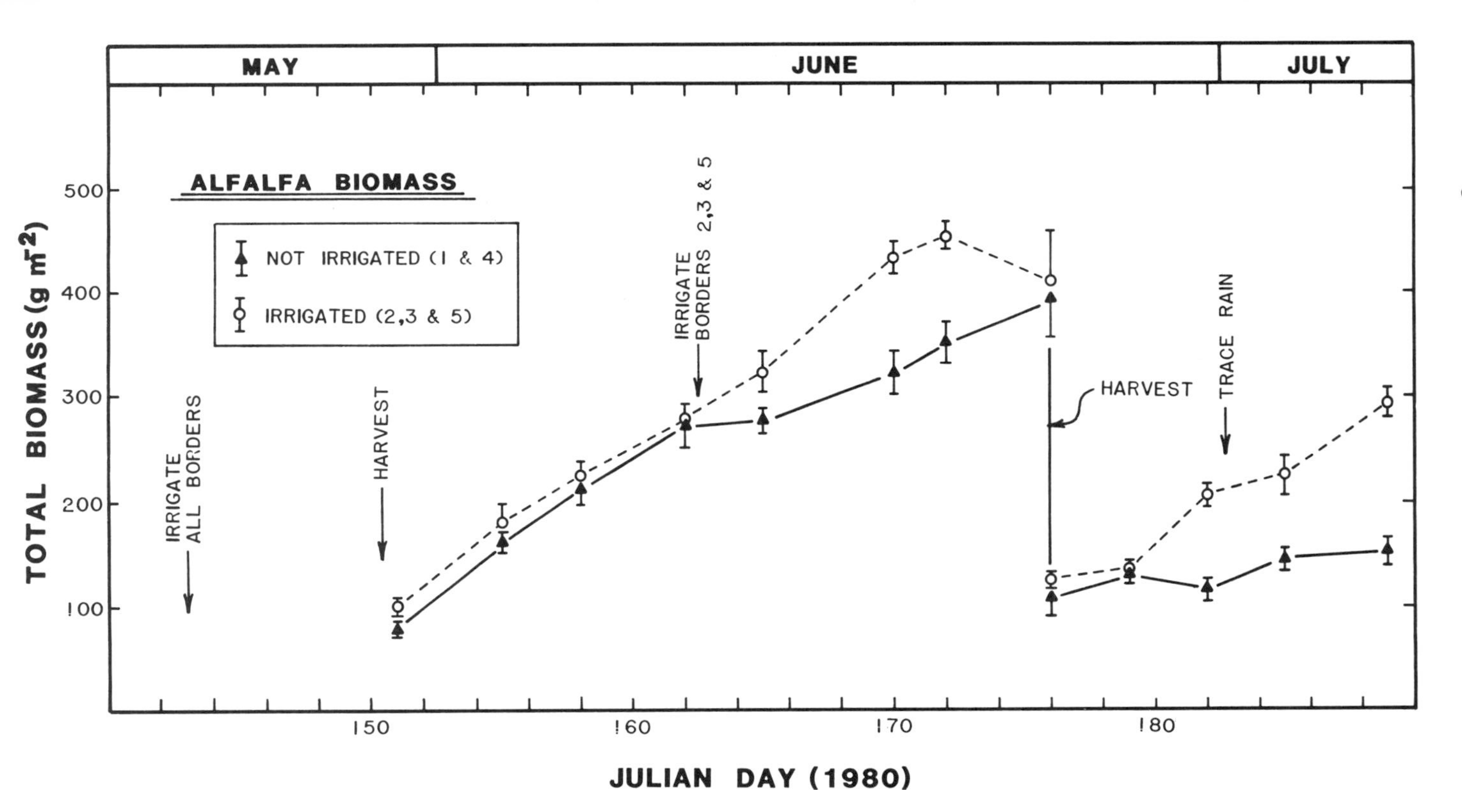

Fig. 4. The dry biomass of differentially irrigated alfalfa borders at Phoenix, AZ. Vertical bars represent $\pm$ standard error of 12- and 8- 0.25m^2 samples of aboveground vegetation in the irrigated and dry treatments, respectively.

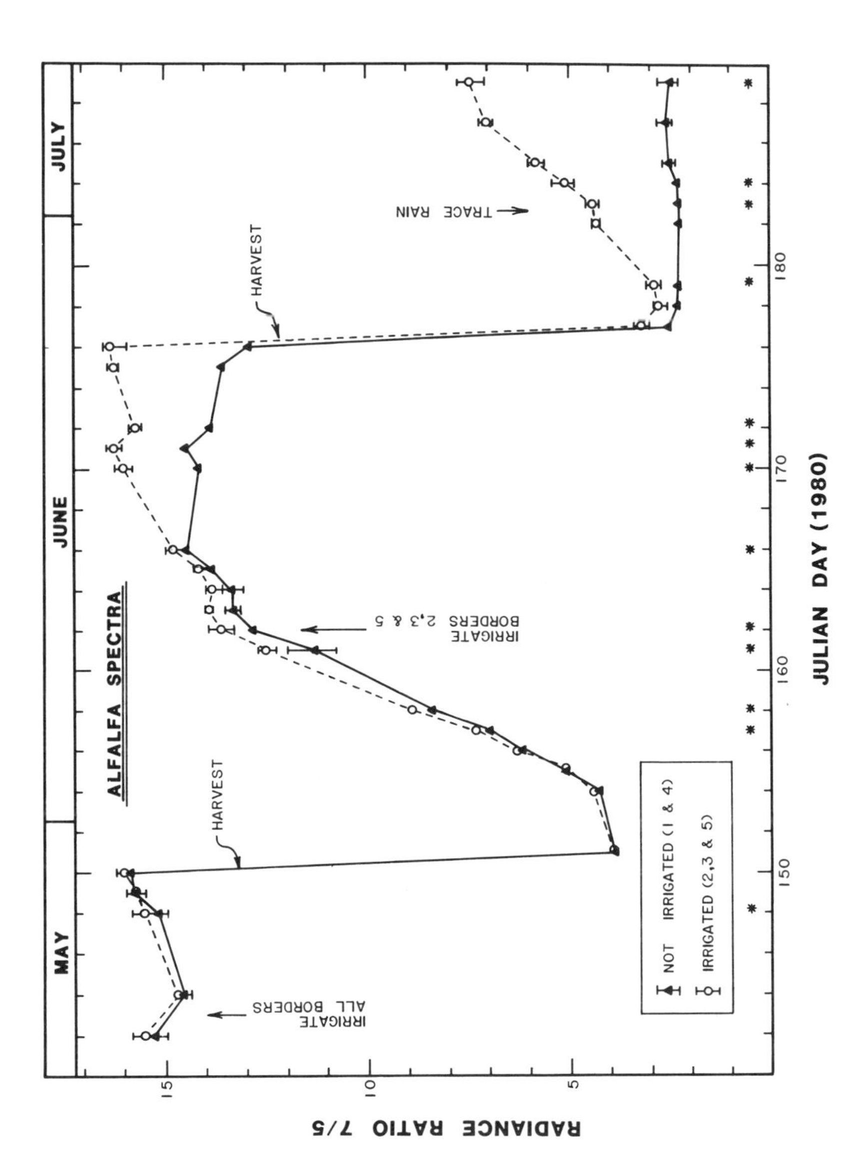

Fig. 5. The spectrally derived vegetation index (MSS7/MSS5) as a function of time for the irrigation treatments. Vertical bars represent ranges for treatment means which were derived from 75 to 125 measurements per transect. Where not shown, error bars were smaller than graph symbols.

to day 162. Following the differential treatment five to eight days elapsed before either visual observations or the spectral data indicated differences between treatments. The spectral index for the wet and dry borders continued to diverge up until day 176, then showing the abrupt effect of harvest, clearly distinguished the differences in regrowth between the two treatments. The data presented in Fig. 5 thus mirror the biomass data quite well, yielding useful information on the development of water stress, cultural operations, and rates of crop growth. However, note that the same ratio value is associated with both moderate biomass levels and drought stress. This is partly responsible for Tucker et al. (1980a) finding that spectral reflectance measurements are not good predictors of biomass when alfalfa becomes drought stressed. Thus to separate water-stress induced effects from those caused by biomass, the data must be interpreted within a multitemporal framework (as in Fig. 5) or with additional remote sensing information such as crop canopy temperature, a technique which will be explored in Section III, E and IV.

The transect data are unique in that windrowed, drying vegetation as well as clipped stubble and regrowth are included in the average spectral index value for the day. This means that these data are similar to those which aircraft and satellite platforms would observe in commercial fields (atmospheric effects aside). In addition, the spectral data were collected on 13 cloudy days when the sun was partially or completely obscured (asterisks in Fig. 5). Apparently one need not be strictly bound by clear-sky conditions for accurate interpretation of ratioed radiance data.

Potential uses for expanded coverage spectral data of this nature in IPM are bound only by the imagination of the user. For example, the cutting of alfalfa triggers migration of certain adult insects into adjacent crops where control measures are sometimes necessary and causes increased mortality among others unable to escape the abruptly inhospitable microclimatic conditions (Pinter et al., 1975). Knowledge of dates and patterns of cutting could minimize the impact of many economic insects on crops grown on a region wide basis.

Likewise, information on the extent of canopy development is a necessary input for algorithms which transform macrometeorological weather station data into temperature, humidity, and light parameters that are more meaningful on a microscale for crop growth, disease and pest simulation efforts. Mean daily temperature is perhaps the single most important input into many simulation models yet it is notoriously difficult to come by for remote areas. If a relationship between growing degree days and the rate of change in a spectral index of a nonstressed crop were established, then clearly a mean daily temperature could be derived from multidate spectral observations. These approaches are valid yet to my knowledge none have been tried. They certainly merit serious consideration for use in IPM.

III. THE THERMAL INFRARED

The nature of many fundamental meteorological processes can be characterized by the spatial and temporal variability of the earth's surface temperature. Infrared thermometry provides a noncontact method for documenting these patterns over wide geographical regions in a cost-effective manner. It is based on the principle that the thermal radiation emanating from a surface is proportional to the fourth power of its absolute temperature. This seemingly straightforward relationship is complicated by several factors which make it difficult to infer absolute temperatures from apparent radiant temperatures (Fuchs and Tanner, 1966; Perrier, 1971; and Sutherland et al., 1979). Fortunately relative apparent temperatures are sufficient for many applications of thermal IR data (e.g., comparison of targets with similar emissivities and under like sky conditions). When the emissivity of a surface is known [many vegetation surfaces range from 0.95 to 0.99 (Idso et al., 1969)] absolute temperatures can be estimated with an accuracy equal to the performance specifications of most sensors. As an example, an apparent crop temperature of 30 C measured with an 8-13 μm radiometer under clear sky conditions (i.e., 200 W m^{-2}, 8-13 μm sky irradiance) would translate into a 31.4 C absolute temperature for a crop with an emissivity of 0.97 (Fuchs and Tanner, 1966).

A. Meteorological Conditions

Meteorologists rely on frequent thermal IR images of cloud patterns from geostationary weather satellites for short and medium term mesoscale weather predictions. Although the resolution of these data are relatively coarse, they are useful in delineating and tracking long-term surface temperature patterns indicative of climatic trends. With improved resolution that is possible from aircraft or orbital satellites, patterns of cold air drainage and north-south slope temperature differences become evident (Vogler et al., 1974; Nixon et al., 1974; Stewart et al., 1978; and Schneider et al., 1979). Growers could take advantage of those areas where early season soil temperatures are conducive to germination, and avoid those with high freeze, frost or dew potential. Once the spatial patterns of temperature are defined, intelligent decisions can be made to ensure that meteorological stations are sited in representative areas. Similar data could identify the optimum location for orchard heaters or wind machines; infrared radiometers switching them on or off as local conditions dictate. One particularly promising but neglected aspect is the direct input of IR derived temperatures into biological models. In many cases these data would be more relevant than those from a distant meteorological station. Pinter and Butler (1979) established the magnitude of errors involved when predicting insect development within irrigated fields using National Weather Service temperatures instead of microclimatic parameters. Giddings (1976) reported on an algorithm that predicted daily air temperatures over a broad geographical region from meteorological satellite data. These data were intended as inputs for a population growth model in the screwworm eradication program in the southern United States and Mexico.

B. Soil Moisture

The complexities which govern the transfer of heat within and from the soil enable predictions concerning relative water content, the proximity of subsurface ground water, soil composition, texture, slope and aspect to be made from repetitive IR sensing to be made (Myers et al., 1970). Some potential applications of IR thermometry in soil moisture detection have been reviewed by Idso et al. (1975). They demonstrated that wet soils remained cooler than those with a lower moisture content. In fact, where both incident solar radiation and evaporative demand are high, differences of 30°C are not uncommon. The amplitude of daily and annual surface temperature oscillations reveals subsurface moisture, soil depth and physical properties which are not apparent from once-a-day measurements. This type of "thermal inertia" approach is useful in detecting different soil moisture levels as well as geological formations (Pratt and Ellyet, 1979). Dalsted et al. (1979) found a thermal IR technique superior to CIR photography in the detection of incipient, saline seeps. Other related uses include locating shallow groundwater sites and blocked field drainage tiles.

C. Crop Water Stress and Yield

The energy balance of a crop canopy is largely a result of the micrometeorological conditions to which it is exposed. These, of course, are integrated within the context of adaptive physiological and morphological mechanisms as well as the availability of microenvironmental resources such as water and nutrients. Stresses which interfere with the partitioning of energy fluxes within the plant canopy often result in elevated temperatures. This forms the basis for the utility of IR thermometry in assessing crop vigor, scheduling irrigations and predicting yields.

Many well-watered agricultural crops have midday radiant canopy temperatures at or below that of the ambient air. With ample water, evapotranspiration (ET) proceeds at near potential levels and cools foliage at a rate proportional to the evaporative demand of the atmosphere. The amount of cooling appears to be a function of air vapor pressure deficit (VPD) in cotton (Ehrler, 1973), wheat, alfalfa, soybeans, and kidney beans (Idso et al., 1981a; Walker, 1980). In fact, Idso et al. (1981a) showed that well-watered alfalfa was 3°C cooler than air under relatively moist (2 kpa VPD) conditions in Arizona; under dry (5 kpa VPD) they found a 8-10°C depression below air temperatures.

As moisture in the root zone becomes limiting, ET decreases, and canopy temperatures rise relative to nonstressed crops and air temperature (Tanner 1963; Wiegand and Namken, 1966). Severely stressed alfalfa and soybeans had apparent canopy temperatures several degrees warmer than ambient air (Idso et al., 1980b). This relationship suggests a method to evaluate the water status of entire canopies that has obvious advantages over the conventional, time consuming approaches currently used (Bryne et al., 1979). Indeed, evidence is emerging which establishes IR thermometry as a reliable surrogate for leaf water potential observations (Ehrler et al., 1978; Idso et al, 1981b). In addition to the ease and rapidity with which measurements can be taken, it is nondestructive and remote, permitting repetitive measurements over the same crop day after day without interferring with its growth. Furthermore, it is an averaging technique and not as subject to the sampling variability associated with single plant or leaf measurements.

Previsual warning of stress is especially important when management options exist to alleviate the condition before yield reduction occurs. Thus, it follows that irrigation scheduling is a logical application of IR thermometry. Hiler et al. (1974) proposed that a "stress day index" be employed for irrigation timing purposes. Although they developed the index using plant and soil water potential, they suggested that a canopy-air temperature differential could be used in a similar fashion. Subsequently, Jackson et al. (1977) pursued a thermal approach for scheduling water application in wheat. Their technique was to sum the positive stress degree days (SDD is the difference between radiant canopy and air temperature) beginning with an irrigation, continuing until the accumulated SDD's reached a predetermined level and then applying water once again.

Noting that the SDD parameter, of itself, bears no simple relationship to available soil moisture, several refinements were proposed which take the evaporative demand of the atmosphere into account. The first refinement is mostly empirical, expanding upon the aforementioned relation between the canopy-air temperature differential of a well-watered crop versus VPD and also the upper temperature limit observed for a severely water stressed non-transpiring crop which is independent of VPD (Idso et al., 1981a). It appears that both these parameters are crop specific and, because a major determinant of evaporative demand (i.e., VPD) is included, independent of geographic location. A dimensionless crop stress index is defined within this context and is shown to characterize water stress better than the conventional SDD.

A second approach is based upon theoretical considerations of a canopy energy balance model (Jackson et al., 1981b). For a given set of ambient air temperatures, VPD, wind and net radiation parameters, the domain over which the canopy minus air temperature can be expected to vary solely due to extremes in latent heat exchange is calculated. A water stress index is then derived as the ratio of the difference between the observed canopy minus air temperature and that calculated for a well-watered crop, to the range of possible extremes. Figures presented in Jackson et al. (1981b) show correspondence between the water stress index and total extractable soil water. Both the Idso et al. (1981a) and Jackson et al. (1981b) approach have considerable merit in scheduling irrigations.

The utility of crop temperatures in the prediction of final yield has been established for a number of agricultural crops (Idso et al., 1977; Reginato et al., 1978; Walker and Hatfield, 1979; Idso et al., 1980b; Gardner, 1980). In general, crops which experience minimal stress throughout their growth have cooler temperatures and higher yields than those encountering less favorable conditions. The yield prediction technique uses a temperature index (either the SDD or, alternatively, the difference in temperature between a stressed crop and a well-watered crop measured under the same macroclimatic conditions) which is summed over critical periods of crop development. The summations are inversely correlated with seed and/or biomass yield.

Walker (1980) assessed the validity of this approach for yield modeling. He emphasized that the intervals over which the index is accumulated, and the techniques used for obtaining infrared temperature have a significant bearing on the final results. He draws the distinction between canopy temperatures and foliage temperatures even though both are obtained by viewing the plants at an oblique angle. The former includes both plant and background soil temperatures when plant densities are low. The latter, however, represents temperatures of only plant elements. Walker's results demonstrated that canopy temperatures were superior for yield predictions in kidney beans, probably because they provided indirect information on plant density which influenced yield but was not contained in foliage temperatures.

D. Other Plant Stresses

Fox (1978) defines previsual detection as the "detection of vegetation damage through remote sensing before a person using their eyes, at close range, could detect the damage." He points out that the remote sensing of temperature changes accompanying plant water stress is the only well-documented example of previsual detection. In theory, at least, any stress of physical or biological origin which interferes with the movement of water through the soil-plant-atmosphere continuum should be detectable previsually by using a thermal approach.

Reports in the literature support this hypothesis. For example, Myers et al. (1970) indicate a positive correlation between soil salinity and cotton leaf temperature. They found that cotton plants growing in salty soils were as much as 5°C warmer than those in low salt soils. They further state that salinity-induced differences will be enhanced under low soil moisture conditions and high atmospheric evaporative demand. Canopy temperatures are useful for discriminating between certain diseased and healthy plants (De Carolis et al., 1974 and 1976; Olson, 1977); however, it is not always possible to discern from these reports whether the detection was previsual. Kumar et al. (1978) present results showing that radiant temperatures of corn canopies increased with the severity of southern corn blight disease. Since contact temperatures of blighted lesions and healthy foliage revealed no systematic differences, they concluded that the higher radiant canopy temperatures resulted from less dense canopy cover conditions where blight levels were more severe. Hence, the radiometer was viewing a higher proportion of hot sunlit soil under those conditions.

1. Root rot in sugar beets

In an effort to determine the feasibility of using IR thermometry to distinguish different levels of biological stress before they became obvious visually, Pinter et al. (1979) investigated sugarbeets infected with a root rotting, soil borne fungus, _Pythium aphanidermatum_ Edson (Fitz). This organism, which is indigenous to Arizona, has an optimum temperature range of 31-37°C for germination and vegetative growth (Stanghellini and Russell, 1973). As a consequence, infection of mature sugarbeet roots often begins in late May and increases rapidly with time as soil temperatures become warmer. Uninfected beets continue to increase in sugar content until harvest which normally occurs in

June or July, but the recoverable sugar in infected beets declines sharply as the decay progresses. Although completely decayed beets are left in the field, partially infected beets are harvested. Processing of them, however, not only yields less sugar, but also reduces the efficiency of the extraction process. Thus, this disease represents a significant loss to the processors as well as the individual growers. At present, a method for control of this disease is not available. Above ground symptoms are not visible until extensive root rot has occurred. The beet is then worthless from a sugar standpoint. If a method for timely previsual detection of disease incidence and rate of infection were available, harvests could be scheduled earlier to maximize economic return.

Hand-held infrared thermometers were used to measure the radiant leaf temperatures of plants in commercial fields located south of Phoenix, Arizona. Then, because above ground symptoms of disease were not evident, we pulled each plant from the ground and two observers classified the root according to degree of root involvement with the fungus. Three disease categories were established: 1) healthy, no evidence of disease; 2) slightly diseased, less than 10% of the root volume was damaged; and 3) moderately diseased, 10 to 60% of the root volume destroyed.

Results (Table 2) showed that moderately diseased sugarbeets had leaves several degrees warmer than healthy plants. An important finding was that this differential persisted despite progressively drier soil conditions on each sampling date, even on 6 July when wilting occurred to all plants. Slightly diseased plants could not be distinguished thermally from healthy plants implying that the disease had not progressed to the point where transpiration was reduced.

The radiant temperatures shown in Table 2 reflect the plant's integration of atmospheric evaporative demand, available soil moisture and physiological status as evidenced by substantial changes in temperature over the three week period for each disease

Table 2. Mean leaf temperature (±1SE) of sugarbeets classified according to root damage by *P. aphanidermatum*. Data were collected south of Phoenix during 1978 from live green leaves. Sample size in parenthesis. Ambient air temperatures and vapor pressures were calculated from an aspirated psychrometer positioned 1 m above the soil surface.

		Radiant Leaf Temperatures (°C)				
Date	Time (MST) (h)	Healthy noninfected plants	Slightly diseased plants	Moderately diseased plants	Ambient air temp. (°C)	Vapor pressure (mb)
19 June	1120-1230	31.5 ± 0.36	31.5 ± 3.35	34.5 ± 1.13	37.3	18.5
		(27)	(3)	(5)		
29 June	1200-1400	33.1 ± 0.49	32.4 ± 2.00	35.7 ± 1.46	35.9	18.9
		(11)	(2)	(5)		
6 July	1200-1400	38.5 ± 0.41	37.6 ± 0.80	42.1 ± 0.52	38.6	7.5
		(33)	(7)	(13)		

category. Attempts to evaluate the disease parameter using infrared thermometry alone may be frustrated unless the soil moisture and evaporative demand are also known. In this study, this problem was largely circumvented by examining plants adjacent to one another, and thus exposed to the same soil and macroclimate. However, once baseline data which relate the air-foliage temperature differential to VPD become available for nonstressed sugarbeets then temporal and spatial variability patterns may permit separation of disease from water stress effects.

2. Root rot in cotton

The soil-borne fungus, _Phymatotrichum omnivorum_ (Shear) Dug., manifests its symptoms in cotton during the hottest periods of the season, striking the roots of an otherwise healthy-appearing plant, causing it to wilt severely and die within several days after the onset of the first visible above-ground symptoms. Mainly because of the change in apparent canopy cover, aerial photographic (Henneberry et al., 1979) and probably thermal scanner techniques would delineate infested areas within fields very well. However, in order to determine whether a thermal approach would give a previsual indication of disease, radiant temperature measurements were taken of cotton plants in commercial fields near Marana, Arizona (Pinter et al., 1979). The apparent temperatures of three fully expanded leaves near the top of each plant were measured using a hand-held infrared thermometer. Then each plant was extracted from the soil and assigned to one of three vigor categories based on the appearance of the root system: 1) healthy; 2) slightly diseased with discoloration of the root vascular structures; or 3) moderately diseased with root decay, sloughing of the outer root cortex and a network of fungal mycelia on the surface of the main root. Dead plants were avoided in this study, also.

Table 3 shows results similar to those in the sugarbeet disease investigation. Moderately diseased plants had leaves that

Table 3. Mean leaf temperatures (±1SE) of cotton plants classified according to root damage by *P. omnivorum*. Data were collected near Marana, Arizona, during 1978 from live green leaves. Sample size in parenthesis. Ambient air temperatures and vapor pressures were calculated from an aspirated psychrometer positioned 1 m above the soil surface.

		Radiant Leaf Temperatures (°C)				
Date	Time (MST) (h)	Healthy noninfected plants	Slightly diseased plants	Moderately diseased plants	Ambient air temp. (°C)	Vapor pressure (mb)
18 August	1104-1145	29.3 ± 0.29 (21)	37.5 ± ... (1)	32.9 ± 1.03 (8)	32.0	24.3
18 August	1215-1228	31.2 ± 0.34 (15)	30.9 ± 0.82 (6)	36.5 ± 1.03 (9)	33.9	22.7
24 August	0930-1000	29.2 ± 0.51 (13)	31.0 ± 0.97 (13)	33.0 ± 1.02 (16)	30.2	24.5
24 August	1100-1114	29.2 ± 0.13 (28)	30.2 ± 0.40 (9)	34.1 ± 0.85 (3)	31.2	24.5
24 August	1440-1448[a]	30.5 ± 0.57 (5)	29.9 ± ... (1)	33.8 ± 1.08 (4)	35.0	20.2

[a] Data for this time period only were from Pima cotton (*Gossypium barbadense* L.). Remaining cotton data were from Deltapine cotton (*G. hirsutum* L.).

were 3 to 5°C higher than those on adjacent healthy plants. Even though leaves from slightly diseased plants were warmer than healthy plants, they were also more variable precluding their identification as a separate class. Interestingly, the difference between healthy and moderately diseased plants prevailed on 18 August several days after an irrigation when visual clues to disease (i.e., wilting) were not reliable. However, the moderately diseased plants were the first to wilt as the soil began to dry and although the thermal data still enabled clear separation of the extreme disease classes, it failed to provide additional information that was not visually available.

These results demonstrate the utility of infrared thermometry in detecting soil-borne fungal diseases of sugarbeets and cotton. It is likely that the principles involved will be applicable to other plant stresses caused by disease, nematodes, and insects. Within this context, plant physiologists and pathologists will likely find infrared thermometry a valuable nondestructive diagnostic research tool. Plant breeders could use this technique to improve the efficiency of genetic screening. Realistic adjustments to plant growth models to account for yield losses due to disease or pest infestation, could be made if quantitative information on the nature and extent of biological stresses were available over broad regions. Finally, since the goal of investigations such as these is to assist the grower, previsual detection (even with a ground-deployed, handheld instrument) would enable more effective crop management through selective culling, replanting, or treatment of unhealthy individuals.

E. Water Status of Alfalfa

In Section II, B, 4, a ratio of near-IR visible reflected light was used as a vegetation index that permitted discrimination between differentially irrigated alfalfa plots, but only after differences in biomass had developed between the two treatments (Figs. 4 and 5). It is reasonable to assume that prior to a change in biomass accumulation or a major canopy architectural change, a reduction in crop evapotranspiration would occur. If this proved true, then the thermal IR would provide an earlier warning of impending water stress. That this is indeed the case is shown in Fig. 6, where apparent radiant canopy temperatures are plotted versus Julian day for the same alfalfa experiment presented in Figs. 4 and 5.

Note that the temperature differences between the treatments were negligible immediately before and after the May harvest, implying the rates of crop evapotranspiration were very similar during this period. On day 162, the differential irrigation had a more pronounced effect in the thermal IR than it had on the spectrally derived vegetation index. In fact, almost immediately following the irrigation, temperatures were depressed several degrees below those observed in the dry treatment. By the following day, these differences were significant and continued to increase until by day 175, just before harvest, they reached a maximum of 9°C. Subsequent to the June harvest, the irrigated treatment maintained a 3-4°C temperature differential due largely at first to evaporation of residual surface soil moisture then later to the rapid regrowth of alfalfa (see also Fig. 4) and its aasociated higher rate of transpiration. Thus, it appears that the thermal infrared provides an earlier and perhaps less ambiguous warning of impending stress than reflected visible and near-IR light. Additionally, it provides supplementary information on soil moisture which is important for regrowth potential.

IV. NEED FOR AN INTEGRATED APPROACH FOR STRESS DETECTION

In many circumstances, a single remote sensing technique might not be sufficient to either identify a target or separate a particular agronomic parameter from a crop stress. Although important clues are often present in the spatial and temporal structure of the data, it is very likely that several different approaches will be required to resolve ambiguities. A case in

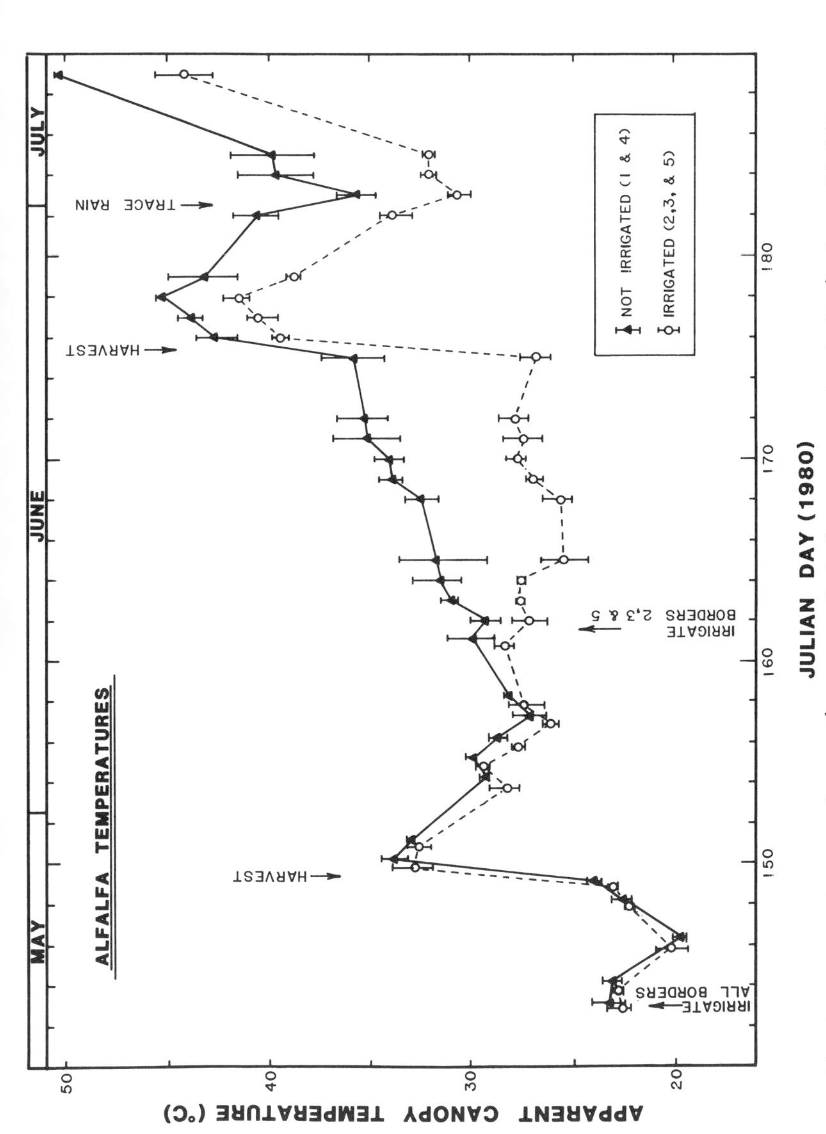

Fig. 6. The radiant canopy temperatures (uncorrected for crop emissivity) for alfalfa grown at Phoenix, AZ, and exposed to the two irrigation regimes from May to July 1980. Data were collected at approximately 1400h by viewing the canopy at an oblique angle so that predominantly vegetation was observed. Vertical bars represent the range of means observed for each treatment.

point involves the thermal data presented for alfalfa in Section III, E, which characterize the short term, almost immediate, response of the canopy to an irrigation. Unfortunately other stresses beside drought are also capable of causing an increase in canopy temperature. Examples include diseases which impair water transport, lodging and skeletonizing of leaflets by herbivorous insects. Furthermore, harvest reduces the full canopy to stubble as often as once every four weeks in some regions with the result that "canopy" temperatures rise very abruptly and are decoupled from the available soil water. One simple way to determine whether elevated temperatures are a result of a recent harvest or water stress is to interpret the thermal data within the context of the canopy biomass information provided by the vegetation index shown in Fig. 5 (Section II, B, 4). Even though the vegetation index did not by itself permit discrimination between moderate biomass and water stress conditions, it certainly gave an unambiguous characterization of canopy development from one cutting to the next.

The partial canopy problem which arises during stand establishment or regrowth is one that has made the interpretation of thermal data from aircraft very difficult because the downward-pointed sensor perceives a composite of soil and vegetation temperatures. Millard et al. (1980) compared aircraft thermal scanner data with simultaneously acquired ground-based canopy temperatures of experimental plots with varying canopy densities. There was good agreement between the two techniques provided the green canopy cover exceeded 75%. Under lower percent cover conditions, aircraft data were 5-6°C warmer than those measured on the ground due to the influence of the hot dry soil. Partial canopy problems can be minimized by only accepting data beyond a certain scanner look angle from aircraft (Jackson et al., 1979) or by deploying a ground-based radiometer at a glancing or oblique angle so that mostly vegetation is viewed. Kimes et al. (1980) proposed that a canopy density or probability gap function be used in a simple model to transform composite temperatures acquired from a nadir-pointed sensor into temperatures more characteristic of foliage alone. It is likely that a spectrally derived vegetation index may serve the same purpose (Cihlar, 1980). Reflectance measurements from satellites can provide seasonal estimates of leaf area index for entire growing regions, an essential input for meteorologically based evapotranspiration and crop modelling approaches (Wiegand et al., 1979).

An integrated approach will also be required before the large scale application of thermal data for disease detection in sugarbeets (Section III, D, 1) can be realized. At close range with a hand-held instrument it was simple to measure the temperature of single green leaves with a narrow field-of-view radiometer. However, by the time major symptoms of root rot in sugarbeets begin to show up in Arizona, most fields have passed their peak of green leaf area index and have less than 100% canopy cover. Interpretation of data from a thermal scanner is virtually impossible because data pixels represent various pure and mixed combinations of green plant parts, senescent leaves and soil targets. With sunlit soil temperatures reaching 60 C, only a small portion of soil need be included within a pixel to shift its apparent temperature by +3°C which is indicative of a diseased plant. This problem can probably be handled by using a technique referred to as a layer classifier (Bartolucci et al., 1975). In this approach, thermal scanner data are registered precisely with multispectral scanner data in the visible and near IR. Then, the pure green vegetation pixels are sorted from the mixed pixels by their low visible and high near IR reflectance, and temperatures from only these pixels are used in the analysis. A similar approach may be a requirement for data interpretation from row crops which have incomplete canopies. In these situations, flight direction, scan direction, sun angles and row orientation all interact to form a confusing composite of temperatures from shaded and unshaded components. Without additional information on canopy geometry provided by multispectral techniques, temperature data could not be interpreted.

It is naive to believe that remote sensing will be a panacea for all our agricultural resource management needs. Indeed a number of authors have pointed out that we have been guilty of overselling remote sensing in the past, promising more than we could deliver with our present level of technology (Sharp, 1979; Beaubien, 1979; and Dellwig and Bare, 1978). Many instances where remote sensing approaches have proven useful in the detection of plant stresses were treated on a unique case-by-case basis, thus it has been difficult to transfer resulting technology to an applied program. Hopefully, we are now past the "Gee whiz, look what remote sensing can do for you" syndrome described by Heller (1978) and can begin to tackle issues which are fundamental to the basic understanding of remotely sensed parameters and then apply them to solve some of our more difficult agricultural management problems.

ACKNOWLEDGMENTS

Appreciation is extended to Mr. Ronald Seay, Mr. Paul Flowers, and Ms. Karen Adamson for technical assistance in the field and to Drs. Ray D. Jackson, Sherwood B. Idso, Robert J. Reginato, and Jerry L. Hatfield for helpful discussions during field experiments and manuscript preparation.

REFERENCES

Aase, J. K., and Siddoway, F. H. (1980). Agron. J. 72:149-152.

Aase, J. K., and Siddoway, F. H. (1981). IEEE Trans. Geosci. and Remote Sensing (in press).

Allen, R. F., Jackson, R. D., and Pinter, P. J., Jr. (1980). Agric. Eng. 61:12-14.

Bartolucci, L. A., Swain, P. H., and Wu, C. L. (1975). IEEE Trans. Geosci. Electron., GE-14(2). 16 p.

Bauer, M. E. (1975). In "Advances in Agronomy," 27:271-304. Academic Press, New York.

Bauer, M. E., Swain, P. H., Mroczynski, R. P., Anuta, P. E., and MacDonald, R. B. (1971). Proc. Symp. Remote Sensing Environ. 7th. Ann Arbor, Michigan. pp. 693-704.

Beaubien, J. (1979). Photogr. Eng. and Remote Sensing 45:1135-1144.
Beer, T. (1980). Remote Sensing of Environ. 9:65-85.
Byrne, G. F., Begg, J. E., Fleming, P. M., and Dunin, F. X. (1979). Remote Sensing of Environ. 8:291-305.
Chiang, H. C., Meyer, M. P., and Jensen, M. S. (1976). Entomologia Experimentalis et Applicata 20:301-303.
Cihlar, J. (1980). Int. J. Remote Sensing 1:167-173.
Colwell, R. N. (1956). Hilgardia 26:223-286.
Colwell, J. E., Rice, D. P., and Nalepka, R. F. (1977). Proc. Symp. Remote Sensing Environ. 11th. Ann Arbor, Michigan. pp. 1245-1254.
Curran, P. J. (1979). Remote Sensing of Environ. 8:249-266.
Dalsted, K. J., Worcester, B. K., and Brun, L. J. (1979). Photo. Eng. and Remote Sensing 45:285-291.
De Carolis, C., Baldi, G., Galli de Paratesi, S., and Lechi, G. M. (1974). Proc. Symp. Remote Sensing Environ. 9th. Ann Arbor, Michigan. pp. 1161-1170.
De Carolis, C., Conti, G. C., and Lechi, G. M. (1976). Proc. Symp. Remote Sensing Environ. 10th. Ann Arbor, Michigan. pp. 1219-1229.
Deering, D. W., Rouse, J. W., Jr., Haas, R. H., and Schell, J. A. (1975). Proc. Symp. Remote Sensing Environ. 10th. Ann Arbor, Michigan. pp. 1169-1178.
Dellwig, L. F., and Bare, J. E. (1978). Photogr. Eng. and Remote Sensing 44:1411-1419.
Drewett, R. J. (1975). Proc. Symp. Remote Sensing Environ. 10th. Ann Arbor, Michigan. pp. 1123-1131.
Eastman Kodak. (1972). "Applied Infrared Photography." Publication M-28, Rochester, New York. 88 p.
Ehrler, W. L. (1973). Agron. J. 65:404-409.
Ehrler, W. L., Idso, S. B., Jackson, R. D., and Reginato, R. J. (1978). Agron. J. 70:999-1004.
El-Baz, F. (1978). Photogr. Eng. and Remote Sensing 44:69-75.
Fox, L., III. (1978). In "Symposium on Remote Sensing for Vegetation Damage Assessment," (P. A. Martha, ed.), pp. 53-64. Am. Soc. of Photogr., Falls Church, Virginia.
Fritz, N. L. (1967). Photogr. Eng. 33:1128-1138.
Fuchs, M., and Tanner, C. B. (1966). Agron. J. 58:597-601.
Gardner, B. R. (1980). "Plant Canopy Temperatures in Corn as Influenced by Differential Moisture Stress," Agricultural Meteorology Progress Report 80-1, pp. 119. Center for Agric. Meteor. and Climatol., University of Nebraska, Lincoln, Nebraska.
Gates, D. M. (1970). In "Remote Sensing with Special Reference to Agriculture and Forestry," pp. 224-252. Nat. Acad. of Sci., Washington, D.C.
Gausman, H. W., Escobar, D. E., Rodriquez, R. R., Thomas, C. E., and Bowen, R. L. (1978). Photogr. Eng. and Remote Sensing 44:481-485.
Gautier, C., Diak, G. and Masse, S. (1980). J. Appl. Meteor. 19:1005-1012.

Giddings, L. E. (1976). "Remote Sensing for Control of Tsetse Flies," Lockheed Electronics Co., Inc., Houston, Texas. Technical Memorandum Job Order 92-105. (National Technical Information Service - HC A03/MF A01 CSCL 06F.

Green, L. R., Olson, J. K., Hart, W. G., and Davis, M. R. (1977). Photogr. Eng. and Remote Sensing 43:1051-1057.

Hart, W. G. and Meyers, V. I. (1968). J. Econ. Entomol. 61:617-624.

Hart, W. G., Ingle, S. J., Davis, M. R., Mangum, C., Higgins, A., and Boling, J. C. (1971). In "Proceedings Third Biennial Workshop on Color Aerial Photography in the Plant Sciences. Gainesville, Florida. pp. 99-113.

Heller, R. C. (1970). In "Remote Sensing with Special Reference to Agriculture and Forestry," pp. 35-72. Nat. Acad. of Sci., Washington, D.C.

Heller, R. C. (1978). In "Symposium on Remote Sensing for Vegetation Damage Assessment," (P. A. Murtha, ed.), pp. 231-252. Am. Soc. of Photogr., Falls Church, Virginia.

Henneberry, T. J., Hart, W. G., Bariola, L. A., Kittock, D. L., Arle, H. F., Davis, M. R., and Ingle, S. J. (1979). Photogr. Eng. and Remote Sensing 45:1129-1133.

Hielkema, J. U., and Howard, J. A. (1976). "Pilot Project on the Application of Remote Sensing Techniques for Improving Desert Locust Survey and Control." Final Report, Food and Agriculture Organization, Rome. 70 p.

Hiler, E. A., Howell, T. A., Lewis, R. B., and Boos, R. P. (1974). Trans. Am. Soc. Agric. Eng. 17:393-398.

Horler, D. N. H., Barber, J., and Barringer, A. R. (1980). Int. J. Remote Sensing. 1:121-136.

Idso, S. B., Jackson, R. D., Ehrler, W. L., and Mitchell, S. T. (1969). Ecol. 50:899-902.

Idso, S. B., Jackson, R. D., and Reginato, R. J. (1975). Am. Scientist 63:549-557.

Idso, S. B., Jackson, R. D., and Reginato, R. J. (1977). Science 196:19-25.

Idso, S. B., Pinter, P. J., Jr., Jackson, R. D., and Reginato, R. J. (1980a). Remote Sensing of Environ. 9:87-91.

Idso, S. B., Reginato, R. J., Hatfield, J. L., Walker, G. K., Jackson, R. D., and Pinter, P. J., Jr. (1980b). Agric. Meteor. 21:205-211.

Idso, S. B., Jackson, R. D., Pinter, P. J., Jr., Reginato, R. J., and Hatfield, J. L. (1981a). Agric. Meteor., 24:45-55.

Idso, S. B., Reginato, R. J., Reicosky, D. C., and Hatfield J. L. (1981b). Agron. J., (in press).

Jackson, R. D., Reginato, R. J., and Idso, S. B. (1977). Water Resour. Res. 13:651-656.

Jackson, R. D., Reginato, R. J., Pinter, P. J., Jr., and Idso, S. B. (1979). Appl. Opt. 18:3775-3782.

Jackson, R. D., Pinter, P. J., Jr., Reginato, R. J., and Idso, S. B. (1980). "Hand-Held Radiometry." U.S. Dept. of Agric., S.E.A. ARM-W-19. 66 p.
Jackson, R. D., Salomonson, V. V., and Schmugge, T. J. (1981a). Proc. Second National Irrigation Symp. Am. Soc. of Agric. Eng., Lincoln, Nebraska. October 20-23, 1980., pp. 197-212.
Jackson, R. D., Idso, S. B., Reginato, R. J., and Pinter, P. J., Jr. (1981b). "Canopy temperature as a crop water stress indicator." Water Resour. Res., (in press).
Jackson, R. D., Jones, C. A., Uehara, G., and Santo, L. T. (1981c). "Remote detection of nutrient and water deficiencies in sugarcane under variable cloudiness." Remote Sensing of Environ., (in press).
Kauth, R. J., and Thomas, G. S. (1976). Proc. Symp. Machine Processing of Remotely Sensed Data 4B:41-51.
Kimes, D. S., Idso, S. B., Pinter, P. J., Jr., Jackson, R. D., and Reginato, R. J. (1980). Appl. Opt. 19:2162-2168.
Knipling, E. B. (1970). Remote Sensing of Environ. 1:155-159.
Kumar, R., Silva, L. F., and Bauer, M. E. (1978). "Effects of systematic and nonsystemic stresses on the thermal characteristics of corn." Publication N78-29537 (National Technical Information Service, Springfield, Virginia).
Lillesand, T. M., Eav, B. B., and Manion, P. D. (1979). Photogr. Eng. and Remote Sensing 45:1401-1410.
Lindqivst, S., and Mattsson, J. O. (1975). Proc. Symp. on Remote Sensing of Environ. 10th. Ann Arbor, Michigan. pp. 123-130.
MacDonald, R. B., Bauer, M. E., Allen, R. D., Clifton, J. W., Landgrebe, D. A., and Erickson, J. D. (1972). Proc. Symp. on Remote Sensing Environ. 8th. Ann Arbor, Michigan. pp. 157-190.
MacDonald, R. B. and Hall, F. G. (1978). Proc. of the LACIE Symp., Document No. JSC-14551. NASA Johnson Space Center, Houston, Texas. pp. 17-48.
Millard, J. P., Reginato, R. J., Goettelman, R. C., Idso, S. B., Jackson, R. D., and Leroy, M. J. (1980). Photogr. Eng. and Remote Sensing 46:221-224.
Morgan, K. M., Morris-Jones, D. R., Lee, G. B., and Kiefer, R. W. (1979). Photogr. Eng. and Remote Sensing 45:769-774.
Myers, V. I., Heilman, M. D., Lyon, R. J., Namken, L. N., Simonett, D., Thomas, J. R., Wiegand, C. L., and Woolley, J. T. (1970). In "Remote Sensing with Special Reference to Forestry and Agriculture. Nat. Acad. of Sci., Washington, D.C. pp. 253-297.
Nalepka, R. F., Colwell, J. E., and Rice, D. P. (1978). Proc. Symp. on Remote Sensing Environ. 12th. Ann Arbor, Michigan. pp. 1149-1154.
Nixon, P. R., Phinney, D. E., Arp, G. K., and Wiegand, C. L. (1974). J. Rio Grande Valley Hort. Soc. 28:86-90.
Novak, C., and Young, M. (1977). Proc. Symp. on Remote Sensing Environ. 11th. Ann Arbor, Michigan. pp. 1589-1598.
Olson, C. E., Jr. (1977). Proc. Symp. Remote Sensing Environ. 11th. Ann Arbor, Michigan. pp. 933-944.

Perrier, A. (1971). In "Plant Photosynthetic Production Manual of Methods," (Z. Sestak, J. Catsky and P. G. Jarvis, eds.), pp. 632-671. Dr. W. Junk N. V. Publishers, The Hague.
Pinter, P. J., Jr., Hadley, N. F., and Lindsay, J. H. (1975). Environ. Entomol. 4:153-162.
Pinter, P. J., Jr., and Butler, G. D., Jr. (1979). Environ. Entomol. 8:123-126.
Pinter, P. J., Jr., Stanghellini, M. E., Reginato, R. J., Idso, S. B., Jenkins, A. D., and Jackson, R. D. (1979). Science 205:585-587.
Pinter, P. J., Jr., Jackson, R. D., Idso, S. B., and Reginato, R. J. (1981). Int. J. Remote Sensing, (in press).
Pratt, D. A., and Ellyett, C. D. (1979). Remote Sensing of Environ. 8:151-168.
Rango, A. (1975). "Operational applicationas of satellite snow cover observations" Scientific and Technical Information Office, NADA, Washington, D.C. 426 p.
Reeves, R. G. (1974). "Manual of Remote Sensing." Am. Soc. of Photogr., Church Falls, Virginia. 2144 p.
Reginato, R. J., Idso, S. B., and Jackson, R. D. (1978). Remote Sensing of Environ. 7:77-80.
Richardson, A. J., and Wiegand, C. L. (1977). Photogr. Eng. and Remote Sensing 43:1541-1552.
Schmugge, T. J. (1980). Photogr. Eng. and Remote Sensing 46:495-507.
Schneider, S. R., McGinnis, D. F., Jr., and Pritchard, J. A. (1979). Remote Sensing of Environ. 8:313-330.
Sharp, J. M. (1979). Photogr. Eng. and Remote Sensing 45:1487-1493.
Sprigg, W. A. (1978). Proc. Symp. on Remote Sensing Environ. 12th. Ann Arbor, Michigan. pp. 513-522.
Stanghellini, M. E., and Russell, J. D. (1973). Phytopathol. 63:133-137.
Stewart, R. B., Mukammal, E. I., and Wiebe, J. (1978). Remote Sensing of Environ. 7:187-202.
Sutherland, R. A., Bartholic, J. F., and Gerber, J. F. (1979). J. Appl. Meteor. 18:1165-1171.
Talerico, R. L., Walker, J. E., and Skratt, T. A. (1978). Photogr. Eng. and Remote Sensing 44:1385-1392.
Tanner, C. B. (1963). Agron. J. 55:210-211.
Thompson, D. R., and Wehmanen, O. A. (1979). Photogr. Eng. and Remote Sensing 45:201-207.
Thompson, D. R., and Wehmanen, O. A. (1980). Photogr. Eng. and Remote Sensing 46:1087-1093.
Tucker, C. J. (1979). Remote Sensing of Environ. 8:127-150.
Tucker, C. J. (1980). Remote Sensing of Environ. 10:23-32.
Tucker, C. J., Elgin, J. H., Jr., McMurtrey III, J. E., and Fan, C. J. (1979). Remote Sensing of Environ. 8:237-248.
Tucker, C. J., Elgin, J. H., Jr., and McMurtrey III, J. E. (1980a). Int. J. Remote Sensing 1:69-75.

Tucker, C. J., Holben, B. N., Elgin, J. H., Jr., and McMurtrey III, J. E. (1980b). Photogr. Eng. and Remote Sensing 46:657-666.
Ulaby, F. T., Batlivala, P. P., and Bare, J. E. (1980). Photogr. Eng. and Remote Sensing 46:101-105.
Vickery, P. J., Hedges, D. A., and Duggin, M. J. (1980). Remote Sensing of Environ. 9:131-148.
Volger, K., von Hesler, A., and Bartels, H. (1974). Proc. Symp. on Remote Sensing Environ. 9th. Ann Arbor, Michigan. pp. 1669-1678.
Walker, G. K. (1980). Relations between crop temperature and the growth and yield of kidney beans (Phaseolus vulgaris L.). Ph.D. Dissertation, University of California, Davis. 203 p.
Walker, G. K. and Hatfield, J. L. (1979). Agron. J. 71:967-971.
Wallen, V. R., and Philpotts, L. E. (1971). Photogr. Eng. 37: 443-446.
Wiegand, C. L. and Namken, L. N. (1966). Agron. J. 58:582-586.
Wiegand, C. L., Richardson, A. J., and Kanemasu, E. T. (1979). Agron. J. 71:336-342.
Younes, H. A., Abdel-Aal, R. M., Rhodair, M. M., and Abdel-Samil, A. G. (1974). Proc. Symp. on Remote Sensing Environ. 9th. Ann Arbor, Michigan. pp. 1105-1125.

MODIFICATION OF THE MICROCLIMATE VIA MANAGEMENT

J. L. Hatfield
Department of Land, Air and Water Resources
University of California
Davis, California

Pests that invade crops either develop within the microclimate or are transported into the microclimate. They live, grow, and reproduce within the microclimate. Carlson described in Chapter 1 the physical features of the microclimate and developed an energy balance for evaluating the relationship between microclimatic components. If we describe the physical relationship between factors, we can evaluate the possibility of modifying the microclimate by management and the impact of such modification on pest populations.

As shown in Chapter 1, the microclimate can be treated as a volume, and there are several components of energy exchange. We can expand Eq. 1 in Chapter 1 to include a more detailed approach, as shown in Eq. 1.

$$(1-\alpha)\ St + Ld + D = \varepsilon\sigma T_c^{4} + \rho_a C_p \frac{(T_c - T_a)}{r_a} + \rho_a d \frac{(e_s(T_c) - e_a)}{r_a + r_c} + G + P \quad (1)$$

This expanded form of the energy balance approach includes factors that describe both canopy features and atmospheric parameters.

ISBN 0-12-332850-0

Because this complex interaction creates the microclimate, we can state that a canopy often creates its own microclimate. In Eq. 1, two terms often neglected must be considered if we are to fully understand the microclimatic response: the horizontal advection of energy (D) and the net photosynthesis term (P). The advection term allows for the possibility of evaluating the microclimatic response given a horizontal influx of sensible heat or water vapor, and the net photosynthesis provides an avenue for energy that is incorporated into plant material, the obvious result of growth and thus a realistic approach if one considers an annual or perennial crop. We are not evaluating an hour or a day but rather weeks, months, and, in the case of perennial crops, seasons. Because we need to consider a longer time span, we must first discuss the seasonal microclimates within crop canopies and then determine how the microclimate is modified by various management schemes.

I. SEASONAL MICROCLIMATE OF CANOPIES

Important parameters of the microclimate for pests are radiation, both quality and irradiance, temperature of both air and soil, water vapor, windspeed, and soil moisture. The behavior of these parameters can be determined by examining their profiles within the canopy. The midafternoon profiles of the microclimatic parameters of sorghum and bean canopies are given in Figs. 1 and 2, respectively. These idealized profiles have been generated by considering Eq. 1 and the canopy size and shape at mid to late season. Saugier (1977) has shown similar profiles for a corn crop. However, the question arises as to temporal and spatial variability of these parameters. The profiles in Figs. 1 and 2 are good for generalization to evaluate the gross microclimatic differences between canopies, but they do not allow the evaluation of daily or seasonal differences within a given crop, which is

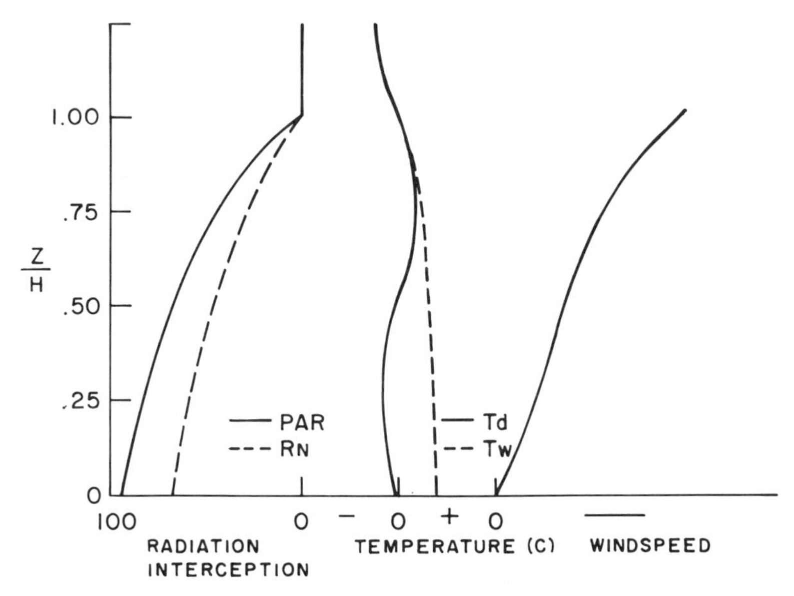

Fig. 1. Generalized microclimate profiles within grain sorghum canopies at midday for full ground cover.

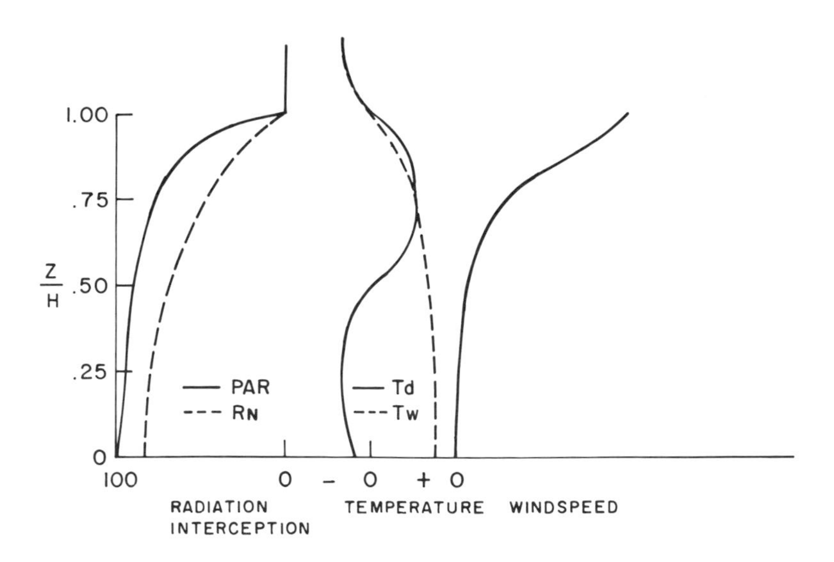

Fig. 2. Generalized microclimate profiles within a soybean canopy at midday for full ground cover.

extremely important, because often subtle changes in environmental factors can influence diseases, insects, and weeds. Dew may persist longer, the soil may remain wetter longer, or soil temperatures may be reduced to nearer the optimum for germination and emergence of weed seeds, or temperature and humidity may become more favorable for disease or insect growth as the canopy develops.

During a growing season, the microclimate of crops grown under normal management undergoes many changes. These changes are particularly evident in dry-bulb temperature profiles. Examples for grain sorghum and soybean canopies are shown in Figs. 3 and 4, respectively. For both crops, the ambient air temperatures decreased throughout the season due to the late planting in this experiment. For an earlier planting we expect the ambient temperatures to increase because of seasonal warming. The shape of the air-temperature profile changes throughout the season and depends on canopy morphology. In the soybean canopy with a high leaf-area index, the dry-bulb temperatures were depressed more within the canopy and at the soil surface, whereas in the grain sorghum canopy a more uniform temperature profile resulted from a more open canopy structure. In both canopies, the soil surface temperature declined because of reduced energy available at the soil. Because of reduced mixing within the canopy, wet-bulb temperatures increased during the season due to transpiration by the leaves, which caused higher relative humidities and a longer duration of dew. Also, as the canopies closed, the number of days that the soil surface remained near field capacity following an irrigation increased. This canopy works as a mulch in reducing evaporation. Unger and Parker (1976) found that any crop residue on the surface reduces the rate of evaporation by 20 to 80%.

Research in California has shown that in grape canopies of light and medium foliage density, canopy air temperatures are warmer than shelter temperatures; in heavy canopies, canopy temperatures are cooler than the air temperature (Sall and

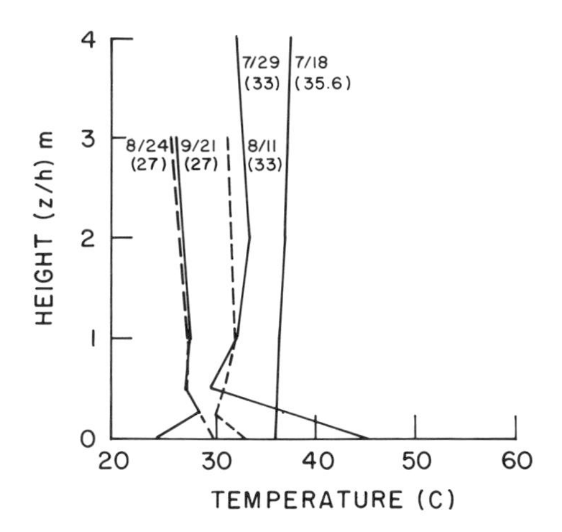

Fig. 3. Seasonal dry-bulb temperature profiles for grain sorghum canopies at midday. Ambient air temperature from a nearby weather station and leaf area values given in parenthesis.

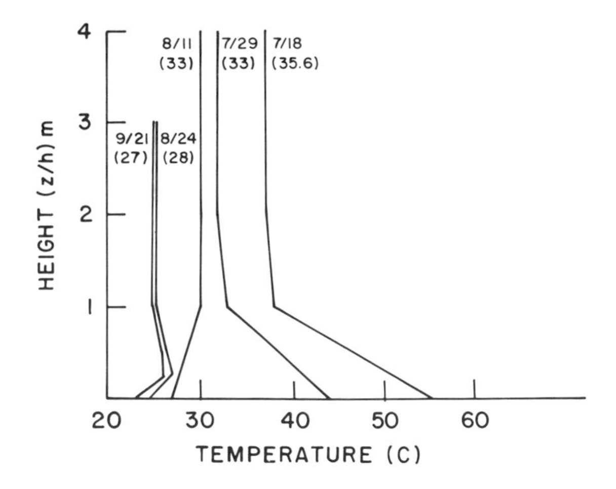

Fig. 4. Seasonal dry-bulb temperature profiles for soybean canopies at midday. Ambient air temperature from a nearby weather station and leaf area values given in parentheses.

Hatfield, unpublished data). These data show that heat units calculated from shelter temperatures diverge from those based on canopy temperature as the season progresses and foliage density changes. Sall and Hatfield (unpublished data) suggest that these differences in heat units have implications for the prediction of insect emergence or powdery mildew development.

The albedo of the surface also changes throughout a growing season, which offsets the energy balance. The seasonal trend in albedo is dependent on the stage of crop development, soil background, soil water content and any crop residue on the surface. Idso et al. (1977), Hatfield (1979), and Aase and Siddoway (1980) have shown that seasonal progression is mainly dependent on the base soil albedo, since crop albedos range from 0.15 to 0.20 when the crops fully cover the soil and reach that value by accumulating leaf area and covering the soil. Thus, the seasonal albedo may increase or decrease from the bare soil background.

Ambient weather conditions also change. Macroclimatic conditions may cause changes in the microclimate, even if the crop canopy remains the same throughout the season. Figure 5 shows how the microclimate and macroclimate are related and the magnitude of the seasonal change in the macroclimate for 1978. Figure 5 corresponds to the growing season for the data shown in Figs. 3 and 4. If we contrast these data for Davis with those for other locations, the temperature patterns should be similar but the radiation patterns should depend on cloud patterns. For any integrated pest management program, macroclimatic conditions should be characterized for the particular location.

In West Virginia, Bennett and Elliott (1972) showed that disease severity was related to slope for the same crop. They found the fungus (Sclerotinia trifoliorum) was more severe on five forage species growing on north-facing slopes then those on south-facing slopes. Differences between the slopes are related to temperature and humidity.

If we accept that the microclimate within a crop canopy has a particular daily and seasonal behavior dependent on the canopy

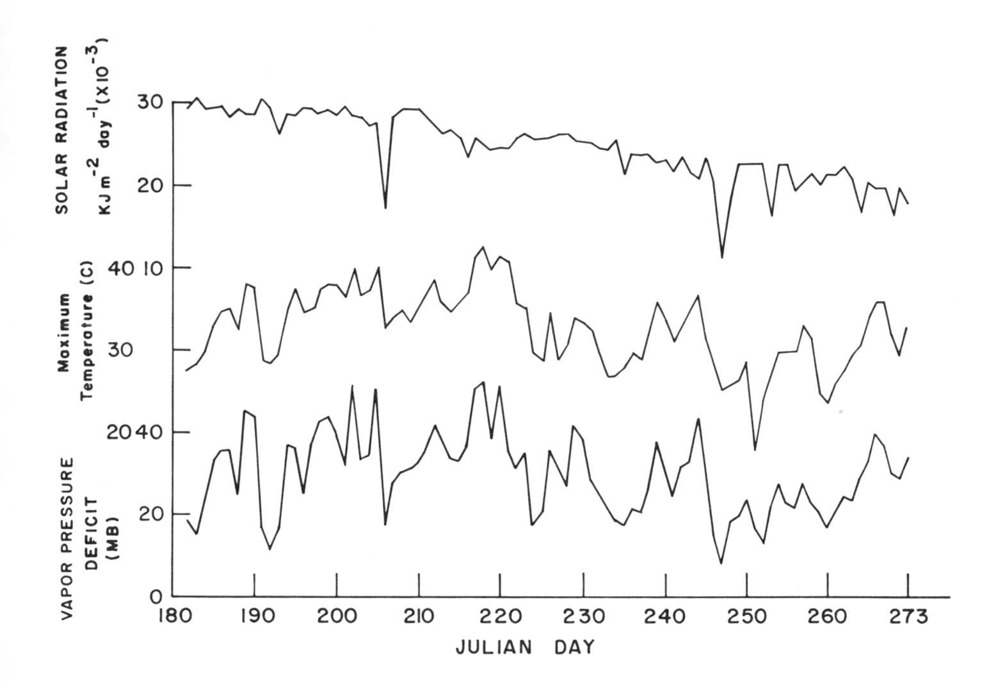

Fig. 5. Ambient weather conditions for maximum dry-bulb temperature, maximum vapor-pressure deficit, and solar radiation for the 1978 growing season for data given in Figs. 3 and 4.

morphology and macroclimate, then to what extent is the microclimate modified by management decisions? For example, what are the effects of crop residue, irrigation, row direction, and other cultural practices and windbreaks on the microclimate within canopies? Each of these factors will be discussed in detail and examples on pest behavior are given where possible.

II. IRRIGATION

European corn borer populations have been positively related to irrigation in Nebraska, particularly during dry summers (Chiang et al., 1961; Hill et al., 1967). In this study conducted in the Midwest, irrigation promoted a favorable environment for borer development. Steadman et al. (1973) and Blad et al. (1978) showed that white mold in beans was more severe in canopies irrigated more frequently. Disease infection by irrigation does not have to be related to the amount of irrigation but can be related to management practices. Thomson and Allen (1976) proposed that Phytophthora parasitica on citrus is contained in tailwater collected from infected orchards and suggested that this practice be avoided to prevent disease transmission. Hipps (1977) proposed that irrigation effects be summarized as shown in Fig. 6.

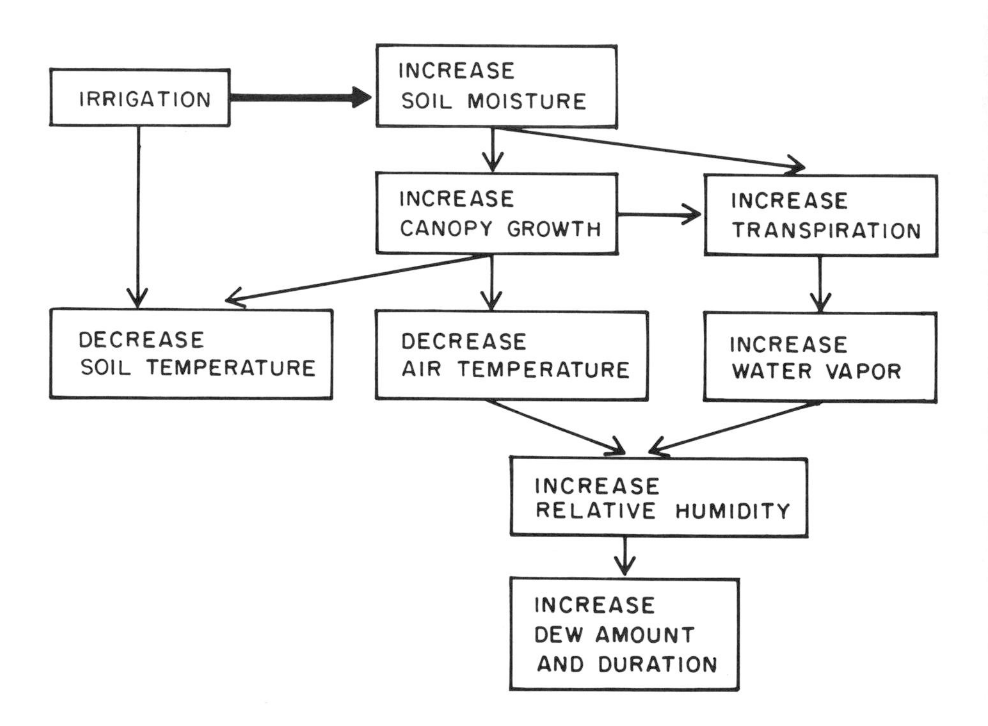

Fig. 6. Effects of irrigation on plant growth and microclimate. Adopted from Hipps (1977).

Overirrigation leading to standing water is often associated with disease problems. Easton and Nagle (1977) found increased tuber rot via Phytophthora infestans with ponding of irrigation water. Miller and Burke (1977) showed that beans were more severely injured via _Fusarium solani_ f. sp. _phaseoli_ when roots were subjected to near-zero oxygen potential in the soil atmosphere and that they were similarly injured in the field by excessive wetness during furrow irrigation. Yield was reduced 30% in infested soils but only 10% in uninfested soils (Miller and Burke, 1977). Higher soil moisture contents aggravate pea seed and seedling rot (_Phythium ultimum_ and _Fusarium solani_ f. sp. _pisi_) (Short and Lacy, 1976). Mircetich and Matheron (1976) and Mircetich et al. (1977) found that periodic standing water or wetting of trunks contributed to trunk diseases in cherries and oak. Mircetich and Matheron (1976) found that infestations of _Phytophthora cambivora_, _P. megasperma_, and _P. drechsleri_ were more severe in cherry orchards in poorly drained soils than in orchards in well-drained soils. They also found an interaction between rootstock and susceptibility to infestation. Mircetich et al. (1977) said the effects that contributed to _Phytophthora_ trunk canker in oak were overirrigation of poorly drained soils, depressions or watering basins left around the trunk for several years, soil around the trunks, and trunks frequently irrigated by lawn or shrubbery sprinklers throughout the summer. They suggested that this disease was not a problem until the sprinkling of lawns and shrubs around trees became common.

Irrigation promotes a denser canopy, which affects the microclimate within the canopy. Downey and Caviness (1973) found that cm air temperatures 15 cm above ground were 4 to 9°C cooler in irrigated soybean canopies than in unirrigated canopies. This reduction in temperature caused the relative humidity to be 16% higher in irrigated plots. Crandell et al. (1971) also found that mist irrigation lowered air temperature by 3°C immediately following the irrigation, although relative humidity increased from 50 to 80%.

Promoting cooler canopy temperatures may affect disease. Gilbert et al. (1975) found that aflatoxin-producing strains of

Aspergillus flavus resulting in an increased percentage of bright greenish-yellow fluorescence of locks and seeds increased as the duration of a daily maximum temperature of 30C increased and as the number of diurnal maximum temperature cycles of 30C increased.

Canopy temperatures are reduced by both overhead and furrow irrigation because both methods create a denser canopy and higher transpiration rate. Rotem et al. (1962) found that relative humidities were different only during sprinkling and the two to three hours after irrigation had ceased. In a later study, Rotem and Cohen (1966) found that furrow irrigation did not significantly change the microclimate in the canopy, but sprinkler irrigation lowered the air temperature and increased the relative humidity. They also found that the effect of sprinkler irrigation on the microclimate was more pronounced when the macroclimate had high temperatures and low relative humidities. The effect of irrigation on the microclimate persisted longer when the macroclimate was cooler and more humid, which was shown by drying rate on outer and inner leaves of the canopy (Rotem and Cohen, 1966). Relative humidities in tomato canopies were different after sprinkler irrigation up to five days (Raniere and Crossan, 1959). Differences between irrigated and unirrigated plots were as large as 20% relative humidity up to five days later and as much as 10% relative humidity up to nine days later.

Lomas and Mandel (1973), in comparing above- and within-canopy sprinkler irrigation treatments, found that the above-canopy treatment decreased air temperature by 7°C, whereas below-canopy irrigation treatments increased it by only 2°C. In their studies on avocado orchards, they found that the above-canopy treatment also increased relative humidity by 27% yet below-canopy treatment increased it only 8%.

Downey and Caviness (1973) also found that humidity levels were higher within irrigated soybean canopies than within unirrigated canopies. This difference was due to cooler temperatures rather than to an increase in dewpoint temperatures. They found the dewpoint temperatures within the canopy were best approximated by the dewpoint temperature of the prevailing air mass.

Weiss et al. (1980) showed that heavy furrow irrigation (5.5 cm every five days) yielded cooler air, leaf, and soil temperatures than the normal furrow-irrigation treatment (5.5 cm every ten days). They also found that the difference in the microclimate was greatest on hot days just before the normal irrigation. This is consistent with the findings in Israel of Rotem and Cohen (1966), who found the canopy structure affected the severity of the white mold disease: the denser canopies had a cooler air temperature than the more open canopies, and within a canopy type. The more open canopies exhibited less difference in air temperature between the normal and heavy irrigation treatments. The presence of the canopy and the resultant microclimate modify the macroclimate so that the white mold disease can develop (Steadman et al. 1973; Weiss et al., 1980).

Ross (1975) found that sprinkler-irrigated soybeans had greater incidence of brown spot (*Septoria glycines*) and benefited from fungicide treatments. He showed that yields from irrigated plots were higher than those from the unirrigated areas only if fungicides were applied. The yield differences between irrigated and unirrigated plots where negated by heavy August rainfall, but the timing of rainfall had little effect on the fungal outbreaks. He suggested that September rainfall has more of an effect on disease occurrence.

Most researchers have found that irrigation promotes a favorable microclimate for disease; however, some researchers have found that in potatoes, sprinkler irrigation decreased tuber infestation by the potato tuber moth (*Phythorimaea operculella* Zell.) and common scab (*Streptomyces scabies*) (Foot 1974; Lapwood and Lewis, 1967). Foot (1974) found that irrigation reduced tuber moth damage because the cracks and soil spaces were reduced: that precludes moth and larvae entry, which are repelled by damp conditions. He also that noted that irrigation can cause compaction and cracking if applied incorrectly and that irrigation was more effective than any other control measure when timed to allow soil moisture to be maintained above 30% moisture content by weight. Irrigation increased potato yields but increased population of foliage miners per unit area with a reduction in intensity because of increased foliage density (Foot, 1974). He suggested that because of population increases under irrigation, other control measures are necessary to insure high yield under continuous production of potatoes. Lapwood and Lewis (1967) reported that common scab in potatoes was reduced by maintaining the soil near field capacity throughout the season and by late irrigation after tubers were 8 mm in diameter, as opposed to early irrigation or no irrigation.

Davis et al. (1976), in later research on potatoes, found that less scab was present at -0.45 or -0.63 bars of soil moisture tension than at -0.96 or -1.60 bars. Lesions on tubers at -1.60 bars were significantly deeper than lesions on tubers grown at -0.96 or -0.65 bars. They found that production of potatoes was greatest when irrigation levels were at -0.65 bars. Lowered soil-moisture tension increased calcium uptake by the tubers. Results of those experiments suggest that irrigation is beneficial to control of tuber diseases in potatoes; however, Foot (1974) and Lapwood and Lewis (1967) report that foliage densities are increased in irrigated plots and thus the microclimate should be cooler.

Sprinkler irrigation promotes diseases in canopies more than furrow irrigation does (Menzies, 1954; Duvdevani et al., 1946; Rotem and Chorin, 1961). Sprinkler irrigation appears to affect foliage most by simulating dew. Duvdevani et al. (1946) showed that when dew was absent, only sprinkler irrigation caused outbreaks on cucumbers of _Peronoplasmopara cubensis_ but when dew was present, the disease was found in both sprinkler- and furrow-irrigated plots. Menzies (1954) showed that mint rust (_Puccinia menthae_) was more prevalent in sprinkler- than furrow-irrigated fields. Rotem et al. (1962) found that the dew was only slightly more prevalent in sprinkler-irrigated than furrow-irrigated plots. In a later study, Rotem and Cohen (1966) obtained the same result. Raniere and Crossan (1959) found in comparing sprinkler-irrigated plots of tomatoes with unirrigated plots that the dew periods were more intense and prolonged in the irrigated plots and caused more disease damage by Colletotrichum phomoides. They stated that the initial effect was caused by the splashing drops from sprinkler irrigation with the increased humidity; dew was the second, equally important effect. Dew lasted three to five hours longer in the irrigated plots than in the unirrigated plots (Raniere and Crossan, 1959).

Rotem et al. (1976) suggested that the limiting and optimum temperatures and wetting periods of sporulation of Phytophthora infestans in potatoes vary and are affected by other factors. They suggested that optimal spore production and survival under higher temperatures require less leaf wetness, either continuous or interrupted, and under lower temperatures require longer wet periods. Thomas (1977) found that the period of dew was the predominant factor in the onset of downey mildew (Pseudoperonospora cubensis) on cantaloupes: five to six hours was required. Spotts et al. (1976) found that when apples were misted to delay bloom fire blight (Erwina amyloroa) was nine times more severe then on nonmisted trees. Misting favored both dissemination and infection.

As shown in Fig. 6, irrigation promotes greater growth of the canopy, which reduces air temperature. Reicosky et al. (1980) found that the difference between irrigated and unirrigated soybeans for the air temperature 20 cm above ground was largest between 1000 and 1600 hours. During this period, the differences were as much as 8°C, and this difference persisted for as long as five to six hours each day. When the 120- and 20-cm-high air temperatures of the irrigated and unirrigated plots were compared, unirrigated had differences (T_{20} - T_{120}) up to 5.8°C, whereas the irrigated had differences less than 2°C. When the day was cloudy, the differences between above and within the canopy were less than 1°C. Hipps and Hatfield (unpublished data) showed that for grain sorghum and soybeans grown in various irrigation treatments, differences in temperature within the canopy were large. At 1500 hours, the differences in grain sorghum canopies grown under various irrigation treatments in the average 25-50-cm-high temperature were only 2°C different, as shown in Fig. 7. The irrigated plots were cooler than the stressed plots throughout the canopy; however, the differences in the soybean canopy were 5°C for the same period (Fig. 8). The differences between the canopies were due to leaf-area density. The soybean canopies were more dense and had less exchange with the atmosphere above the canopy and thus had cooler temperatures in the irrigated treatments but warmer temperatures than the sorghum canopies. These data indicate that irrigation management can cause differences in canopy structure, which would influence the wet- and dry-bulb temperatures within the canopy. The presence of dew was more pronounced than in the unirrigated plots. Similar differences between sorghum and soybean canopies were shown by Chin Choy and Kanemasu (1974). The differences shown in Figs. 7 and 8 are due to water stress imposed at different times during the growing season; stress during the vegetative stage reduced the height and leaf-area index primarily because of a reduction in average leaf size. The result is that the exchange with the atmosphere is greater in the canopies with reduced leaf size, and hence the air temperature profiles are not as cool as in the well-watered plots (well-watered vs vegetative stress in Figs. 7 and 8). These data suggest that the canopy can be modified by irrigation management and the resultant changes should be enough to affect pest development.

Downey and Caviness (1973) found that air temperatures within unirrigated soybean fields were best approximated by temperatures from a weather shelter. They stated, however, that the use of shelter temperatures to predict canopy air temperatures should result in an overprediction of the thermal regime in irrigated soybean canopies. Data from Davis on fully developed grain sorghum and soybeans show that shelter temperatures from a grass plot during the afternoon overpredict the irrigated soybean canopy air temperature by 2°C and underpredict that in a unirrigated plot

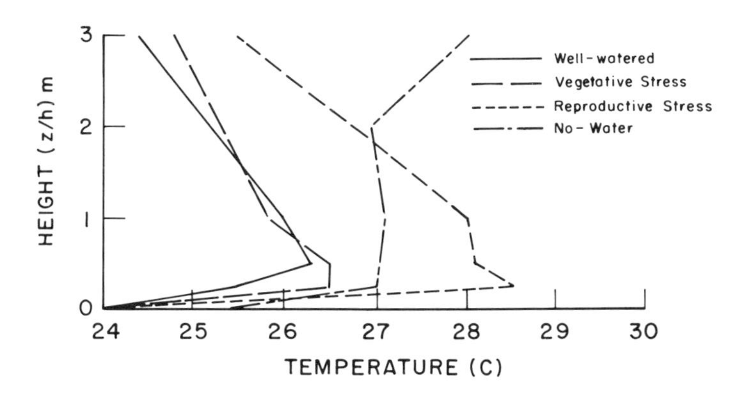

Fig. 7. Irrigation treatment effects on dry-bulb air temperatures above and within grain sorghum canopies measured during mid afternoon.

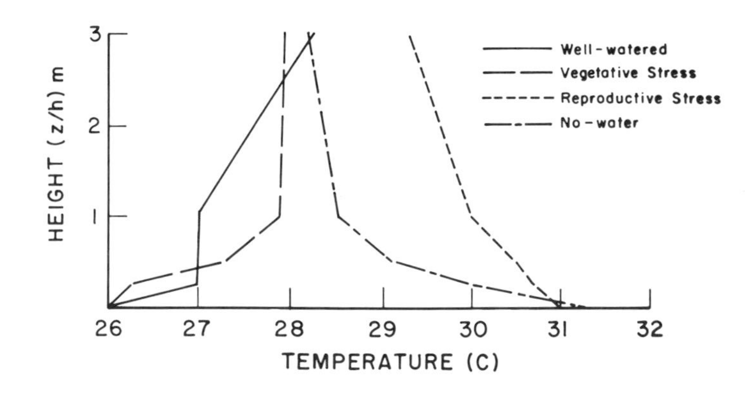

Fig. 8. Irrigation treatment effects on dry-bulb air temperatures above and within soybean canopies measured during mid afternoon.

by 3°C. In grain sorghum canopies, the shelter temperatures estimated the canopy air temperatures to within 1°C for irrigated canopies, but underestimated the unirrigated temperatures by 2°C. These data suggest that attention be given to the stage of development, because early in the growing season little difference existed, and to the length of time subsequent to an irrigation.

Bertrand et al. (1976) found that French prune trees subjected to post-harvest moisture stresses developed significantly larger cankers following inoculation with mycelium of *Cytospora leucostoma*. Adequately irrigated trees did not exhibit these symptoms, which were related to a decrease in moisture in the bark. They suggested that improper irrigation management results in increased disease problems. It is not possible to provide a set of exact criteria for irrigation effects on diseases, but anyone attempting the practice of pest management should be aware of the effect of irrigation practices on the microclimate.

III. ROW SPACING AND DIRECTION

Both row spacing and direction influence the microclimate within canopies and have been shown to contribute to disease. Crandell et al. (1971) found that closer row spacing allowed the canopy to close more quickly in the season and have a higher incidence of disease. A closer row spacing then hastens the canopies ability to promote a cooler, more humid environment. The effect is to simulate a denser canopy; the result is similar to that shown by Weiss et al. (1980).

Row direction effects depend on the direction of the prevailing winds. Rows planted perpendicular to the wind have greater mixing than those parallel to the wind. Crandell et al. (1971) found that relative humidity was lower in plots with rows perpendicular to the prevailing wind and found the difference was greater when row spacings were wide. They also noted that canopy air temperatures were affected very little with a difference of less than 2C between parallel and perpendicular rows. The pattern in canopy air temperatures was the same at the 18-cm and 36-cm height. Although these differences appear to be minor, they are sufficient to influence disease and insects within a canopy and the rate of weed growth.

Haas and Bolwyn (1973) obtained results contrary to those reported by Crandell et al. (1971) and showed that the incidence of white mold disease was more pronounced in rows perpendicular to the prevailing winds. The difference between the results of these researchers can be attributed to the windspeed, row spacing, and humidity of the ambient air.

IV. CANOPY EFFECTS ON SOIL TEMPERATURE

Soil temperatures are affected by the presence of a plant canopy, and the diurnal soil temperature range decreases at the surface. Hay (1976) and van Wijk et al. (1959) showed that the soil under crop cover had lower maximum and higher minimum temperatures than bare soil. Soil moisture affected soil temperature: the diurnal range decreased when the soil was near field capacity and increased as the soil surface dried. These changes affect the rate of emergence of weeds and the growth rate of nematodes or overwintering of disease spores and insects.

Kohl (1973) quantified the effects of sprinkler irrigation on soil temperature under potatoes. Irrigation at five-to-seven-day intervals on silt loam reduced soil temperature by 1-2°C, but a three-to-five-day interval was needed to achieve the same reduction on a loamy-fine sand. These results are for full cover and warm conditions; the reduction was smaller in cooler weather. Daily irrigation had no more effect on the 10-cm-soil temperature of silt loam soil than nine days of irrigation had, but the temperature difference of the upper 10 cm at nine days was 6C for the bare soil, 4.5°C for 50% cover, and 4°C for full cover. In loamy-fine sand, the bare soil and 45% cover reached the maximum difference six days after irrigation: they were 4°C and 3.5°C warmer, respectively, then plots irrigated daily. The 85% cover plot did not exhibit the maximum difference until nine days after irrigation: it was 3°C warmer than the irrigated daily plot (Kohl, 1973). These data suggest that irrigation timing has a large effect on soil temperature and interacts with the amount of ground cover.

Bhusman et al. (1971) found that the size of aggregates in the field influenced the soil temperature at 8 cm. As the size of the clods decreased, the soil temperature decreased; the largest effect was on the amplitude. That is to be expected, because the bulk density (g cm^{-3}) increased with the decreasing clod size. Diurnal amplitude of temperatures also differed: the largest amplitude was in the largest aggregates (Bhusman et al., 1971), which was attributed to a lower albedo, a rougher surface, and higher thermal diffusivity. Average daily temperatures, however, increase only as soil size of the aggregate decreases. These data suggest that attention be given to aggregate size when evaluating soil temperature input for pest models.

Soil temperature plays a large role in the development of all pests that live in the soil for part or all of their life cycle. Short and Lacy (1976) found the incidence of pea seed and seedling rot to be greater when growth-chamber temperatures simulated field conditions rather than being constant. Fluctuating temperatures favor both *Pythium ultimum* and *Fusarium solani* f. sp. *pisi* because of their different optimum temperatures for development. Carter (1975) showed that soreshiun of cotton from combined infections of *Rhizoctonia solani* and *Meloidogyne incognita* was more severe when critical soil temperatures and inoculum concentrations were present. Optimum soil temperatures were between 18 and 21C. Prasad et al. (1976) found that the effects of temperature and soil moisture were not consistent between barley varieties or infection by *Helminthosporium gramincum*: conditions promoted by microclimatic changes may not be favorable for all varieties.

V. MULCHES AS A MICROCLIMATE MODIFICATION

Mulches as a covering of the soil surface have a dramatic effect on soil and crop microclimates. Of considerable importance are the effects on insects, diseases, and weeds, which arise from changes in air temperature, soil temperature, and soil moisture caused by covering the soil surface. Mulches range from organic (straw or manure) to inorganic (plastic or paper); the results depend on the material. Davies (1975) summarizes the effects of various mulches in a comprehensive review.

Organic mulches tend to reduce soil temperatures during the summer and increase them during the winter. Kohnke and Werkhoven (1963) found that a straw mulch caused the 10-cm-depth temperatures to be 1°C warmer in the winter, 1°C cooler in the spring, and as much as 2°C cooler in the summer than bare soil temperatures. Mulch caused a damping of the diurnal temperature range at all depths. Hanks et al. (1961) found that organic mulches created cooler temperatures down to 150 cm, whereas temperatures were warmer under plastic mulch. van Wijk and Derksen (1961) provide a methodology for evaluating the effect of soil temperature. This procedure is also given by van Wijk (1966) along with procedures for calculating soil temperature.

Mulches reduce soil evaporation and conserve soil moisture. Organic and inorganic mulches act as barriers to water vapor movement; they are most effective when temperatures are cool and humidities high. Hanks et al. (1961) showed that all mulches reduce soil evaporation, but moisture saved is less than 2.5 cm over an entire year. Mulch is most effective in reducing soil evaporation when the soil surface is wet; however, the difference is negligible when comparing a dry, bare soil with a dry, mulch-covered soil. Unger and Parker (1976) found that any crop residue on the surface will reduce soil evaporation by 20 to 80%. Aase and Siddoway (1980) also showed that stubble increased the catch of snow and the soil water recharge.

Several incidences of mulches interacting with disease and insects have been reported. Black plastic mulch caused fungus (Rhizoctonia solani) to increase and nematode (Pratylenchus penctrans) populations to decrease (Miller and Waggoner, 1963). This effect was attributed to the cooler soil temperatures that resulted under the black plastic; this effect was lacking under a clear plastic, which created warmer soil temperatures. It was also noted that under an organic hay mulch, the fungus decreased without a decrease in nematode populations, although the mulch created lower soil temperatures. This result was believed to be due to a lack of nematode-attacking fungi. Other workers found that white and black plastic mulch that caused lettuce heads to come into contact with soil reduced slime rot (Pseudonomas sp. as the causal agent) from 53% to 16% and 18%, respectively (Hilborn et al., 1957).

Aase and Siddoway (1980) found in a stubble height experiment that disease was a major factor in yield each year and was more prevalent in taller stubble. In their experiments, the severity of the disease increased as the height of the stubble increased. This aspect needs further research in evaluating no-fill farming effects on the microclimate and pest populations.

In general, organic mulches do not reduce weed populations, but inorganic mulches, which act as inpenetratable barriers, do reduce emergence. In some cases the soil temperature may be modified to be more favorable for weed-seed germination.

VI. WINDBREAKS

Windbreaks or shelterbelts reduce the velocity of wind in the sheltered area but have some implications that may affect pest management. Air temperature and relative humidity are higher in the sheltered area during the day than in the unsheltered area, and relative humidity is higher and air temperature lower at night (Brown and Rosenberg, 1972). These effects can be quantified by Eq. 1, noting that r_a is increased, T_a is increased during the day, C_a increases in the sheltered area, S_p is the same as in an unsheltered area, and r_c is lower for a sheltered crop. van Eimern et al. (1964) summarized the effects of shelterbelts on the microclimate and yield of various crops.

Air temperature in sheltered areas has a larger diurnal range but generally exhibits changes within 2°C, and the change is dependent on the individual case (van Eimern et al., 1964). For a number of crops, relative humidity changes were found to depend on soil moisture, air temperature, and crop, but humidity generally increased. Soil temperatures increase because of decreased soil evaporation, but once the crop covers the soil, the temperature differs little between sheltered and unsheltered areas. Again, the difference is less than 2°C in most instances (van Eimern et al., 1964).

Coyne et al. (1974) found that white mold disease was more severe in dry beans grown next to a windbreak than in those grown without a windbreak. They attributed that to the vigorous nature of the canopy growth caused by the shelter. Their data showed that outbreak of the disease was not uniform over the entire crop sheltered but rather was more localized, which suggests that subtle microclimate changes can dramatically influence disease outbreak.

Shelterbelts encouraged fungus diseases in strawberries. This effect was primarily associated with increased humidity. van Eimern et al. (1964) reported an increase in *Phytophthora* on potatoes in fields sheltered by rye compared with those in the open.

van Eimern et al. (1964) reported several effects of shelterbelts on animal populations because of the ideal environment. They also noted that several harmful insects can develop in shelterbelts; however, they did not provide any specific examples.

VII. EVALUATING THE EFFECTS OF MANAGEMENT ON THE MICROCLIMATE

For implementation of integrated pest management models to be successful, the magnitude of the microclimatic change must be quantified. Most of the effects created by management decisions are in the canopy growth or ground cover; those, in turn, modify air temperature, windspeed, radiation, and relative humidity within the canopy. The basis for quantifying these effects is Eq. 1 and the factors modified are given in Table 1. Barfield and Gerber (1979) provide a comprehensive treatise on microclimate modification by a number of methods, including measurement techniques.

Table 1. Factors of the Energy Balance that are Modified Management Practices.

Management Practice	Energy Balance Factor Initially Modified
Mulching	ρ, G
Reflectants	ρ
Irrigation	G, r_a, r_s, T_a, C_a
Row Direction	r_a
Windbreaks	r_a, u
Trellising	r_a, ρ, G, u, r_s

The procedures outlined by Norman in Chapter 2 provide the basis for quantifying microclimatic changes and incorporating management factors. Further research is needed on the role of management factors in the microclimate for perennial crops, irrigation practices and schedules in all climates, and modification of canopy energy exchange via pruning or trellising.

REFERENCES

Aase, J. K., and Siddoway, F. H. (1980). Agric. Meteor. 21: 1-20.

Barfield, B. J., and Gerber, J. F. (1979). "Modification of the Aerial Environment of Crops," ASAE, St. Joseph, Michigan. 538 p.

Bennett, D. L., and Elliott, E. S. (1972). Plant Dis. Reptr. 36:371-375.
Bertrand, P. F., English, H., Uriu, K., and Schick, F. J. (1976). Phytopathol. 66:1318-1320.
Bhusman, L. S., Varade, S. B., and Gupta, C. P. (1971) J. Agric. Engy. 11:20-24.
Blad, B. L., Steadman, J. R., and Weiss, A. (1978). Phytopathol. 68:1431-1437.
Brown, K. W., and Rosenberg, N. J. (1972). Agric. Meteor. 9: 241-263.
Carter, W. W. (1975). J. Nematol. 7:229-233.
Chiang, H. C., Jarvis, J. C., Burkhardt, C. C., Fairchild, M. L., Weekman, G. T. and Triplehorn, C. A. (1961). North Central Regional Publ., No. 129, Res. Bull. 775. 95 p.
Chin Choy, E. W., and Kanemasu, E. T. (1974). Agron. J. 66: 98-100.
Coyne, D. P., Steadman, J. R., and Anderson, F. N. (1974). Plant Dis. Reptr. 58:379-382.
Crandell, P. C., Jensen, M. C., Chamberlain, J. D., and James, L. G. (1971). Hort. Sci. 6:345-347.
Davies, J. W. (1975). "Mulching Effects on Plant Climate and Yield." UMO. Tech. Note. No. 136. 92 p.
Davis, J. R., McMaster, G. M., Callihan, R. H., Nissley, F. H., and Pauck, J. J. (1976). Phytopathol. 66:228-233.
Downey, D. A., and Caviness, C. E. (1973). Bull. 784. Agric. Exp. Sta. University of Arkansas. 43 p.
Duvdevani, S., Reichert, I., and Palti, J. (1946). Pal. Bot. Rehouot. Ser. 5:127-151.
Easton, G. D., and Nagle, M. E. (1977). Plant Dis. Reptr. 61: 1064-1066.
Foot, M. A. (1974). N. Z. J. of Exp. Agric. 2:447-450.
Gilbert, R. G., McMeans, J. L., and McDonald, R. L. (1975). Phytopathol. 65:1043-1044.
Haas, J. H., and Bolwyn, B. (1973). Can. J. Plant Sci. 54: 525-533.
Hanks, R. J., Bowers, S. A., and Bark, L. D. (1961). Soil Sci. 91:233-238.
Hatfield, J. L. (1979). Water, Air, and Soil Pollution 12:73-81.
Hay, R. K. M. (1976). Soil Sci. 27:121-128.
Hilborn, M. T., Hepler, P. R., and Cooper, G. F. (1957). Phytopathol. 47:245.
Hill, R. E., Sparks, A. N., Burkhardt, C. C., Chiang, H. C., Fairchild, M. L., and Guthric, W. D. (1967). North Central Regional Publ. Res. Bull. 223. 100 p.
Hipps, L. E. (1977). "Influence of irrigation on the microclimate and development of white mold disease in dry edible beans." M.S. thesis, University of Nebraska. 79 p.
Idso, S. B., Reginato, R. J., and Jackson, R. D. (1977). Nature 266:625-628.
Kohl, R. A. (1973). Agron. J. 65:962-964.

Kohnke, H., and C. H. Werkhoven. (1963). Soil Sci. Soc. Am. Proc. 27:13-17.
Lapwood, D. H., and Lewis, B. G. (1967). Plant Path. 16: 131-135.
Lomas, J. and Mandel, M. (1973). Agric. Meteor. 12:35-48.
Menzies, J. D. (1954). Phytopathol. 44:553-556.
Miller, D. E., and Burke, D. W. (1977). Plant Dis. Reptr. 61:175-179.
Miller, P. M., and Waggoner, P. E. (1963). Plant and Soil. 18:45-52.
Mircetich, S. M. and Matheron, M. E. (1976). Phytopathol. 66: 549-558.
Mircetich, S. M., Campbell, R. N., and Matheron, M. E. (1977). Plant Dis. Reptr. 61:66-70.
Prasad, M. N., Leonard, K. J., and Murphy, C. F. (1976). Phytopathol. 66:631-634.
Raniere, L. C., and Crossan, D. F. (1959). Phytopathol. 49: 72-74.
Reicosky, D. C., Deaton, D. E., and Parsons, J. E. (1980). Agric. Meteor. 21:21-35.
Ross, J. P. (1975). Plant Dis. Reptr. 59:809-813.
Rotem, J., and Chorin, M. (1961). Israel J. Agric. Res. 11: 189-192.
Rotem, J., and Cohen, Y. (1966). Plant Dis. Reptr. 50:635-639.
Rotem, J., Palti, J., and Rawitz, E. (1962). Plant Dis. Reptr. 46:145-149.
Rotem, J., Clare, B. G., and Carter, M. V. (1976). Physiol. Plant Path. 8:297-305.
Saugier, B. (1977). In "Environmental Effects on Crop Physiology" (J. J. Landsberg, and C. V. Cutting, eds.), p. 39-55, Acad. Press, New York.
Short, G. E., and Lacy, M. L. (1976). Phytopathol. 66:188-192.
Spotts, R. A., Stang, E. G., and Ferrec, D. C. (1976). Plant Dis. Reptr. 60:329-330.
Steadman, J. R., Coyne, D. P., and Cook, G. E. (1973). Plant Dis. Reptr. 57:1070-1071.
Thomas, C. E. (1977). Phytopathol. 67:1368-1369.
Thomson, S. V., and Allen, R. M. (1976). Phytopathol. 66: 1198-1202.
Unger, P. W., and Parker, J. W. (1976). Soil Sci. Soc. Am. J. 40:938-942.
van Eimern, J., Karschon, R., Razumova, L. A., and Robertson, G. U. (1964). "Windbreaks and Shelterbelts," WMO. Tech. Publ. 59. 188 p.
van Wijk, W. R. (1966). "Physics of Plant Environment," North Holland Publishing Co., Amsterdam. 382 p.
van Wijk, W. R., and Derksen, W. J. (1961). Agron. J. 53: 245-246.
van Wijk, W. R., Larson, W. E., and Burrows, W. C. (1959). Soil Sci. Soc. Am. Proc. 23:428-434.
Weiss, A., Hipps, L. E., Blad, B. C., and Steadman, J. R. (1980). Agric. Meteor. 22:11-21.

OVERALL APPROACH TO INSECT PROBLEMS IN AGRICULTURE

Carl B. Huffaker
Department of Entomological Sciences
Division of Biological Control
University of California, Berkeley
Berkeley, California

I. INTRODUCTION

I will deal here with the general insect problems in crops from an overall ecological, interactive perspective. Certainly, irrigated cropping systems present some special problems and no doubt avoid others. The ability to control water gives an enormous advantage to the grower in producing the crop. Water can be allotted and timed to best advantage for crop growth, used as a vehicle to distribute chemicals, and even to control some pests (e.g., Bottrell, 1979). But irrigation also brings with it some pest problems – e.g., serious human diseases [malaria and types of encephalitis (mosquito-vectored) and schistosomiases (snails as reservoir hosts) – Reeves et al., 1980; Washino, 1980; Garcia and Huffaker, 1979)].

We are concerned with the interrelatedness of insects and other elements in crop ecosystems, which is why we need detailed studies to closely quantify these relationships. This is, of course, to gain the insights needed in crop production management. We cannot easily deal in isolation with insect pests or how they may be controlled by other insects (biological control), with chemicals or whatever tactic. We need to view these pests as only one component, albeit an important one, with the focus on the crop

ISBN 0-12-332850-0

itself. A good common denominator seems to be the flow of carbohydrates and their production through photosynthesis. This latter is represented by Wilson and Loomis' (1962) sketch (Fig. 1), and the former by a sketch of Gutierrez et al. (1980) (Fig. 2).

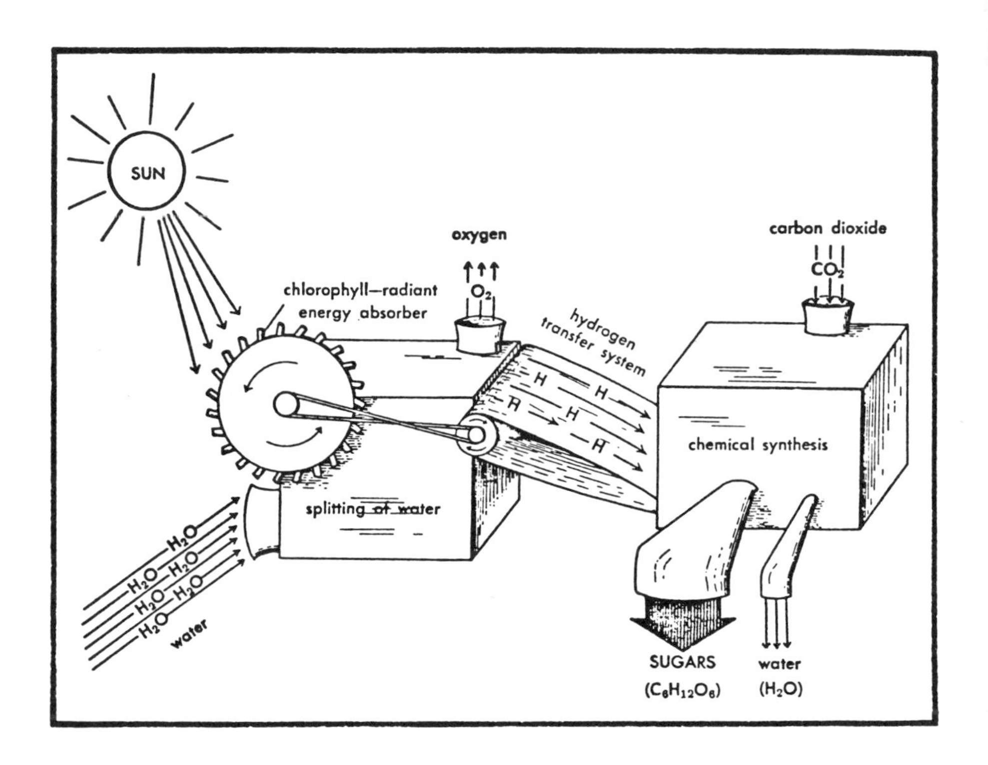

Fig. 1. Schematic representation of photosynthesis. The sun's energy is used to split water in the chlorophyll-containing parts of the leaf into oxygen and energy-charged hydrogen. This energy-charged hydrogen is combined with carbon dioxide from the air to make an energy-storage molecule – sugar; water is released as a by-product. The energy stored in sugar can be used by the plant later in its life cycle for growth, reproduction or maintenance or it may stay in this storage form until the plant is eaten by an herbivore. (From BOTANY, Third Edition, by Carl L. Wilson and Walter E. Loomis. Copyright 1952, (c) 1957, 1962 by Holt, Rinehart and Winston, Inc. Reprinted by permission of Holt, Rinehart and Winston.

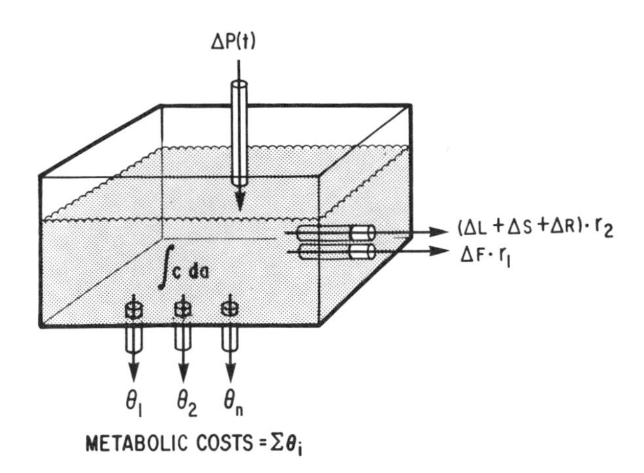

Fig. 2. The metabolic pool model for allocating photosynthate (P), carbohydrate reserves (C) to meet metabolic needs (θ_i) of the plant or growth of plant parts [e.g., leaves (L), stem (S), root (R), or fruit (F)]. Note that the levels of the outflows indicate a priority scheme, and r_1 and r_2 indicate the fraction of the maximum growth achieved during some $\Delta t (0 \leqq r_i \leqq 1)$. (From Gutierrez et al., 1980.)

To a marked degree, light, soil moisture and fertility, and temperature determine crop growth. Crop growth, along with all other factors, combine to determine potentials for growth and reproduction of the other organisms in the crop ecosystem, i.e., the insects, weeds, plant pathogens, and other pests, and also their biological agents of suppression (natural enemies). This system must also include man himself – i.e., his activities in regard to features he imposes on the systems (Rabb, 1978). So, it is essential to get a handle on the influences of such biometeorological factors as temperature, light and moisture on the main biological processes. Although normally these factors have profound concurrent influences, there are times when normal biological processes are arrested and not driven in straightforward fashion by temperature and light. Effects of temperature during such periods, as in insect diapause or seed dormancy, must be reckoned differently than in active growth periods. Light periodism, as well as temperature, is a powerful factor in the timing or phenology of many biological events. But these are things others at this conference will deal with. My objective is mainly to discuss insect control in crop systems in the perspective of integrated pest management (IPM) which itself must be considered in the perspective of crop production as a whole.

What is IPM? Why so much talk about it? Why did President Carter request of CEQ a special report on IPM? Rather suddenly, it seemed to promise a panacea for all pest control without conducting extensive new researches — at least to many of the uninformed. But integrated pest management as a concept and discipline is only a more sophisticated version of integrated control. Its origins lie in work in the late 1800's and early 1900's when pesticides were little used. Such greats in entomology as Stephen A. Forbes, C. V. Riley, J. H. Comstock, C. W. Woodworth and W. D. Hunter come to mind. The principles embodied were recommended in the control of boll weevil on U.S. cotton by U.S.D.A. entomologists W. D. Hunter and B. R. Coad in 1923 (Hunter and Coad, 1923). Pest tolerant cotton varieties and residue destruction were the main means of control, with insecticides used only when a measured incidence (monitored) of weevil damage occurred. Plant pathologists, too, had developed disease management programs utilizing varietal selection and cultural manipulations, along with minimal use of the early fungicides such as Bordeaux mixture. These methods were used well before modern chemical control technology had developed (Bottrell, 1979).

In the late 1800's, Stephen A. Forbes, of the University of Ilinois, had emphasized the application of ecological principles in crop pest control. Yet, though a true pioneer in this respect (and inspiration to C. W. Woodworth who later quietly led California entomologists to develop specific integrated control concepts and methods), Forbes utterly missed the ecological significance of the natural enemies of the pests themselves, claiming they had no role in depressing ultimate pest numbers.

Use of chemicals for pest control greatly expanded in this century, at first slowly and then, following the very successful launching of DDT, with precipitous momentum. The President's Council on Environmental Quality (Bottrell, 1979) reported that in 1979, one-third to one-half the world's use of chemical pesticides were used in this country. Production of synthetic organic pesticides had increased from less than one-half a million pounds in 1951 to about 1.4 billion pounds in 1977 – or about 3,000 times as much.

A major change took place in recent decades in the kinds of pesticides used. While 30 years ago, insecticides constituted 75% of total pesticide use in the U.S., now herbicides constitute more than 50% and insecticides less than 40% (Train, 1972; USDA, 1978). Much of this change is associated with a near total shift in recent years from mechanical to chemical means of weed control.

By the 1950's and 1960's, ecological methods of controlling insects had given way largely to insecticides. Even during this replacement, serious weaknesses in practicing sole reliance on insecticides had surfaced. At the same time, the benefits in lives saved, improved human and livestock health, and increases in food and fiber from the use of these materials became strong medicine against advocacy of a return to a broader, ecologically based approach.

The concept of integrated pest control (or management) received a major thrust in the formal definition, statement of principles, and description of the necessary techniques in its accomplishment by a group of California workers (Stern et al., 1959). In that paper, the consequences of a policy of sole reliance on pesticides were in effect predicted and its weaknesses inherent to crop protection itself were illustrated. This work had itself been foreshadowed by development of practical integrated control programs by A. D. Pickett in Nova Scotia for apples, and by Dwight Iseley for cotton in Arkansas.

Stern et al. (1959) stressed that the destruction of natural enemies by use of chemicals can have a two-fold effect: 1) in reducing the natural suppressive effects of the enemies on the target pest species, rapid target pest resurgence may occur; and 2) the reduced action of natural enemies on other innocuous or minor pests can unleash these to become major pests. But an even more significant factor forcing consideration of other methods was the widespread development of resistance to the insecticides being used. Previously a few entomologists had in fact foreseen the possibility that soon we might be unable to control some, or many, insects by use of chemicals (Smith, 1941).

Along with these purely economic entomological concerns there was an awareness growing of the dangers to man, other animals, and the crops themselves from such unrestrained use of biocides. Rachel Carson's "Silent Spring" (Carson, 1962) not only alluded to these crop-centered problems but also dared to view the broader implications of unrestrained use of pesticides to human health, and to all organisms on earth. Events moved swiftly after "Silent Spring" caused the public, and even Congress, to look at the dangers inherent to an unrestrained release of deadly chemicals in the environment, and not just pesticides.

Regarding pest control itself, several efforts had been launched to realign our thinking and practices. Two opposing concepts were emphasized. One was that of eradication of a pest species over a whole region of occupancy on an island or continent. But this method goes beyond and is not consistent with the basic goal of integrated pest management, which is simply to contain the pests, not to eradicate them or completely prevent their establishment. Secondly, an ecologically centered concept (IPM) was launched in a major way in the early 1970's by the U.S.D.A. and the National Science Foundation (joined later by the Environmental Protection Agency). The approach was to concentrate on a given crop or cropping system and to seek, first of all, a thorough holistic understanding of the system, especially in regard to crop growth, pest impact and cost/benefit relationships. This, in essence, was simply the launching of a widescale investigation of what integrated pest control specialists had been saying since 1959 (Stern et al., 1959). These efforts initiated work to developed technologies of systems analysis, engineering, and computer science.

In this context, I would define IPM as the discipline that deals with analysis of the production system as related to pest impact, and of the biological and physical factor interactions and their interactions with production practices that bear upon pest impact and the combining of appropriate tactics and strategies to optimize net benefits of pest control to the producer, society, and the environment. It is no longer acceptable to simply consider profit to the producer regardless of other adverse effects. Profitable production remains the primary objective of IPM, but this must be achieved in a socially acceptable way.

In his May 23, 1977, Environmental Message to Congress, President Carter acknowledged that chemical pesticides have been the foundation of agricultural, public health, and residential pest control for several decades. But he expressed concern that of the approximately 1,400 chemicals used in pesticide products, some pose unacceptable risks to human health and the environment. Others pose similar risks under unrestrained use. By presidential directive he ordered encouragement of IPM wherever practicable. Also the State of California has decreed by law use of IPM wherever feasible.

As scientists, it no longer makes sense for the chemist, entomologist, plant pathologist or weed scientist to proceed along his/her own separate path to develop and fully launch a tactic; while serving a limited objective, its employment may be detrimental to society or the land itself. An example of the latter has been the use of atrazine in Illinois to control weeds in corn. This seems to have rendered the soils in about one-fourth of Illinois unsuitable for growing soybeans. This is a major crop in the area and soybean culture offers rotation control, the best method of controlling the main insect pest of corn, i.e., Diabrotica rootworms (Roskamp, 1977). Other examples have been accumulations of lead in apple orchard soils in Washington (Benson, 1968) and copper in Florida citrus soils, with severe consequences (Brown and Jones, 1975).

It has now become clear to our administrators who inherited the old system of strict disciplinary divisions and the filtering down of funds for isolated pieces of work, that such a fundamental change is needed - hence, our statewide IPM project in California and two major federal projects came into existence.

Fortunately, the principal gain to growers from use of IPM, that is, lower costs for equal, better and less risky insect control, also means, usually, less use of insectides with its other attendant benefits. To replace the mode of insect control "insurance" through calendar date chemical treatments, as was formerly common, IPM has been advanced as a more ecologically sound solution. But before any broad, national, or state solutions can be realized, much more needs to be done to increase or improve (1) the research base, (2) the technology of implementation, (3) pest control advising, (4) education of farmers, and (5) regulatory systems. If all this can be done, a longer lasting, more economically and socially profitable, healthful, and environmentally sound system of pest control can be obtained.

In order to optimize their combined use in relation first to the farmer and secondly to our social and environmental responsibilities, integrated pest management requires multidisciplinary research on a variety of direct tactics to suppress the pests as well as on quantitative effects of other related tactics or components of the system. But a system of pest control on a specific farm must be a priori profitable to the producer; otherwise, he will not use it or he may switch to another crop.

The foremost requirement is to understand the system and how the pests and other components interact. The focal point in making these assessments is to determine influences on crop-produced photosynthate and its allocation through the system (Fig. 3). Figure 4 presents a model of the different major components in the system and their breakup into units that can be measured and an analysis made of specific interactions.

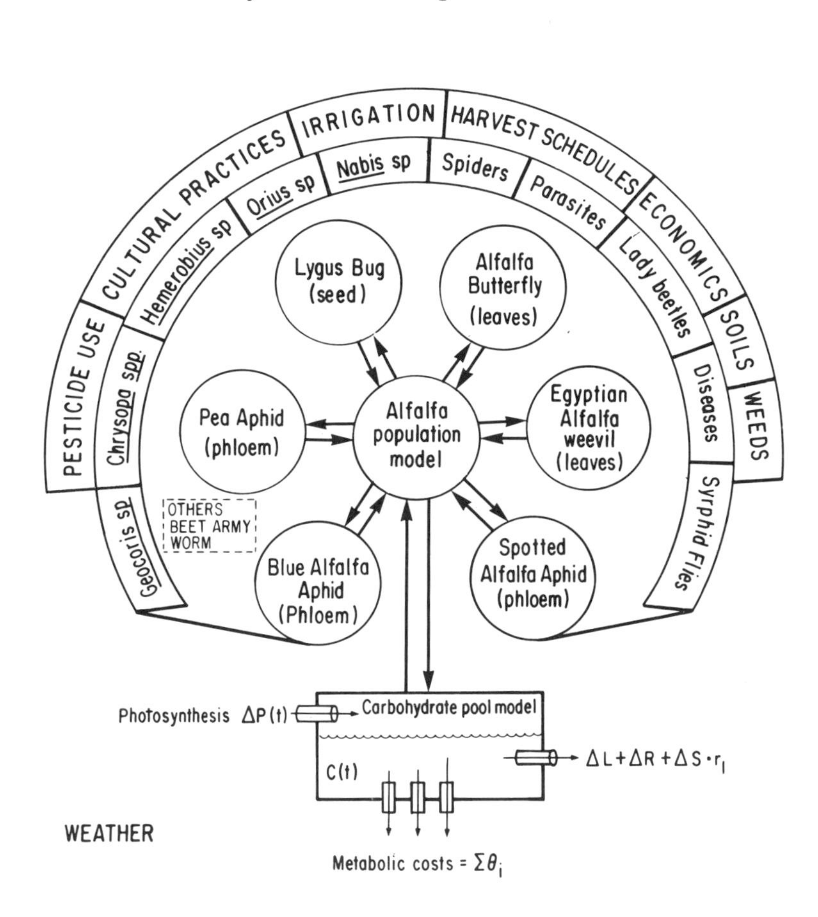

Fig. 3. Some of the potential interactions in an alfalfa ecosystem. Notice that all interactions ultimately impinge on the carbohydrate pool model. (From Gutierrez et al., 1980.)

The 18-university NSF/EPA project, which I directed, concentrated on the insect and mite pests, with only a little

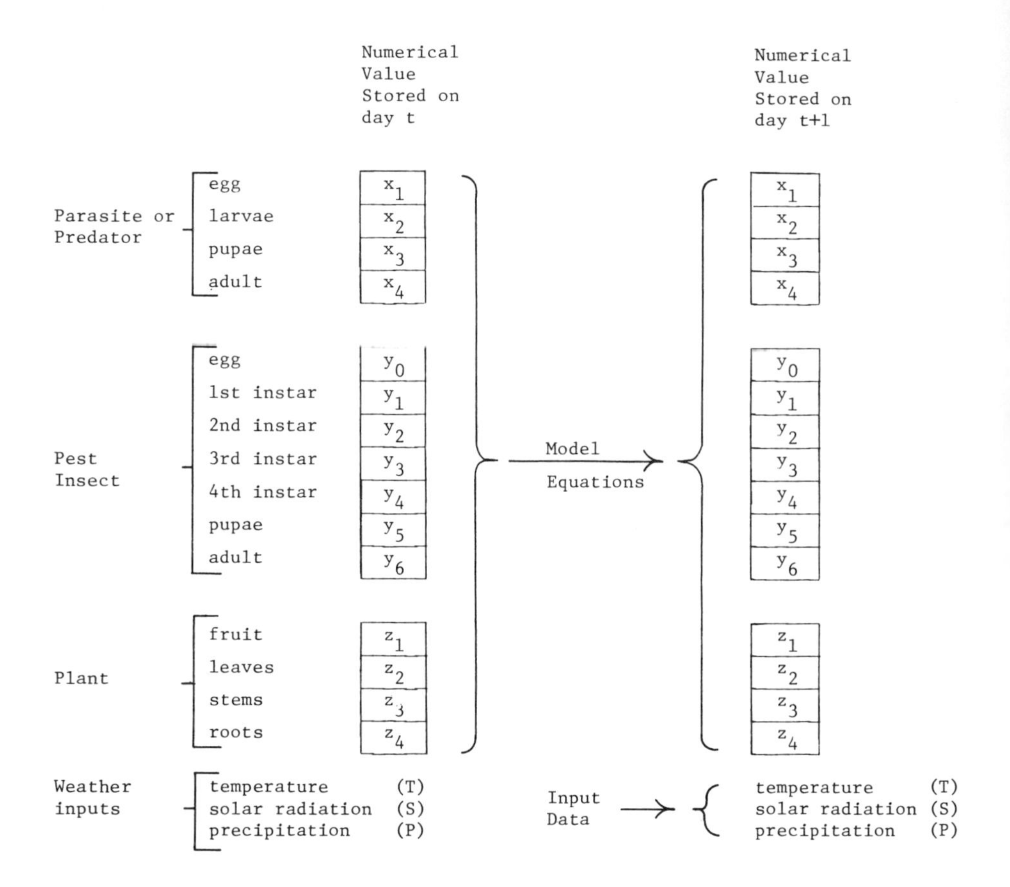

Fig. 4. The structure of a typical pest management simulation model. (From Shoemaker, 1980.)

additional work on plant diseases. It is now time to consider all these features and classes of pests in selecting problems and planning their solution. This is being done in the new national "Adkisson project" and in the statewide IPM project in California. In such efforts, central management must have the power to set priorities and allot funds if it is to keep overall programs on a balanced track.

The securing of support is no easy task for a project, which by its very nature, departs from traditional federal and state funding and administrative routes, and challenges traditional patterns of pest control. It has proved both administratively and politically involved.

II. SPECIFIC STEPS OR GUIDELINES

Because pest problems differ according to type of pest and situation, general guidelines cannot be absolute. Bottrell (1979) modified and condensed guidelines developed by USDA-ARS scientist F. R. Lawson and colleagues and modified by Huffaker and Smith (1972) as the basis of the NSF project. Bottrell specifies four items:

1. Analyze the "pest" status of each of the reputedly injurious organisms and establish economic thresholds (or action levels) for the "real pests."

That is, some organisms are "induced pests" and become pests only through what man does to the environment. Some are only occasionally pests though not induced, and some are regularly troublesome and are "key" pests. In most situations, many potential insect pests are held at characteristic (or "equilibrium") noneconomic densities by natural control factors (Fig. 5). Only a few are key pests and often a single insect is the key insect pest. The way one manages key insect pests alleviates or aggravates problems with otherwise noninjurious or occassional pests. Their "equilibrium" densities may become greatly increased and then suppressed naturally only by severe food shortage from their own destruction of the crop; thus, new or further chemical treatments are required. As one of many examples, Rabb (1971) showed that aphid populations on tobacco increased dramatically when the insecticide carbaryl was used to control other pests.

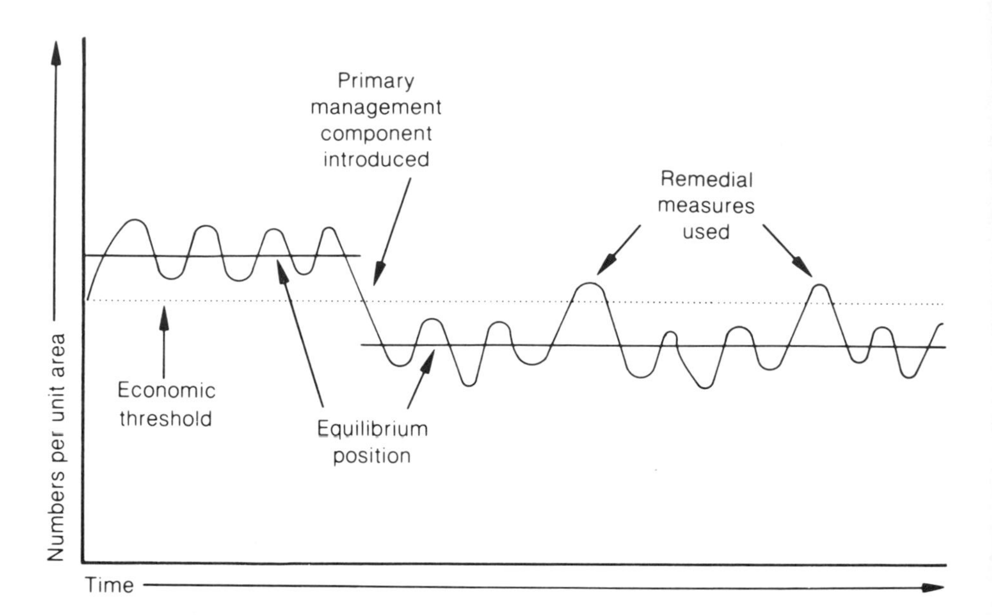

Fig. 5. Lowering the equilibrium position of a pest. (Bottrell, 1979, after Rabb, 1978.)

2. Devise schemes, basic tactics, for lowering equilibrium densities for key pests.

a) Biological control has often been highly successful.

Figures 6 and 7 on olive parlatoria scale and Klamath weed control by introduced parasites and leaf-feeding beetles, respectively, illustrate well the power of natural enemies to lower pest equilibrium densities far below their respective economic injury levels (far greater reductions than shown in Fig. 5).

These two examples offer a glimpse of the general importance of biometeorology in IPM. In the case of the Klamath weed beetles, three species of leaf beetles, _Chrysolina quadrigemina_, _C. hyperici_ and _C. varians_ were introduced into California. The first two were well-adapted to the California climate and readily established and spread, while _C. varians_ requires wet summers

Fig. 6. St. Johnsworth or Klamath weed control by Chrysolina quadrigemina at Blocksburg, California: a) 1948, foreground shows weed in heavy flower while remainder of field has just been killed by beetles; b) 1950, portion of same location when heavy cover of grass had developed; c) 1966, showing degree of control that has persisted since 1950. This story was repeated all over California. (From Huffaker and Kennett, 1969. Photography by J. K. Holloway and J. Hamai.)

Fig. 7. a) Olive tree to left treated with DDT to inhibit action of *Aphytis paramaculicornis* and showing lack of vigor, defoliation and dieback resulting from heavy *P. olea* densities. Trees to right had no treatments with DDT, and *Aphytis* kept the scale under good control with tree growth and condition excellent; b) twig from representative DDT-treated, largely parasite-free tree showing nature of damage by *P. oleae*; c) twig from representative untreated parasite-present tree, showing clean condition. (After Huffaker et al., 1962) (Photography by F. E. Skinner.)

and did not establish. Yet, C. quadrigemina is more responsive than C. hyperici to the fall breaking of the summer drought and it completely displaced C. hyperici when competition for the vastly dwindling supply of Klamath weed became intense (Fig. 6). Moreover, in western Australia this beetle is also quite effective, whereas only variably so in eastern Australia. Only in western Australia is the climatic pattern adequately similar to that in California - i.e., there is a long summer drought and adequate winter rains. The range vegetation there is also different from that in eastern Australia, being again a near duplicate of California in the dominance of specific species of alien grasses and forbs of Mediterranean origin (Huffaker, 1967).

Regarding the olive scale parasites, only Aphytis paramaculicornis (= maculicornis) proved very effective in California. In this case, due to the hot dry summers, even it was handicapped; a supplementing species, Coccophagoides utilis which readily survives the summer, was introduced and together they accomplish remarkable control of the pest (as good as any insecticide) (Huffaker and Kennett, 1966).

Early season releases of Pediobius epilachnae, in Florida and adjacent southern states, have been highly effective against the Mexican bean beetle. This insect does not overwinter and must be reared and released each spring; it breeds and spreads phenomenally fast (Sailer, 1972).

b) Use of pest-resistant or pest-escaping varieties.

There has been substantial research in recent years to develop crop varieties resistant to various pests. This tactic has also been highly successful. Work on grain insects and diseases has been highly successful and includes work with viruses, bacterial, and fungal pathogens with a number of insects, including Hessian fly, corn earworm, European corn borer, greenbug and chinch bug (Klassen, 1979). Another point of interest is the great variety of characters available in the germ plasm of a single crop (e.g., cotton) that can be used to develop resistance against insects (Beck and Maxwell, 1976).

Extensive research in the soybean IPM project in the South has been conducted to develop a variety of commercially acceptable soybean resistant to Mexican bean beetle. This was accomplished but the cultivar so far developed has a lower yield – sufficiently lower that chemical treatment of higher yielding but more susceptible varieties is preferable. Success with *Pediobius* promises to make continuation of this breeding effort unnecessary. This is an illustration of how one line of research may be terminated by success with another. This latter was true also when the success with Klamath weed beetles in California in the early 1950's made obsolete further studies for its control using grazing patterns and herbicides.

A specially developed short-season dwarf cotton variety was used in a production system for South Texas designed to avoid prolonged exposure to boll weevil damage; by harvesting the crop early, weevil overwintering would be greatly reduced. This is combined with selective insecticide treatment (below) of weevils destined to overwinter (in diapause) so that the weevil population is slow to develop the following season. Thus, summer treatments can be delayed and natural enemies of bollworms and tobacco budworms (*Heliothis* sp.) can be active and keep those pests under control. Such a system has not only restored biological control of the devastating budworm and also bollworms, but the costs of insecticides have been reduced. Yields increased 30%, pesticide use decreased 27%, cost per pound of lint decreased 29%, energy use per pound of lint decreased 48%, and profit increased from $12.40 to $104.97 per acre, an 846% increase (Huffaker and Lacewell, 1979). There are similar successful programs in other regions of Texas (Lacewell et al., 1977; Walker, 1980), but this example serves to illustrate the point.

c) Modification of the environment to hinder the pest and/or to favor natural enemies.

Space will allow only a listing of these tactics, among which there are many possibilities in: i) removal or destruction of breeding refuges and overwintering breeding; ii) destruction of alternate hosts or volunteer plants; iii) crop (or pasture) rotation; iv) tillage; v) trap crops; vi) habitat diversification; and vii) water management.

3. During (or to prevent) emergency situations, seek remedial measures that cause minimum ecological disruption.

The best combination of natural enemies, resistant varieties, and environmental modification may eliminate need for action against many pests. Essentially, permanent control of key pests of some agricultural crops has been achieved by integrating such cultural practices as plowing and timing of irrigation with pest-resistant crop varieties and conservation of natural enemy populations (Bottrell, 1979). When key pests require it or secondary pests are out of control, remedial measures must be taken; pesticides may be the only recourse. In integrated pest management programs, selection of the pesticide, dosage, and treatment time are carefully coordinated to minimize ecological disruption and other problems. Economic threshold analysis serves to identify when and where such measures are justified.

There are a number of items that should be mentioned here as direct suppressive measures.

i) Eradication - This method is not normally considered IPM or as causing minimal disturbance, but its success against a key pest could greatly ease or simplify IPM of other pests of a crop.

ii) Use of behavioral chemicals - Certain of these materials are used mostly for monitoring but they present direct suppression possibilities. Repellents have been widely used in the medical and veterinary field; work in developing antifeedants for phytophagous insects offers some promise.

iii) Selective chemical insecticides - The effort here has been to enhance the selectivity of pesticides by altering dosage, timing, coverage and the like to give a selective effect. This has been employed in IPM of most crops, e.g., in apple spray programs and for soybeans (Asquith et al., 1980; Newsom et al., 1980).

In Washington State and Michigan, reduced dosages (e.g., of Guthion) for codling moth have allowed predators of the spider mites to be effective. IPM programs were developed centering on spider mite control by pesticide resistant *Metaseiulus occidentalis* in Washington, *Amblyseius fallacis* in Michigan, and the coccinellid *Stethorus punctum* in Pennsylvania. There have been reductions in insecticides and acaricides of about 50% in Washington and 20-30% in Michigan and Pennsylvania (Huffaker and Croft, 1978; Asquith et al., 1980).

For Michigan apples, if the ratio of predatory mites to spider mites is very low, biological control cannot be relied upon and a chemical control will have to be used for that season. If this ratio is high, biological control will be effective for the season, but if intermediate, another sample a week later is taken to determine if the condition moves into either the biological control or chemical control regions of a guide chart used in decision-making (Croft, 1975).

4. Devise monitoring techniques to assess environmental, pest, and crop condition.

Insects grow and reproduce and they move about. To keep abreast of their ever-changing status in a crop, we must monitor events. The movements of insects from place to place is perhaps the least understood of any major facet of their ecology. Many techniques are used to follow insect movements (even radar) and use of pheromone traps to monitor "activity" is becoming a major tool in IPM. But ordinary sampling of pest and crop development and monitoring of weather factors which drive biological processes are required. But one always needs to ask, "How much monitoring can be afforded by growers"? It must all be boiled down to affordable levels. This is seldom given enough attention.

Three methods of control of alfalfa weevils are possible: biological control, cultural control (early harvesting), and chemicals. Resistant varieties have not as yet proved feasible. Insecticides and also early harvesting ("cutting") kill parasites needed in the next growth crop. Development of the crop, weevils, and parasites are functions of temperature accumulation (day-degrees, D°).

In Illinois, for example, if in the first spring sampling 44 weevil larvae are taken and the alfalfa is 3" high, a chart developed by IPM specialists (Wedberg et al., 1977) says not to spray, but to resample 50 D° later. But if

47 or more larvae are taken, the chart says to spray. If 50 D° later the alfalfa is 6" high, then 55 or more larvae would be required to indicate need to spray. A second chart shows the decision process at 540 D° to time of harvest, during which early harvesting may be better than spraying despite some loss in yield.

In Arkansas the main pest of cotton is bollworm. Using simple developmental information and by use of careful monitoring of all stages of bollworms and application of insecticides only as needed, it has been possible to reduce use of conventional insecticides from ten treatments to one in 1977 and from 14 to two in 1978. One additional microbial treatment was made in 1977 while two were used in 1978. These programs were completely adopted in a 60 sq. mile community area in 1977 and over a region ca. 200 sq. miles in 1978, with almost total grower adherence. There was a reduction in conventional insecticides of 87% and overall savings, including costs of the microbial application, of $20 per acre. Conventional costs were $28 per acre compared to $8 for this program. Growers are clamoring for expansion of the program over the whole Arkansas cottonbelt (Phillips, personal communication; Phillips and Nicholson, 1979).

REFERENCES

Asquith, D., Croft, B. A., Hoyt, S. C., Glass, E. H., and Rice, R. E. (1980). In "New Technology of Pest Control" (C. B. Huffaker, ed.), pp. 249-317. Wiley, New York.

Beck, S. F., and Maxwell, F. G. (1976). In "Theory and Practice of Biological Control" (C. B. Huffaker and P. S. Messenger, eds.), pp. 615-636. Academic Press, New York.

Benson, N. R. (1968). Wash. State Hortic. Assoc. Proc. pp. 109-114.

Bottrell, D. G. (1979). "Integrated Pest Management." Counc. on Environ. Qual., U.S. Government Printing Office, Washington, D.C. 120 p.

Brown, J. C., and Jones, W. E. (1975). Commun. in Soil Sci. and Plant Anal. 6:421-438.

Carson, R. (1962). "Silent Spring," Houghton Mifflin, Boston. 368 p.

Croft, B. A. (1975). Michigan State University, Coop. Ext. Serv. Bull. E-825.

Garcia, R. and Huffaker, C. B. (1979). Agro-Ecosystems 5:295-315.
Gutierrez, A. P., Demichele, D. W., Wang, Y., Curry, G. L., Skeith, R., and Brown, L. G. (1980). In "New Technolgoy of Pest Control," (C. B. Huffaker, ed.), pp. 155-186. Wiley, New York.
Huffaker, C. B. (1967). Mushi 39 (Suppl.) pp. 51-73.
Huffaker, C. B., and Croft, B. A. (1978). Calif. Agric. 32(2): 6-7.
Huffaker, C. B., and Kennett, C. E. (1966). Hilgard. 37:283-335.
Huffaker, C. B., and Kennett, C. E. (1969). Can. Entomol. 10:425-447.
Huffaker, C. B., Kennett, C. E., and Finney, G. L. (1962). Hilgard. 32:541-636.
Huffaker, C. B., and Lacewell, R. D. (1979). Proc. Western Soc. Weed Sci. 32:109-119.
Huffaker, C. B., and Smith, R. F. (1972). Proc. Tall Timbers Conf. on Ecol. Animal Control by Habitat Managmt. 4:219-236.
Hunter, W. D., and Coad, B. R. (1923). U.S. Dept. Agric. Farmers' Bull. 1329. 30 p.
Klassen, W. (1979). In "Crop Protection" (W. Ennis, ed.) Am. Soc. Agron. Crop Soc. Am., Madison. pp. 402-442.
Lacewell, R. D., Casey, J. E. and Frisbee, R. E. (1977). Texas Agric. Exp. Sta. Dept. Tech. Rep. 77-4. 215 p.
Newsom, L. D., Kogan, M., Miner, F. D., Rabb, R. L., Turnipseed, S. G., and Whitcomb, W. H. (1980). In "New Technology of Pest Control" (C. B. Huffaker, ed.), pp. 51-98. Wiley, New York.
Phillips, J. R., and Nicholson, W. F. (1979). Proc. 1979 Beltwide Cotton Prod.-Mech. Conf. pp. 39-41.
Rabb, R. L. (1978). Bull. Entomol. Soc. Am. 24:55-61.
Rabb, R. L. (1971). In "Biological Control" (C. B. Huffaker, ed.), pp. 294-311. Plenum, New York.
Reeves, W. C., Milby, M., and Hardy, J. L. (1980). Calif. Agric. 34(3):6-7.
Roskamp, G. (1977). 19th Illinois Custom Spray Operator Training School: Summaries of Presentations. University of Illinois, Coop. Ext. Serv., Urbana. pp. 174-177.
Sailer, R. I. (1972). Agric. Sci. Rev. 4th Quart. pp. 15-28.
Shoemaker, C. S. (1980). In "New Technology of Pest Control" (C. B. Huffaker, ed.), pp. 25-49. Wiley, New York.
Smith, H. S. (1941). J. Econ. Entomol. 34:1-13.
Stern, V. M., Smith, R. F., van den Bosch, R., and Hagen, K. S. (1959). Hilgardia. 29:81-101.
Train, R. E., Chmn. (1972). "Integrated Pest Management," Council on Environmental Quality. U.S. Government Printing Office, Washington, D.C. 41 p.
United States Department of Agriculture. (1978). "Pesticide Review, 1977" Agric. Stab. and Conserv. Serv. 44 p.
Walker, J. K. (1980). 1980 Proc. Beltwide Cotton Prod. and Res. Conf., Nat. Cotton Counc. Am. pp. 153-155.

Washino, R. K. (1980). Calif. Agric. 34(3):11-12.
Wedberg, J. L., Ruesink, W. G., Armbrust, E. J., and Bartell, D. P. (1977). University of Illinois, Coop. Ext. Serv. Circ. 1135. 7 p.
Wilson, C. L., and Loomis, W. B. (1962). "Botany" Holt, Rinehart and Winston, New York.

INSECT MOVEMENT IN THE ATMOSPHERE[1]

R. E. Stinner and K. Wilson
Department of Entomology
North Carolina State University
Raleigh, North Carolina

C. Barfield
Department of Entomology and Nematology
University of Florida
Gainesville, Florida

J. Regniere
Great Lakes Research Station
Sault Ste. Marie
Ontario, Canada

A. Riordan and J. Davis
Department of Geosciences
North Carolina State University
Raleigh, North Carolina

I. INTRODUCTION

If there is any single problem in ecology which requires the inputs of many diverse views, it is our lack of understanding of insect movement. This complex set of processes and behaviors could be the focal point for bringing together the often divergent opinions of applied ecologists, experts in population dynamics and genetics, and behavioral and evolutionary ecologists. Indeed, the added complexity introduced by movement may well be one of the

[1] Paper No. 8126 of the Journal Series of the North Carolina Agricultural Research Service, Raleigh, NC 27650.

ISBN 0-12-332850-0

major reasons for some of the different philosophies among these groups.

We are all aware that populations can change by births, deaths, or movement. Often we assume, in theoretical studies, that we are dealing with the entire population and thus net movement is zero and can be ignored. What a tragedy to population dynamics that we have taken this view! It totally ignores the importance of spatial patterns. Why should such patterns be any less important than temporal changes? Such a view ignores genetic differences among sub-populations and the evolutionary significance of spatial heterogeneity.

In practice, we simply cannot look at whole populations and ignore gene flow, even if net movement were near zero. The many documented cases of insecticide resistance spread should stand as a constant reminder of the importance of this process.

Why then has there been so little work in applied ecology and agricultural entomology on insect movement? There are a number of reasons or at least rationalizations. First, our historical information is generally anecdotal; no one has developed a truly satisfactory conceptual framework. Perhaps such a single framework does not exist and our approach should concentrate on examining, in detail, selected species to develop an in-depth understanding of the role of movement in the dynamics, genetics and potential evolution of specific targets.

Secondly, with most species, movement is difficult to measure directly. Since biologists in general dislike extrapolation, deductive reasoning, or circumstantial evidence, it is extremely difficult to prove hypotheses relating to movement to the satisfaction of the biological community.

Thirdly, there has been a temptation to look at movement as a process apart from the resource structure within which that movement occurs. Species vary not only in their behavioral modes with regard to movement, but perhaps more basically in their ability to perceive resources and their propensity to move. One of our first tasks must be to define for given species the grain or patchiness of the ecosystem (in our case, agroecosystem) as recognized by the organism. We also must be cognizant that there is a gradient of movement responses from trivial flight to long-distance migration. In the same individual, specific internal and external cues may interact to cause movement of quite different magnitudes and for distinctly different "purposes." For practical reasons, we will examine a sampling of on-going and potential research dealing with the two ends of this movement gradients: local or resource-finding movement; and migration or long-distance displacement.

II. SHORT-RANGE MOVEMENT WITHIN THE AGROECOSYSTEM

One must first, however, have a conceptualization which includes not only displacement, but also the resource structure and other factors which impinge upon this movement. We will, therefore, present examples of three distinct species to demonstrate how a single agroecosystem structure can be quite differently perceived by species of varying vagility.

In North Carolina, Heliothis zea Boddie is a multivoltine, highly polyphagous species which overwinters in the soil as diapausing pupae. Adults are highly mobile and crop hosts (maize, soybean, tobacco, cotton, among others) provide the major food resources. Given the average North Carolina crop field size of 2-3 hectares and the random nature of field location, the adults move freely from field to field, ovipositing in a pattern that reflects both basic preferences for and availability of specific hosts and maturity states of those hosts. For any given host, the adults prefer to oviposit in fields of reproductive, but not senescing, plants (Johnson et al., 1975). Although discussion of this species at this point is restricted to localized movement, it is important to note that Heliothis is also a long-distance disperser, being found as far north as Canada every year (Kogan et al., 1978; Rabb and Stinner, 1979). Presently, studies of differences among populations from various parts of South and North

America are being conducted (Gould et al., 1980). Included in these studies are several hypotheses concerning movement propensity and agroecosystem structure.

Development of every stage of this species, even under constant temperatures is highly variable (Stinner et al., 1975). Larvae, when confined, are highly cannibalistic (Barber, 1936). Due to the abundance of maize in North Carolina (over 800,000 hectares) and early planting and flowering dates, over 90% of the population in the first two generations is "funneled" through this crop. On reproductive maize, larvae are typically confined within the fruit (ears) and intraspecific predation (cannibalism) effectively places a ceiling on the population of 40-50,000 larvae/hectare (Stinner et al., 1977a). This competition, however, also affects the "apparent" generation time. Since larvae from the first-laid eggs and the fastest developing individuals survive at the expense of slower-developing and later-born individuals, subsequent adult emergence peaks can be as much as a week earlier than peaks predicted on the basis of physiological development alone. Additionally, the inherent variability in development strongly affects this decrease in "apparent" generation time; the greater the variability, the shorter the time interval between adult peaks at higher larval densities. Since the choices available to ovipositing adults, in terms of available host phenologies, change rapidly and their ability to move is great, with respect to field size, this shift in generation time can greatly alter the spatial pattern of resultant progeny (Stinner et al., 1977a, 1977b).

During usual weather patterns, the bulk of the *Heliothis* population remains in maize until August. As maize dries up, the population shifts (via adult movement and oviposition) to late-planted, open canopy soybeans (as well as other hosts). Under these same conditions, early planted soybeans escape utilization by *Heliothis*, since maize is more attractive than these soybeans during their bloom phase. In 1977, much of North Carolina experienced a prolonged drought in July. A large proportion of the maize either matured faster or simply dried up and was unattractive to *Heliothis* adults. Blooming in soybeans is largely photoperiodically controlled, so maturation in this crop was not greatly accelerated, although there was a significant decrease in canopy development. The result of this early loss of maize as a preferred oviposition resource, and the concurrent reproductive state of early-planted soybeans led to a high oviposition on these soybeans, which normally escape *Heliothis*. Thus, we have an example of weather variation indirectly affecting the spatial distribution of a species in a subsequent generation, through influence of patterns, movement, and oviposition.

Looking at our second species, the Japanese beetle (*Popillia japonica* Newman), we find a univoltine, polyphagous species which has a much lower vagility than that exhibited by *H. zea*, but is still quite mobile. This species overwinters as a grub in the soil and requires perennial hosts, most of which, at least in North Carolina, are not crop hosts (Deitz et al., 1976). The spatial structure of importance to this species is not the crop field arrangement but rather the size and spatial arrangement of weed species and soil moisture gradients conducive to adult feeding and immature survival (Regniere et al., 1981b). The agroecosystem, in a simplistic view for this insect, can be partitioned into (1) intensive production sites, (2) marginal production sites, (3) "migration" alleys, and (4) transient feeding sites (Regniere et al., 1981b).

(1) Intensive production sites are locations with an abundance of preferred hosts for adults (smartweed, *Polygonum pennsylvanicum*; blackberry and related *Rubus* sp.; elderberry, *Sambucus canadensis*; and wild grape, *Vitis* sp.) and in or near which high densities of grubs are present. Such sites are often located along creeks, and drainage or irrigation canals. In these sites, biotic agents, such as diseases and parasites (Fleming,

1968; Regniere and Brooks, 1978) usually are found. Furthermore, because of high grub densities, it is likely that intraspecific competition also is acting upon immature populations (Regniere et al., 1981a). In such sites, the gregarious adults from neighboring sites converge to feed and mate. Oviposition in or near aggregation sites is normally intensive, beginning at the onset of the emergence period and continuing as long as adults are present. Usually, environmental conditions in the soil of these intensive production sites are favorable to immature survival.

(2) Marginal production sites are areas which become suitable for oviposition and survival in a more sporadic manner. They are often located at the margin of intensive producton sites, or in more remote areas where preferred hosts are not abundant. In these sites, immature populations are lower due to lower ovipositional activity and more frequent occurrence of environmental conditions unfavorable for immature survival. Biotic mortality factors play a less important role in population regulation, although they may be present. Under unfavorable conditions (e.g., even limited drought periods), marginal production sites may be dramatically reduced in area, and immature populations may suffer high mortality. Adults emerging from these sites migrate to aggregation sites soon after emergence to feed and mate.

(3) Migration "alleys" are areas where environmental conditions are usually unfavorable to oviposition and immature survival. Grub density in migration "alleys" is low. Major causes of mortality are abiotic. Adults present in these areas are transient.

(4) Feeding sites are "islands" of host plants in migration "alleys" where transient beetles may aggregate. In North Carolina, field edges of soybean or corn may be utilized by beetles as such feeding sites. Usually, aggregations of beetles in feeding sites are unstable, changing frequently in location, giving an observer the impression that the insects have a highly unpredictable behavior.

In such a system, limitation of population growth is associated with the temporal variation in the size and occurrence of intensive and marginal production sites. This variation is largely temperature and rainfall related, but man plays an important role with his weed and water management actions. Oviposition is determined by the availability and quality of food, the suitability of oviposition sites and the spatial separation of feeding and oviposition foci (Regniere et al., 1979).

Finally, although this species is univoltine, soil moisture greatly affects both survival and timing of immature development. Egg development can be greatly delayed and survival decreased by soil moisture that are either too high or too low (Regniere et al., 1981c). The Japanese beetle, thus, can impinge upon crop hosts, but its dynamics are not strongly tied to the crop fields themselves.

In contrast, the Mexican bean beetle (*Epilachna varivestis* Mulsant) is a multivoltine, oligophagous species whose population dynamics are highly dependent on the availability of garden beans (genus *Phaseolus*) and soybeans (*Glycine max* L.) (Deitz et al., 1976). Additionally, adult fecundity and larval survival are reduced severely by moderately high temperatures and low humidities. This sensitivity to weather is exaggerated when the beetles feed on soybean foliage, a marginal yet far more abundant host. Also, the phenological stage of the soybean host also influences its nutritional adequacy (Lockwood et al., 1979). Under favorable physical conditions, both survival and fecundity are significantly lower on prebloom soybean foliage compared to postbloom foliage (Lockwood et al., 1979; Kitayama et al., 1979; Wilson, 1979). Since a dense canopy can moderate ambient temperatures and increase relative humidity, the microclimate experienced by the insect in the crop canopy can be quite different from the ambient conditions (Sprenkel and Rabb, 1981). An understanding of the interactions of these environmental factors is beginning to clarify this beetle's response to the heterogeneous agroecosystem of the southeastern United States (Wilson and Stinner, personal communication). The presumed area of origin of the Mexican bean beetle is the Central Plateau of Mexico (Landis and Plummer, 1935) where Phaseolus is the predominant host and the climate during the growing season is characterized by moderate temperatures and daily rainfall. This may partially explain the patchy spatial dynamics of this beetle in more northern American regions where the climate and the most abundant host are marginal. In the arid Southwest, these beetles traditionally reach pest proportions only in irrigated fields (Sweetman, 1932).

The Mexican bean beetle overwinters as an adult in forest leaf litter so at least local movement must occur in the spring and fall to and from feeding sites (Deitz et al., 1976). Although little is known about its flight potential, this beetle appears to be a moderate flyer and the availability of accessible overwintering sites does not seem to be a limiting factor in North Carolina, where fields are small and woodlots common.

Though planted at about the same time as soybeans (mid-May), garden beans (Phaseolus) usually senesce in late July, when soybeans are blooming and are at peak suitability as a host. Thus, localized movement from woodlots to small patches of garden beans for first generation survival to large fields of soybeans and back to woodlots for overwintering is an annual movement cycle and represents an important aspect of this insect's population dynamics.

Although difficult to assess, movement in the Mexican bean beetle is presently being addressed through two avenues of research. One study area involves lab and field experimentation to determine the cues which lead to movement out of a resource by examining intrinsic factors such as the physiological and reproductive state of individuals, and extrinsic factors such as nutrition (host species and maturity), photoperiod, moisture, temperature, and wind. The second avenue is the development of alternative mathematical models using various hypothesis which include both the behavior of the organism and the resource field structure. For example, one can assume (1) that the population searches directionally, randomly, or both and (2) that "waiting time" in a given field is or is not a function of host maturity. Combining the above with various assumptions concerning resource distribution and patch size with respect to different movement x distance probability curves can lead to a wide array of simulated population patterns, only a limited number of which are actually observed. Thus, the models (i.e., hypotheses) become a tool in designing field experimental situations and vice versa.

III. LONG-DISTANCE DISPLACEMENT

So far we have discussed only local movement. When one examines long-distance displacement, several quite different difficulties arise. In addition to behavioral and physiological studies which define when an organism is ready or able to move, information on general weather patterns, fronts, low level jet streams, etc., over wide areas is required (Muller, 1979; Riordan, 1979). The effort to study such movement has to be regional, often international. Such a program is presently in the planning stage in the southeastern United States for studying movement of the fall armyworm (*Laphygma frugiperda*) and velvetbean caterpiller (*Anticarsia gemmatalis* Hübner). Previous studies of insect movement have generally dealt in behavioral-physiological studies of species with a clearly defined oogenesis flight syndrome (e.g., Dingle, 1979; Rankin and Rankin, 1979) or have looked at movement from a relatively empirical view (e.g., Taylor, 1979; Haile et al., 1975). The proposed work on fall armyworm and velvetbean caterpillar adult movement arose as the result of a symposium on movement of highly mobile *Lepidoptera* (Rabb and Kennedy, 1979). Presented is our initial approach to this problem in terms of the hypotheses raised and how we propose to test these hypotheses. Most of the ideas expressed are the result of a number of meetings among state and USDA entomologists from all the southeastern states with Carl Barfield (University of Florida) as Chairman, ecologists Hugh Dingle (University of Iowa) and Mary Ann Rankin (University of Texas at Austin), and meteorologists Art Muller (Louisiana State University) and Al Riordan (North Carolina State University).

As a result of these meetings, a proposal is being developed which will include a holistic approach to the study of long-distance displacement within the context of the meteorological and agroecosystem structures. Some of the research described is already underway.

The first objective is to determine seasonal patterns of appearance and geographic distribution of fall armyworm and velvetbean caterpillar. Subobjectives include:

(a) compilation of existing data through comprehensive literature search and

(b) determining patterns of initial appearance.

This objective is to be met by a large network of survey sites using pheromone traps, light traps and in-field larval sampling. Temperature and host information will be used in developmental models to estimate adult arrival from the in-field larval sampling. All three techniques are necessary, since a pheromone is not available for velvetbean caterpillar and light traps are not necessarily efficient capture devices for either species. In addition, "trap crops" for both species will be planted and monitored over much of the region.

The second objective is to determine overwintering (either quiescence or continuous breeding) habitats of fall armyworm and velvetbean caterpillar in the continental United States, with subobjectives:

(a) to determine the ability to overwinter under various temperature regimes with initial emphasis in the Gulf Coast states, and

b) to determine crop and noncrop hosts and the distribution and abundance of these plants in overwintering areas.

Laboratory experimentation (where missing or inadequate) will focus on delineating physical environmental tolerances for fall armyworm and velvetbean caterpillar. These same experiments will be conducted on the principal host plants of fall armyworm and velvetbean caterpillar in an attempt to separate physical environment from unavailability of host plants as affecting "winter" survival. Detailed studies will focus on placing fall armyworm and velvetbean caterpillar larvae and pupae at sites along the Gulf Coast to monitor survival over several years. Larvae of both species should be placed also on key noncrop hosts to determine ability of these hosts to maintain low-level moth populations during periods when major crop hosts are either unavailable or seemingly nonpreferred. A survey of a wide variety of habitats should be made during "winter" to determine whether the fall armyworm or velvetbean caterpillar quiescent stages are present or where continuous breeding populations of these species occur. Key questions to be answered are: (1) are fall armyworm and velvetbean caterpillar populations overwintering in the continental USA? (2) If so, where and in what densities? (3) Can these densities produce sufficient populations to account for our fall armyworm and velvetbean caterpillar pest problems?

The third objective involves developing methods for characterizing more localized populations of fall armyworm and velvetbean caterpillar as needed to determine the origins of these highly mobile individuals. Much "homework" is needed to determine the applicability of current technology to determining origins of mobile arthropod species. Personnel possessing pertinent analytical equipment must be sought for potential cooperation in this project. All fall armyworm and velvetbean caterpillar life stages collected under the first objective will ultimately be retained and forwarded to predesignated locations for analyses, not in the first years but, rather, only after analysis details are resolved.

Specimens from Mexico and other Central American and Caribbean countries will be treated similarly. Eventually, the following ten techniques should be investigated thoroughly to determine the validity and shortcomings of each as tools to determine fall armyworm and velvetbean caterpillar origins: (1) rare alleles, (2) x-ray of trace elements, (3) isoenzymes, (4) pollen analysis, (5) phoretic mites, (6) biotypic moth races, (7) insecticide resistance, (8) differential response to pheromone, (9) artificial application of some trace element (e.g., strontium) to a given habitat, and (10) determination of developmental thresholds for immatures of moths collected under Objective 1. From existing information, x-ray analysis and phoretic mites seemed to be the most promising and these will be investigated first. Techniques to best utilize these tools will then be assessed and regional centers established where fall armyworm and velvetbean caterpillar stages can be sent for determination of origins.

The fourth objective is to determine which weather variables are needed to characterize specific flight altitudes, and to relate arrival of fall armyworm and velvetbean caterpillar populations at specific sites with probable flight paths at various altitudes. (The USDA has already committed funds to this area.)

(a) To utilize radar techniques to determine flight altitudes (couple airplane collections with techniques).

(b) To expand synoptic weather types for the entire region.

(c) To integrate various weather parameters currently measured over the region (surface and low altitude) and utilize these with data from objective 1 to develop "back-tracking" capabilities.

Once aloft, fall armyworm and velvetbean caterpillar moths respond (theoretically) to a vector of weather variables including, for example, wind speed and direction, barometric pressure, temperature, and dew point. These and other weather variables vary with height above the earth's surface. Thus, we must know how high fall armyworm and velvetbean caterpillar moths fly and to what weather variables they respond. Methods either to measure or assess data on measurement of these key variables must be developed. If profiles of these "key weather variables" along vertical gradients can be developed, our ability to "backtrack" populations of fall armyworm and velvetbean caterpillar to their probable origins will be much enhanced. Since a mere 100 m shift in altitude can result in a totally different pathway for moving moths, much consideration must be given to flight altitudes and how they shift as a result of major environmental changes (e.g., seasonal implications).

Much research has centered on the use of radar as a tool for monitoring insect movement (e.g., Wolf, 1979). One major drawback is that individual species cannot (currently) be separated from a mass of mixed species on the move. In short, there are insufficient data to evaluate the use of radar as an insect population monitor.

A computerized trajectory program developed by Heffter and Taylor (1975), which has been widely used in air quality research and has recently been improved and simplified (Heffter, 1980), is suitable for application to the insect problem. The program computes trajectories for either fixed levels or levels integrated over the depth of the planetary boundary layer. Although the Eastern United States will be the region of concentration of this objective, if suitable insect data is available from Central America and the Caribbean, the study could be extended to include those areas as trajectory starting points.

Insect sampling data from the cooperative effort of independent investigators in the regional study will provide the input for determining the choice of trajectory starting points. Such field data will be readily available for use in this objective. Early in the season, known overwintering sites would be classed as source regions when emergent moths are first observed, then as migration of populations becomes evident, the new source regions would be included.

The Heffter program (Heffter, 1980) will be used to compute the height of the planetary boundary layer (generally 1 to 2 km

A.G.L.) and compute forward trajectories for separate layers from the surface to the top of the boundary layer, and for the layer-average flow. Air temperature, wind speed, dew point, cloudiness, pressure, and presence of precipitation along the trajectory routes will be included with each trajectory.

Using actual weather data, simulated backward air trajectories at several heights from ground locations of reported new presence of fall armyworm or velvetbean caterpillar as the seasonal population shifts northward will also be computed, to narrow the range of possible source populations. Although it would not be feasible to evaluate each trajectory separately, case studies of effects of trajectory length and transport over "data-sparse" regions will be investigated. Trajectories will be computed for data sets containing random errors representative of the accuracy of meteorological data. These trajectories will be compared with those for the actual unmodified data. Studies, such as those by Hoecker (1977) in which actual paths of tetrons (constant-level balloons) and air trajectories computed by the Heffter model were compared, will also aid in defining the accuracy of computations.

Initially, trajectories could be sorted into classes described as "favorable" and "less favorable" for insect flight based on current knowledge. As new information which quantifies the influence of meteorological factors on fall armyworm and velvetbean caterpillar flight becomes known, it will be used to limit trajectory heights and starting times, and to determine absolute limits of trajectory and points.

This step would incorporate results of insect behavioral studies. For example, temperature or pressure thresholds for migrating moths, as determined by independent investigators, could be used to classify some trajectory routes as being less likely than others. Conversely, if local origins of some migratory populations can be identified and if times of departure and subsequent arrival are known from field sampling, then the trajectory information will provide estimates of heights, distances traveled by individuals, and meteorological conditions in flight.

Finally, the last objective is to determine the physiological and behavioral attributes conducive to initiating and/or terminating fall armyworm and velvetbean caterpillar movement, with subobjectives to:

(a) device a usable laboratory flight test procedure,
(b) determine whether or not an adult diapause occurs,
(c) determine the physiology of reproduction and flight,
(d) determine the effects of larval host and nutrient reserves on flight, and
(e) determine the effects of physical variables on flight.

It should be apparent from the above description of our proposed study of these two species that many biological facets in addition to movement are to be studied. It is our contention that any studies of this process must examine movement in the content of the total life history. To study movement alone would produce an unacceptably inaccurate view of the species dynamics, just as population dynamics studies which have not considered movement as a viable process have done in the past. If we are to develop more optimal control strategies, we cannot permit our work to fall into either category.

REFERENCES

Barber, G. W. (1936). USDA Tech. Bull. 499. 18 p.

Deitz, L. L., Van Duyn, J. W., Bradley, Jr., J. R., Rabb, R. L., Brooks, W. M., and Stinner, R. E. (1976). A guide to the identification and biology of soybean arthropods in North Carolina. N.C. Agr. Exp. Sta. Tech. Bull. 238. 264 p.

Dingle, H. (1979). In "Movement of Highly Mobile Insects: Concepts and Methodology in Research," (R. L. Rabb and G. G. Kennedy, eds.), pp. 64-87. University Graphics, N.C. State University, Raleigh.

Fleming, W. E. (1968). In "Biological control of the Japanese beetle." USDA Tech. Bull. 1383. 78 p.

Gould, F., Holtzman, G., Rabb, R. L., and Smith, M. (1980). Ann. Entomol. Soc. Am. 73:243-250.

Haile, D. G., Snow, J. W., and Young, J. R. (1975). Environ. Entomol., 3:75-81.

Heffter, J. L. (1980). In "Air resources laboratories atmospheric transport and dispersion model." NOAA Tech. Mem. ERL ARL-81. Silver Springs, MD.

Heffter, J. L., and Taylor, A. D. (1975). In "A regional-continental scale transport, diffusion, and deposition model, Part I: trajectory model." NOAA Tech. Mem. ERL ARL-50, Silver Springs, MD.
Hoecker, W. H. (1977). J. Appl. Meteor., 16:374-383.
Johnson, M. W., Stinner, R. E., and Rabb, R. L. (1975). Environ. Entomol. 4(2):291-297.
Kitayama, K., Stinner, R. E., and Rabb, R. L. (1979). Environ. Entomol. 8:458-464.
Kogan, J., Sell, D. K., Stinner, R. E., Bradley, Jr., J. R., and Kogan, M. (1978). In "A bibliography of Heliothis zea (Boddie) and H. virescens (F.) (Lepidoptera: Noctuidae)," INTSOY Ser. No. 17. 242 p.
Landis, B. J., and Plummer, C. C. (1935). J. Agr. Res., 50:989-1001.
Lockwood, D. F., Rabb, R. L., Stinner, R. E., and Sprenkel, R. K. (1979). J. Ga. Entomol. Soc., 14:220-229.
Muller, R. A. (1979). In "Movement of Highly Mobile Insects: Concepts and Methodology in Research," (R. L. Rabb and G. G. Kennedy, eds.), pp. 133-146. University Graphics, N.C. State University, Raleigh.
Rabb, R. L., and Kennedy, G. G. (1979). In "Movement of Highly Mobile Insects: Concepts and Methodology in Research." 456 p. University Graphics, N.C. State University, Raleigh.
Rabb, R. L., and Stinner, R. E. (1979). In "The role of insect dispersal and migration in population processes." NASA Conf. Publ. No. 2070, NASA Wallops Flight Center, Wallops Island, VA. 248 p.
Rankin, M. A., and Rankin, S. M. (1979). In "Movement of Highly Mobile Insects: Concepts and Methodology in Research," (R. L. Rabb and G. G. Kennedy, eds.), pp. 35-63. University Graphics, N.C. State University, Raleigh.
Regniere, J., and Brooks, W. (1978). J. Invert. Pathol., 32:226-228.
Regniere, J., Rabb, R. L., and Stinner, R. E. (1979). Can. Entomol. 3:1271-1280.
Regniere, J., Rabb, R. L., and Stinner, R. E. (1981a). Environ. Entomol. (In press.)
Regniere, J., Rabb, R. L., and Stinner, R. E. (1981b). Can. Entomol. (In press.)
Regniere, J., Rabb, R. L., and Stinner, R. E. (1981c). Environ. Entomol. (In press.)
Riordan, A. J. (1979). In "Movement of Highly Mobile Insects" Concepts and Methodology in Research." University Graphics, N.C. State University, Raleigh. pp. 120-132.
Sprenkel, R. K., and Rabb, R. L. (1981). Environ. Entomol. (In press.)
Stinner, R. E., Butler, Jr., G. D., Jr., Bacheler, J. S., and Tuttle, C. (1975). Can. Entomol. 107:1167-1174.
Stinner, R. E., Jones, J. W., Tuttle, C. and Caron, R. E. (1977a). Can. Entomol. 109:879-890.

Stinner, R. E., Rabb, R. L., and Bradley, Jr., J. R. (1977b). In "Natural factors operations in the population dynamics of *Heliothis zea* in North Carolina." Proc. XV Int. Congr. Entomol. (Washington, D.C., 1976):622.42.
Sweetman, H. L. (1932). Ann. Entomol. Soc. Am. 25:224-240.
Taylor, L. R. (1979). In "Movement of Highly Mobile Insects: Concepts and Methodology in Research," (R. L. Rabb and G. G. Kennedy, eds.), pp. 148-185. University Graphics, N. C. State University, Raleigh.
Wolf, W. (1979). In "Movement of Highly Mobile Insects: Concepts and Methodology in Research," (R. L. Rabb and G. G. Kennedy, eds.), pp. 263-266. University Graphics, N.C. State University, Raleigh.
Wilson, K. G. 1979.. M.S. thesis. University of North Carolina. Raleigh.

NOCTURNAL ACTIVITY OF THE TOBACCO BUDWORM AND OTHER INSECTS

P. D. Lingren and W. W. Wolf*
Western Cotton Research Laboratory
Agricultural Research
Science and Education Administration, USDA
Phoenix, Arizona

I. INTRODUCTION

The adults of many insect pest species are nocturnal and in many cases they are highly mobile (Rabb and Kennedy, 1979). A single adult female of some species can deposit hundreds and even thousands of eggs. For instance a female tobacco budworm, Heliothis virescens (F.), is capable of laying in excess of 2,000 eggs. If all natural enemies were removed allowing all these eggs from even one female to hatch and the larvae to develop, then the economic threshold on high cash value crops, such as lettuce, would be surpassed. Moreover, similar pressure on lower cash value crops, such as cotton, could result in the initiation of control procedures using insecticides.

For the most part, control procedures for Lepidoptera using insecticides have been aimed at the immature forms which cause the actual damage to crops. If we could prohibit or even limit the reproductive capability of adult females to 100 immature forms per

*Current Address; Southern Grains Insects Research Laboratory, Agricultural Research, Science and Education Administration, USDA, Tifton, Georgia.

ISBN 0-12-332850-0

female, then it seems likely that more efficient and effective control measures could be developed. This seems like a very simple matter, but the nocturnal habits and mobility of the group has restricted the gathering of information necessary for the development of sound adult control procedures.

The nocturnal activity of most animals and especially insects is not well understood (Matthews and Matthews, 1978). There are many reasons for this lack of knowledge of which a weak human sensory mechanism (lack of good night vision) and general diurnal human activity behavioral patterns are among the most important. Nevertheless, in recent years several techniques have been developed that allow researchers to delve into the mysteries of nocturnal insect behavior and a beginning has been made toward an understanding of such activities and their importance to the further development of sound IPM programs.

In this chapter, it is our intention to dwell on some of the information that has been developed on nocturnal behavior of the tobacco budworm. Personal observations will be emphasized along with a limited literature review of the subject. Some of the information to be presented is qualitative. Qualatitative observations are the first step toward understanding complex systems. We hope that our presentation will stimulate others to ask questions and initiate investigations which will yield quantitative description of nocturnal insect activities. If this occurs, then one day we will be able to quantify many of the activities to a point where maximum utilization of the knowledge can be applied to IPM programs. We expect to find that many of the more efficient and effective, as well as new, insect control procedures of the future will result from knowledge of insect behavior and the influence of natural external factors on that behavior.

II. TOOLS FOR OBSERVING NOCTURNAL INSECT BEHAVIOR IN THE FIELD

There are a number of tools that can be used to make direct and indirect observations of the nocturnal activity of insects in their natural environment. Visible light in the form of head lamps or flashlights and night viewing equipment, such as night vision goggles (NVG) or monocular viewing devices that utilize natural or infrared light through microchannel-plate image intensification tubes, are very useful for direct viewing. Indirect methods of observation can be made through the use of devices such as pheromone traps, malaise traps, mating tables, genetic markers, tethered females and radar. Various uses of these tools have been discussed in several recent papers (McLaughlin et al., 1972; Mitchell et al., 1976; Lingren et al., 1977 a, b; Lingren et al., 1978; Lingren et al., 1979; Raulston et al., 1979; Sparks et al., 1979; Lingren et al., 1980; Raulston et al., 1980; and Lingren et al., 1981). We refer the reader to these papers for an in-depth look at the tools and will only casually discuss tools other than radar. More recent studies of using radar to measure the nocturnal flight activity of insects made primarily by the junior author will be discussed in a detailed section later in this chapter.

III. IMPORTANT BEHAVIORAL ACTIVITIES

A. Adult Eclosion

Adults of many of the Lepidopteran species, such as the tobacco budworm, emerge at night. Upon emergence the adult tobacco budworm moves a few inches up on a plant or other debris and expands its wings. It then elevates the wings before returning them to a normal resting position. These processes usually take several minutes. Thereafter the adult moves to the top of plants, such as cotton, where it remains for a period of several hours. During this period the adult is very docile and exhibits little tendency for flight. Consequently, newly emerged adults are easy to locate and collect in the field at night. During this period of limited mobility, the adults are likely at their most susceptible state for attack by both natural and artificial control factors. Therefore, it is of paramount importance to have a good understanding of the behavior and dynamics of the adult emergence and generation cycling of economic species.

Observations of the time of adult tobacco budworm emergence in cotton fields are somewhat limited but it appears that emergence begins about 11 pm and peaks between 2-3 am. The night following emergence the adults begin feeding and mating and on the following night mated females begin to deposit eggs. Normally three days pass before the eggs begin to hatch. This means that no damage occurs to the plants for at least five days after adult emergence. Moreover, the young larvae cause limited damage for the first two to three days. Thus, a period of at least seven days passes following adult emergence before damage occurs. Under optimal field conditions the larvae require about 14 days to complete development to the pupal stage in which they remain for about eight days. Therefore, 30 days are required to complete a generation cycle (adult to adult).

B. Generation Cycling and Population Age Structure.

Raulston et al. (1979) have shown that a majority of the emergence of a generation of tobacco budworms in the field occurs over a relatively short period of time (seven days). Females begin to emerge first during the cycle (Lingren et al., 1981). As a result, information on the in-field sex ratio collected over time gives some insight into the age structure of the generation. Also, the initial mating of a majority of the adults occurs very rapidly following the beginning of emergence of a generation. Males are capable of mating every night during their life cycle but females generally do not mate for two to three days following the initial mating (Raulston et al., 1975). In the process of mating, males pass a spermataphore which remains in the female throughout the adult life span. Therefore, good insight into the age structure of a population can be obtained by collecting adult females from the field over time and dissecting them to determine the numbers of spermataphores present.

C. Diel Patterns of Feeding, Oviposition, and Mating

Adult tobacco budworms generally become active in the field at about one hour after sundown after which they exhibit a series of diel patterns of behavior (Lingren et al., 1977a). The first activity appears as a general downwind movement which seems to be more pronounced and directional when wind speeds are in excess of 5 m/s. This type of activity usually lasts less than ten minutes and occurs with, or is followed by, extensive feeding activity which becomes heavily interspersed with ovipositional activity of mated females. However, the adults involved in feeding and oviposition have been observed and photographed moving in an upwind direction toward nectar sources and oviposition sites. In our opinion this type of orientation suggests the presence of airborne feeding and oviposition attractants or stimulants which are being secreted by the plants.

The feeding and ovipositional activity is usually intense for one to two hours but occurs intermittently at varying levels throughout the remainder of the night. Flight activity appears to become limited for 15-30 minutes following the intensive feeding and oviposition period. During this time the population appears to be inactive as if it were a resting period. Toward the end of this period, one begins to notice flight activity which is oriented cross-wind and appears to be faster than that observed earlier in the evening (Lingren et al., 1978). Information gathered from collections of adults on the plants and those captured by nets during the beginning of the fast flight period along with the type of activity that follows suggests that the fast flight is associated with males searching for mates.

Shortly after the beginning of the fast flight activity period females have been observed near the top of plants secreting pheromone (calling). Few observations of calling native female tobacco budworms in the field have been made except when fields had been treated with Virelure (two component mixture of seven component synthetic pheromone of tobacco budworm). However, numerous observations of calling by laboratory-reared females have been made in the field following their release or confinement on mating tables (Lingren et al., 1979). In general, the calling activity in the absence of males occurs intermittently over a three to four hour period, but peak calling activity in a population generally occurs about one to two hours after the initial observations and remains at a high level for one to two hours. Observations of calling females indicate that they move from site to site during the calling period and the movement is more intense when they are unable to attract mates. In other words, they do not appear to be passive in the mate searching process.

Mating pairs are generally not observed in the field until several minutes after the initial observation of fast flight activity. In the cotton field, most mating pairs are located near the top of the plants and their mobility in terms of flight capability is extremely restricted. The mating activity is closely synchronized with calling activity but calling females are usually seen before observations of mating pairs and mating pairs are observed for one to two hours after calling has ceased (probably because time in copulation generally exceeds one hour). Mating generally begins about three hours after sundown, peaks at about two hours later and ceases about two hours after the peak.

Observations of calling females or point sources of synthetic pheromone indicated that the males move cross-wind and contact the pheromone plume, which is concentrated downwind from the point source (Lingren et al., 1978). The males then move upwind to the source. Our observations of such activity suggests that the pheromone plume is of relatively short range (± 40 m). The male mate searching behavior is a circadian activity and is not activated by secretion of pheromone but is closely synchronized to the calling activity of females.

D. Nocturnal Larval Activity

Larvae of the tobacco budworm are also active at night. We have observed larvae moving from the lower bolls to the top leaves of cotton plants and from plant to plant. On several occassions those larvae observed moving to the top leaves were also observed to moult. They ate their cast skins following the moulting process and then remained inactive for several minutes. This type of activity occurred between 11 pm and 3 am and is likely to be a circadian activity. Certainly many other aspects of the larval behavior of the tobacco budworm, such as patterns of eclosion and feeding, are probably of a circadian nature and should be studied in detail. However, the larval behavior of the tobacco budworm from eclosion to pupation has yet to be described. Admittedly the quantitative as well as the qualitative aspects of most larval activities of many important pest species has not been established, but these activities are extremely important to the development of more efficient and effective control procedures. This is true because such activities are important to the timing and placement of materials used for their control.

E. Nocturnal Predator Activity

Numerous arthropod predators are active at night. This includes several of the spiders, assassin bugs, earwigs, ground beetles and many others. Again, their nocturnal activities are not well understood but could be of extreme importance to the development of better control procedures. For instance, during our night studies, we have observed the adults (several thousand) of numerous species of _Lepidoptera_, including tobacco budworms, being captured by spiders, and much of the activity occurs during the mating period of the insects. On several occassions we have observed the adult insects approaching the spiders as if they were attracted to them. This brings to light the possibility that spiders produce attractants to aid them in capturing prey. In fact, Eberhard (1977) has demonstrated that a bolas spider of the genus Mastophora secretes an attractant which aids it in the capture of primarily males of the fall armyworm, Spodoptera frugiperda (J. E. Smith), and Leucania sp. Although the specific attractant produced by the spider has not been isolated and identified, it is likely that it is a sex attractant of the fall armyworm. Of course the extent of chemical mimicry of insect attractants by spiders is not known but could be of great importance in the development of attractants as insect control agents. This would be especially true if some of the spiders produced feeding or oviposition attractants.

IV. RADAR AS A TOOL FOR STUDYING THE NOCTURNAL ACTIVITY OF INSECTS

A. Reports of Recent Studies

An understanding of the flight behavior of insects in the atmosphere and the influence of meteorological parameters on that behavior are extremely critical to the development of sound IPM programs for many species. In recent years a great deal of new knowledge concerning these subjects has been developed through the use of radar (Schaefer, 1976; Vaughn et al., 1978; Riley, 1979; and Greenbank et al., 1980). Results from these studies and others as discussed by Wolf (1978) and Lingren et al. (1981), or referenced by Greneker and Corbin (1978), show that radar is a powerful tool for the study of insect flight activity and that meteorological parameters have a strong influence on that activity.

B. Types of Radar

Current radars used in entomological studies generally consist of modified commerical marine radars. This type of radar was used in our studies (primarily those of the junior author). In its most simplistic form, this radar transmitts a short burst of radiation along a 2° conical beam. Some of the radiation is reflected from a target and is received by the radar. The distance to the target is related to the time it takes to receive the reflected echo. The position of the target is determined by the direction that the antennae is pointing. The range and direction of the target is displayed on a radar screen called a Planned Position Indicator (PPI) scope. This scope shows the target position for each revolution of the antenna. If the target moves then it produces trajectories on the scope. Height obtained above the ground can be calculated from the antennal angular elevation and slant range to the target.

C. Insect Activities

An early evening "take off" is one of the most obvious activities of the nocturnal insect species observed by radar. This type of activity generally begins during the first hour after sundown and can continue for several hours. Figure 1 shows an example of such activity over cotton fields near Rainbow Valley, Arizona. In this study there were very few insects in the air at 25 minutes after sundown as shown by the absence of small clear air echoes in Fig. 1a. At one hour after sundown, the insect activity had increased considerably (Fig. 1b). Most of this activity was within 16 m of the ground but some activity occurred up to 50 m from ground level.

By using time lapse photography the radar operator can show trajectories of individual insects on the PPI scope. The speed and direction of individual insects can be recorded by this method. Wind direction and speed is generally recorded by tracking small helium filled balloons. Differences between the wind vector and insect displacement vector give individual insect air speed and heading. Figure 2 shows a radar display of tracks of individual insects near Tifton, GA., being displaced at an average velocity of 8 m/s.

Insect activity at or near the canopy is difficult to observe with current radars because of interference from other obstacles,

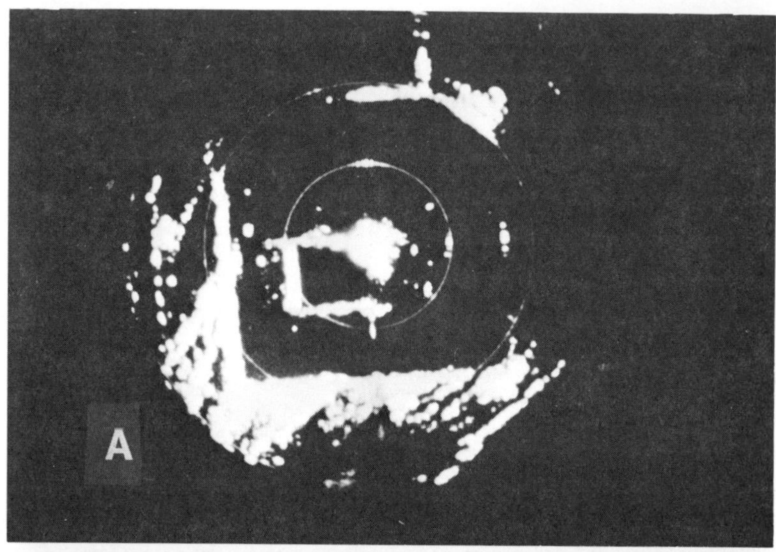

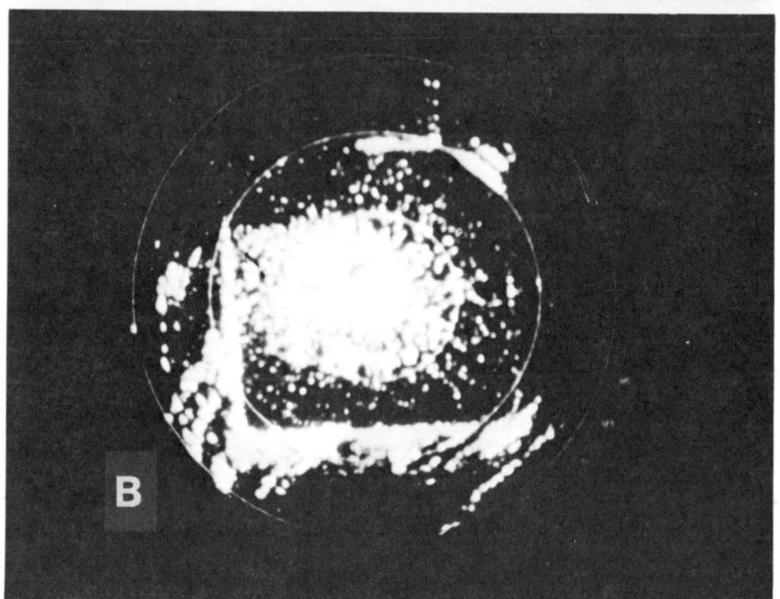

Fig. 1. Early evening activity: a) absence of insects echoes 25 min. after sundown (large area echoes were ground clutter from trees, desert vegetation and irrigation ditches); and b) insect activity one hour after sunset (small dots and large white area around radar represents insects). Radar located at center of photo range circles spaced 460 m (0.25 mi.), radar beam elevation 2.0° (insects at 460 m were 16 m above ground). August 20, 1979, Rainbow Valley, Arizona.

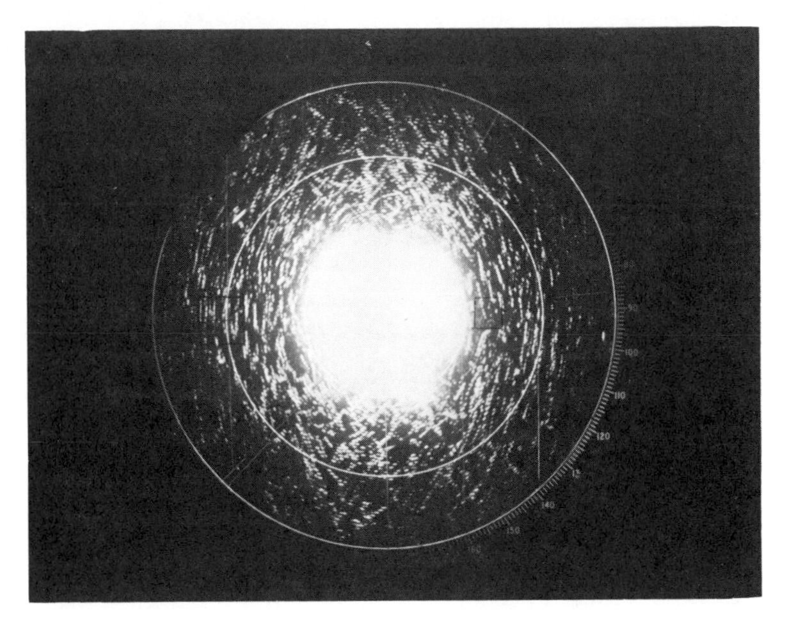

Fig. 2. Radar display of PPI scope showing tracks of flying insects. Average displacement velocity of insects is 8 m/s; range rings 460 m, beam elevation 5°, sweep rate 120°/sec., 5 revolution exposure. August 21, 1980, at Tifton, Georgia.

such as vegetation, dwellings, hills, etc. However, by proper selection of the radar site, insects, such as the tobacco budworm, have been observed between 3 - 9 m above ground (Lingren et al., 1981). Figure 3 shows an example of such movement in which case an observer using night vision goggles correlated visual sightings with those observed by the radar operator. Radios were used to communicate sightings.

Another phenomenon that has been detected by radar is plumes of insects leaving given areas (Schaefer, 1976). Figure 4 shows an example of a plume of insects leaving a cotton field near Phoenix, Arizona. The field is 1400 m NNW of the radar and the plume began about 65 mins after sundown. The surface wind was from the SSW at about 1 m/s. The plume ascended at about a 12° angle with displacement toward SSW and lasted for about

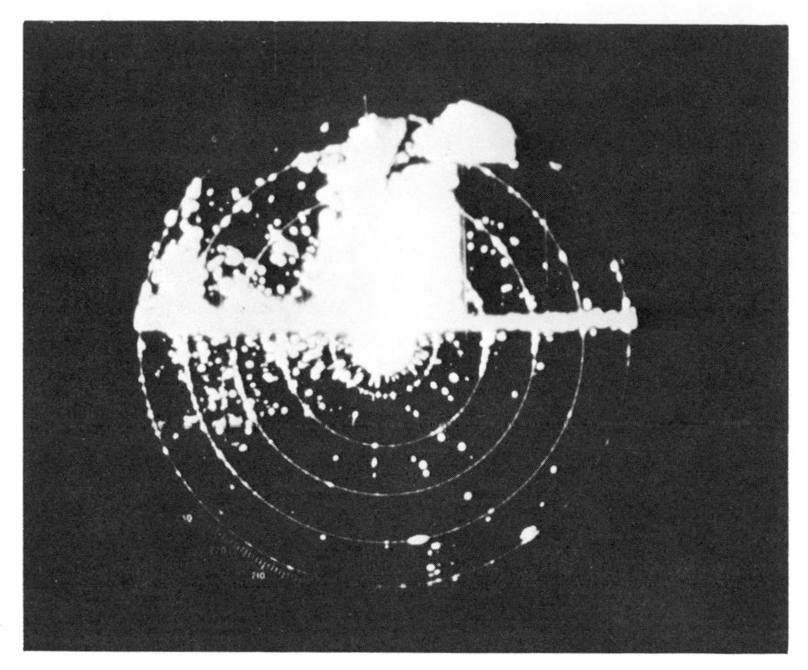

Fig. 3. Insects 3 to 9 m above cotton. An observer 150 m south of the radar correlated visual sightings of insects (using night vision goggles) with targets observed by the radar operator. An orange grove caused large area targets north of radar. Range circles 93 m apart, antenna elevation 1.0°, 2200 hours, July 2, 1979, Phoenix, Arizona.

ten minutes. A balloon was tracked one hour later and the surface wind was less than 2 m/s and wind at 400 m was 4.4 m/s from the NW. Intense studies of such plumes incorporating radar, biometeorology, visual observations of in-field behavior, and plant physiology may eventually lead to a better understanding of the parameters involved in the initiation of insect dispersal.

The convergence of winds can cause excess air to rise. Radar has shown that those insects flying downwind close to the convergence zone will be concentrated as the zone moves along. Figure 5 shows a line concentration of insects formed because of a nocturnal wind shift. We have also observed on numerous occasions that the surface wind direction is frequently much different

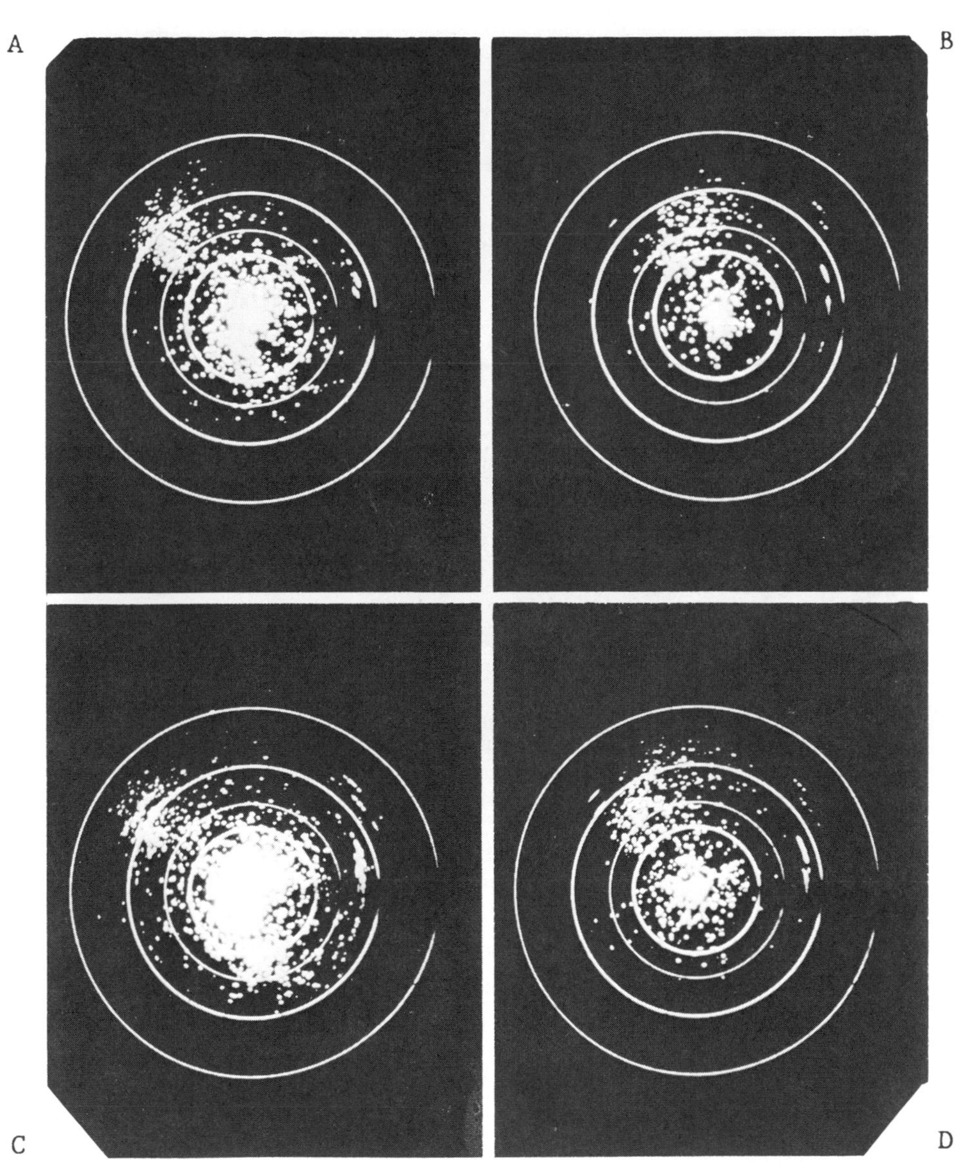

Fig. 4. Plume of insects leaving a cotton field 65 min. after rundown. Field was 1400 m NNW of radar and radar range circles were 460, 560, 930, and 1400 m. Radar beam intercepted the insect plume at coordinates (r = slant range, ϕ = azimuth, θ = elevation of: A) 830 m, 40°, 12°; B) 740 m, 70°, 24° ; C) 930 m, 32°, 8° ; and D) 740 m, 60°, 18°, respectively. The respective altitudes were 130, 170, 230, and 300 m above the ground. 1925 hr, September 13, 1979. Rainbow Valley, Arizona.

Fig. 5. Line concentration of insects due to nocturnal wind shift front. Line located southeast of radar moved toward northwest at 1.2 m/s. During one hour period before and after passage of the front, the surface wind averaged 1.1 ms from west and 0.5 ms from ESE, respectively. Range circles 93 m apart, antenna elevation 1.0°, 2203 hours, July 2, 1979, Phoenix, Arizona.

from that occurring 30 to 60 m overhead. Therefore, one should be careful in relating surface winds to insect activity that is occurring at higher altitudes. However, radar can be used to measure these winds by releasing and tracking small helium filled balloons.

Individual target classification is likely the most difficult measurement to make when using radar to study insect flight behavior. However, parameters such as using wing beat frequencies and amplitude of echoes gives some insight into identification of a specific target (Schaefer, 1976; Riley, 1979). Other devices such as night vision goggles (Lingren et al., 1981), nets attached

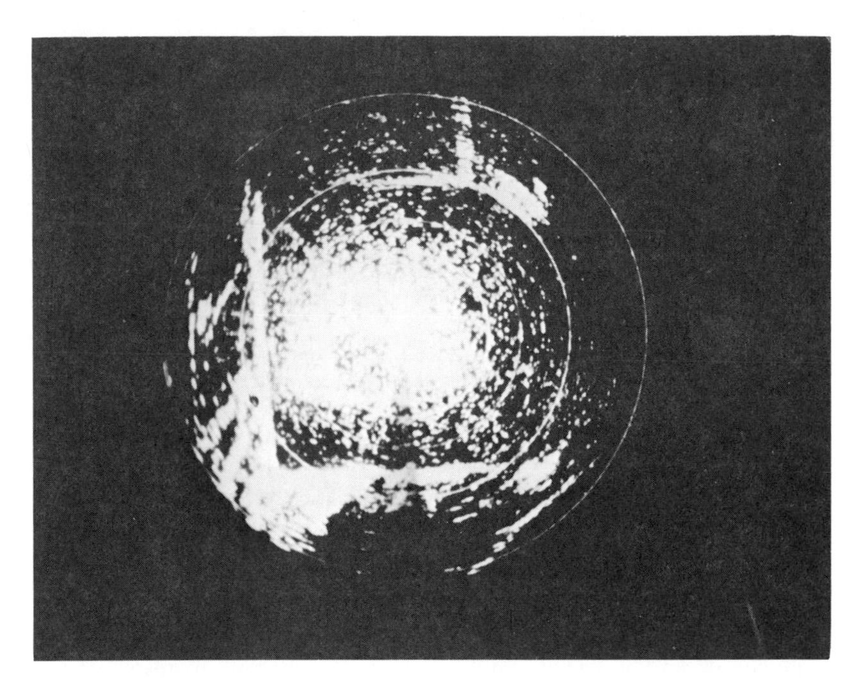

Fig. 6. Absence of insects on radar screen at azimuth 55° and range of 400 m resulted from insects avoiding a bat. The feature (large black hole to the right center of the white mass of insects on the PPI scope) moved across the radar screen toward the southwest at 4.4 m/s, diameter varied from 40 to 140 m and was 14 m above the ground. Surface wind was 0.9 m/s from SSE. Radar screen radius 1400 m, 1934 hours, September 11, 1979, Rainbow Valley, Arizona.

to airplanes and traps on towers (Greenbank et al., 1980) have been used in conjunction with radar observations to classify specific targets. Another possible means of insect classification which may prove to be useful in conjunction with radar observations is ultrasound (Roeder and Treat, 1957; Roeder, 1962). In this case it is well known that tympanate moths, such as the tobacco budworm, detect pulsed ultrasonic cries of bats or similar artificial pulses (Griffin et al., 1960; Agee, 1967; and Payne and Shorey, 1968). Upon detection of the proper ultrasonic pulse, these moths undergo evasive manuevers which in some cases involves diving or looping toward the ground (Agee, 1969). During one of our studies, we observed what appeared to be a bat moving through a concentration of insects. The object is shown as the black hole to the right center of the major insect concentration on the PPI scope in Fig. 6. The feature moved across the area about 14 m above the ground at about 4.4 m/s and caused insects to avoid an air space from 40 to 140 m in diameter, and it changed speed and direction several times as it moved past the radar to the southwest. If the feature was indeed a bat, then it is likely that the insects occupying that air space were moths with tympanic organs. Sticky boards or nets designed to capture those moths as they dove or looped to avoid the bat could have confirmed the identity of the insect targets and night vision goggles could have been used to identify the presence of the bat as well as insect types. Equipment is available for artificial production of ultrasounds to which tympanate moths respond (Webb and Agee, 1969). The use of this equipment along with radar observations and proper capture techniques may prove to be a good method to aid in the study of the nocturnal flight activity of some moth species.

In summary, recent studies (Schaefer, 1976; Riley, 1979; Greenbank et al., 1980) have shown that radar can provide relatively accurate information about insect size, numbers per unit volume, altitude, vertical distribution, flight speed and direction, wing beat frequency, time of flight, rate of climb, flight ceiling, layering during disperal, orientation during flight, and concentrating effects of wind fields. Our studies have confirmed several of these activities and further demonstrated the capability of combining radar and other techniques (Lingren et al., 1981) to define flight behavior of insects near canopy level. Certainly improvements of radars along with development of supporting techniques are possible and further studies will lead to a much better understanding of nocturnal insect flight activity. Eventually, if we are intuitive and persistent in our efforts, then we will determine when, why, where and to what degree some species disperse. That type of understanding of insect populations will lead to better IPM strategies for insect control. Certainly, meteorological parameters influence all of the activities that we have discussed and in many cases this influence are not well understood. Therefore, biometeorological studies will be a prerequisite to any strategies that are developed.

REFERENCES

Agee, H. R. (1967). J. Econ. Entomol. 60:366-369.
Agee, H. R. (1969). Annals Entomol. Soc. of Am. 62:801-807.
Eberhard, W. G. (1977). Sci. 198:1173-1175.
Greenbank, D. O., Schaefer, G. W., and Rainey, R. C. (1980). Memoirs of the Entomol. Soc. of Canada, 110.
Greneker, E. F., and Corbin, M. A. (1978). "Radar Reflectivity of Airborne Insects: A Literature Review." Georgia Institute of Technology, Eng. Exp. Sta., Atlanta, Georgia.
Griffin, D. R., Webster, F. A., and Michael, C. R. (1960). Animal Behavior 8:141-151.
Lingren, P. D., Greene, G. L., Davis, D. R., Baumhover, A. H., and Henneberry, T. J. (1977a). Annals of the Entomol. Soc. of Am. 70:161-167.
Lingren, P. D., Raulston, J. R., and Sparks, A. N. (1977b). Environ. Entomol. 6:217-221.
Lingren, P. D., Sparks, A. N., Raulston, J. R., and Wolf, W. W. (1978). Bull. Entomol. Soc. of Am. 24:206-213.
Lingren, P. D., Raulston, J. R., Sparks, A. N., Proshold, F. I. (1979). Agric. Res. Results W-5, 17 p.
Lingren, P. D., Burton, J., Shelton, W., and Raulston, J. R. (1980). J. Econ. Entomol. 73:622-630.
Lingren, P. D., Raulston, J. R., Sparks, A. N., and Wolf, W. W. (1981). In "Insect Suppression with Controlled Release Pheromone Systems" (G. Zweig, A. F. Kydonieus, and M. Beroza, eds.). CRC Press Inc., Boca Raton, Florida. In press.
Matthews, R. W., and Matthews, J. R. (1978). "Insect Behavior." Wiley, New York.
McLaughlin, J. R., Shorey, H. H., Gaston, L. K., Kaae, R. S., and Stewart, F. D. (1972). Environ. Entomol. 1:645-650.
Mitchell, E. R., Baumhover, A. H., Jacobson, M. (1976). Environ. Entomol. 5:484-486.
Payne, T. L., and Shorey, H. H. (1968). J. Econ. Entomol. 61: 3-7.
Rabb, R. L., and Kennedy, G. G., eds. (1979). "Movement of Highly Mobile Insects: Concepts and Methodology in Research." University Graphics, North Carolina State University, Raleigh.
Raulston, J. R., Snow, J. W., Graham, H. M., and Lingren, P. D. (1975). Annals of the Entomol. Soc. of Am. 68:701-704.

Raulston, J. R., Lingren, P. D., Sparks, A. N., and Martin, D. F. (1979). Environ. Entomol. 8:349-353.

Raulston, J. R., Sparks, A. N., and Lingren, P. D. (1980). J. Econ. Entomol. 73:586-589.

Riley, J. R. (1979). In "A Handbook on Biotelemetry and Radio Tracking" (C. J. Alamander, Jr., and D. W. MacDonald, eds.), pp. 131-140. Permagon Press, Oxford and New York.

Roeder, K. D. (1962). Animal Behavior 10:300-304.

Roeder, K. D., and Treat, A. E. (1957). J. Exp. Zool. 134: 127-158.

Schaefer, G. W. (1976). In "Insect Flight" (R. G. Rainey, ed.), Blackwell Scientific Publ., London. pp. 157-197.

Sparks, A. N., Raulston, J. R., Lingren, P. D., Carpenter, J. E., Klun, J. A., and Mullinix, B. G. (1979). Bull. Entomol. Soc. of Am. 25:268-274.

Vaughn, C. R., Wolf, W., and Klassen, W., eds. (1978). NASA Conf. Publ., 2070.

Webb, J. C., and Agee, H. R. (1969). Trans. of the Am. Soc. of Agric. Engrs. 12:816-821.

Wolf, W. W. (1978). In "Movement of Highly Mobile Insects: Concepts and Methodology in Research" (R. L. Rabb and G. G. Kennedy, eds.), University Graphics, North Carolina State University, Raleigh.

RISK-BASED DESIGN OF METEOROLOGICAL NETWORKS FOR INTEGRATED PEST MANAGEMENT[1]

S. M. Welch and F. L. Poston
Department of Entomology
Kansas State University
Manhattan, Kansas

I. INTRODUCTION

It has long been recognized that weather influences agricultural processes in many ways. Some of the earliest references regarding temperature influence date to the 18th century (Reamur, 1735). By the early 1970's the need for increased monitoring of meteorological data became acute. This was particularly stimulated by the advent of integrated pest management (IPM) models requiring timely abiotic inputs (Ruesink, 1976). Questions soon arose concerning the type and accuracy of data necessary for IPM applications. Haynes, Brandenburg and Fisher (1973) provided an early attempt to answer these questions. A summary of their proposals for specific variables and accuracies is shown in Table 1. This table is primarily based upon the properties of basic agricultural biology. Perhaps more important, however, are questions concerning the economic risks to the decision maker

---

[1] Contribution No. 81-501-A of the Kansas Agricultural Experiment Station. The findings and opinions expressed herein are solely those of the author and do not necessarily reflect those of the Kansas Agricultural Experiment Station or Kansas State University. This work was supported by Hatch Project 998 at Kansas State University.

ISBN 0-12-332850-0

associated with various levels of accuracy. In the final analysis it is the willingness of decision makers to accept such risks which set the lower limit on data quality.

Table 1. Proposed Meteorological Variables with Accuracies Required for IPM

Variable	Accuracy
Air temperature	± 1°C
Soil temperature	± 1°C
Wind Speed	± .5 ms^{-1}
Wind direction	± 10°
Barometric pressure	± 0.067 Pa
Humidity	± 1%
Precipitation	± 2.5 mm
Soil Moisture	± 1%
Irradiance	
Low	± 210 KJ $m^{-2}d^{-1}$
High	± 20%

The central theme of this paper is that these questions can only be answered in the context of a complete systems analysis (including risk analysis) of decision making in the plant protection process. The sections which follow present a plan for this kind of analysis consisting of five steps: (1) a determination of which IPM decisions require meteorological data; (2) an analysis of the relationship in the decision-making process between data errors and the resulting risk of dollar reductions; (3) measurement of the levels of variability in the physical environment; (4) ascertaining the levels of risk acceptable to growers; (5) the integration of the preceding data into designs for weather monitoring systems for IPM.

II. ANALYZING IPM DECISIONS

In any production system there is a series of decisions which the grower must make if he is to produce and/or protect a product. These decisions can often be organized into a chronological sequence based on the annual production cycle (Fig. 1). Viewed in this light, management complexity often seems to stem less from the number of required decisions than from the variety of information each decision ought to incorporate and the wide range of alternative actions available.

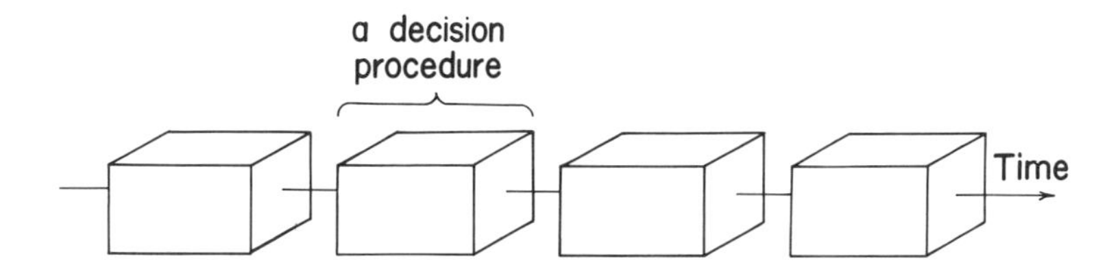

Fig. 1. Schematic of a production system showing a sequential series of management decisions (boxes).

In this step of the analysis, a list of IPM decisions is constructed along with a list of the methods, algorithms, models, etc. by which such decisions are or might be made. An examination of the input variables for such decision procedures will reveal the types (although not the amounts or accuracy) of needed meteorological data. Subsequent steps in the analysis focus on those particular decisions where requirements for meteorological inputs exist.

III. DATA ERRORS VERSUS DOLLAR DEVIATIONS

The objective of this step is to determine the relationship between errors in the data which result from sampling limitations and the resulting income uncertainties.

Figure 2A shows an expanded conceptualization of the decision-making process which is useful in this analysis. The vertical arrow represents the flow of time through the production cycle and the box represents a procedure used to arrive at one particular decision in the course of production. Those decisions requiring meteorological data often combine some form of quantitative model with human interpretation. As shown, the meteorological input is measured at some particular point in the environment.

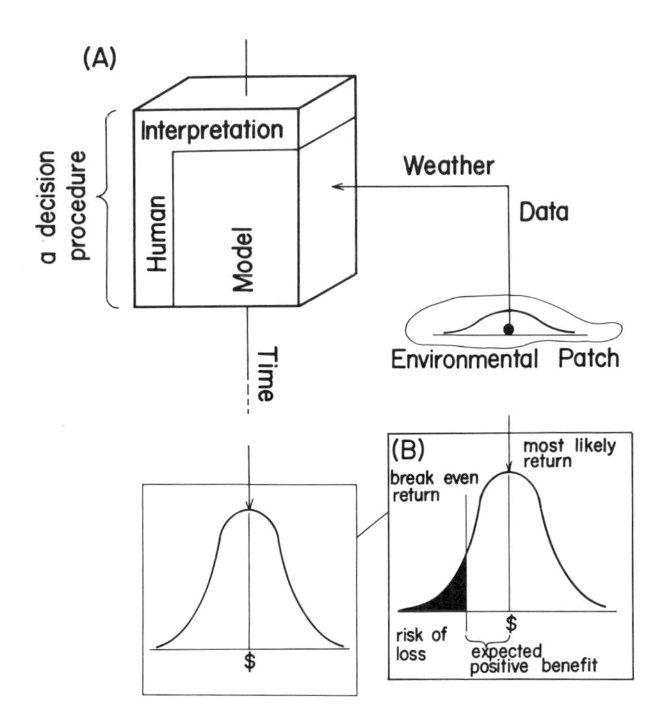

Fig. 2. (A) One management decision expanded to show meteorological inputs. (B) Risks of loss due to imperfect models and inputs.

In practice of course, the results of a particular decision will be applied not only at the single point where the meteorological data was collected but over some extended patch of the environment as indicated. We can, therefore, define a distribution of discrepancies or errors between the particular input values used for decision making and conditions at other specific points in that patch. Such a distribution is indicated schematically in Fig. 2A.

One of the guiding principles of IPM is that one makes those decisions which give the best net dollar return. Thus, associated with the selection of a particular alternative action is a dollar estimate as indicated by "$" in Fig. 2A. On the other hand, because conditions vary over the patch, dollar returns at other points will also vary. Thus, we must talk about a distribution of likely dollar outcomes as shown by the bell-shaped curve at the bottom of Fig. 2A.[2]

Our objective is to relate the distribution of data errors with the distribution of dollar deviations. The following example shows a general approach. Let $\$ = M(T)$ be the dollar benefits resulting from the use of model M (for further detail, see Welch et al., 1981). For simplicity sake, we assume that temperature, T, is the only meteorological input. Letting V denote the variance operator, we desire to find

$$V(\$) = V(M(T))$$

Taking a Taylor series expansion about the temperature at the monitored point, T_u, yields

$$V(\$) = V(M(T_u) + (T-T_u)\, M'(T_u) + \ldots)$$

For small temperature deviations we may neglect high order terms, leaving

$$V(\$) = V(T)\ (M'(T_u))^2$$

---

[2] Effects on the dollar distribution resulting from errors in the model itself are dealt with later in the paper. In the meantime, such effects will be assumed to be constant.

A numerical estimate of $M'(T_u)$ can be obtained from the model by standard sensitivity analysis methods. Standard deviations of dollar returns may be estimated by taking the square root of V($).

Figure 3A shows the dollar deviations as a function of temperature errors. For small temperature errors, the curve is linear with slope $M'(T_u)$. Figure 3B extends this to the case where there are large errors in weather data inputs. In this instance, the linear portions of the function are succeeded by curvilinear responses. It seems likely that the dollar deviations do not increase without limit. This is true because weather data have a finite range of values and therefore a finite range of errors. Figure 3B shows that the curvilinear portion of the function could be either convex or concave. At this point in time it is not clear, at least to this author, which form of curve might predominate. Preliminary work, however, indicates that the concave-down configuration applies in some cases (see discussion

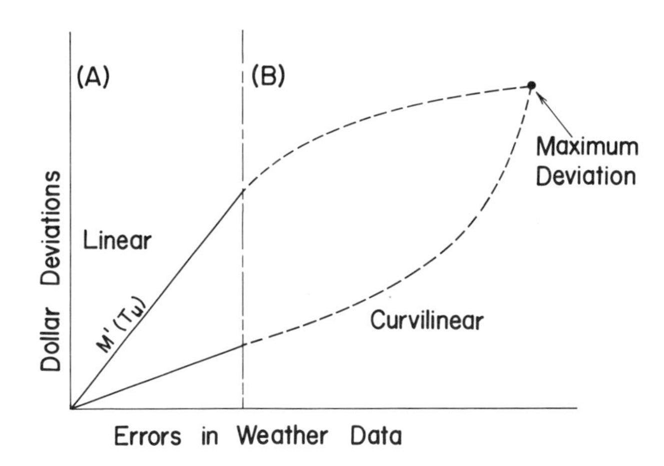

Fig. 3. (A) Linear relation between input errors and dollar deviations. (B) Possible curvilinear relations.

below). If this is a general phenomenon, it would imply that relatively accurate weather data may be needed for pest management applications since larger reduction in monitoring errors would be required to get a given improvement in dollar deviations.

IV. ASSESS NATURAL VARIABILITY

The objective of this step is to determine the level of variability actually present in the environment. Combined with the results of the last section, this will permit estimation of the actual dollar deviations which might be expected from given monitoring system designs.

The most direct approach to assessing variability is simply to establish a large number of monitoring stations in a given area, and make an intensive series of measurements. The stations may be organized in a grid pattern or some other form of arrangement. In some cases, networks will already exist in the environment whose measurements, when condensed into an historical data base, can provide the raw input needed for the analysis. This latter situation is particularly helpful in the case of regional programs. However, such networks usually do not provide the density of monitoring points necessary for microclimate measurements.

It may be difficult, however, to employ this direct approach for variables which are expensive to measure in a field situation. In this case, one may opt for a second approach which is to measure some related set of variables and then infer (often via a model) the degree of variation in the variables of primary interest. A common example of this (Rabbinge, 1976) is the use of leaf-angle distributions and leaf-area index measurements to calculate solar radiation patterns. This eliminates the need to make separate light measurements on a leaf-by-leaf basis.

Both of the above approaches yield a series of time-varying measurements from each of several points distributed in space.

The next question is, how does one analyze them to extract pertinent measurements of variability? Previous analysis efforts in the IPM literature have focused on standard statistical techniques such as correlation or regression (e.g., Fulton and Haynes (1977)). Such studies have been successful in establishing the degrees of relationship between neighboring stations but do relatively little to document distributions of input errors in the sense used in this paper.

To study these distributions, the author examined weather temperature patterns in Kansas. Basic data for the study consisted of daily maximum and minimum temperatures for a three-year period from each of 103 stations distributed over the state of Kansas. Each station was located in a distinct county and only two counties in the state were not represented. For each pair of stations, a simple least-squares univariate regression model was constructed whereby temperatures at one station were predicted from temperatures at the other. This was done so that any first-order deterministic relations between neighboring points could be subtracted from the estimates of intrinsic variability. Since models were generated for both maximum and minimum temperatures, there were, in total, some 21,218 distinct regression models.

The relationship between variation and patch size was then derived by plotting, for each pair of stations, the root mean square errors of their regression models as a function of the great circle distance between them. Figure 4 shows the results of fitting an appropriately shaped curve to this data. The horizontal scale also shows the approximate number of stations in Kansas for given interstation distances.

For projections over short distances, most of the regression slopes and intercepts were in the neighborhood of 1 and 0, respectively. This fact, in conjunction with Fig. 4 might be taken to infer, for example, that on the average the root mean square difference between two stations less than 20 miles apart will be about 1.8°F (1°C). Multiplying by the appropriate value from

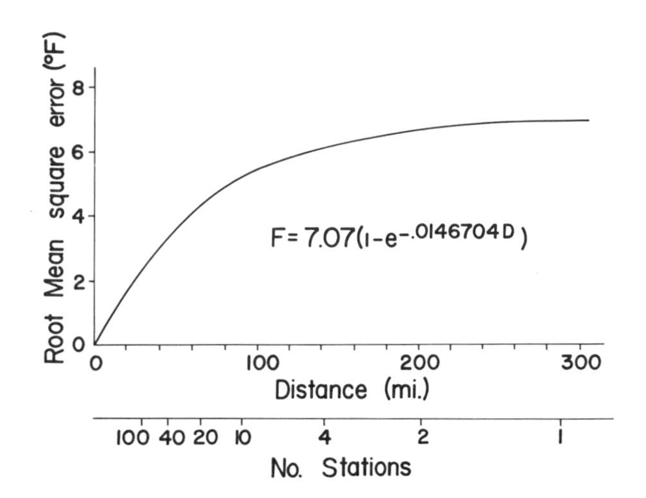

Fig. 4. Relationship between inter-station distances and temperature deviations. The lower horizontal scale shows the number of stations in Kansas at various mesh sizes.

student's t-distribution, we could say that 95% of the time, stations less than 20 miles apart will be within ±3.5°F (2°C) of one another. Of course, in states with more diversified topography than Kansas, these numbers might well be different. What is important here, of course, is the method of analysis by which such error estimates can be obtained.

V. DETERMINE ACCEPTABLE RISK

The topic addressed in this section can best be understood by referring back to Fig. 2B. The distribution of potential dollar outcomes has been modified to show the amount of money a grower must receive in a hypothetical situation in order to break even. In spite of the apparently beneficial effect of the decision, there is, nevertheless, some risk (as indicated by the shaded portion of the curve) that the grower will actually lose money.

The objective of this step in the analysis is to ascertain in exactly what levels of such risks growers are willing to tolerate.

It seems likely that sociologists may be the best equipped to develop assessment procedures for risk tolerance. One method might be simply to ask growers whether they would accept programs which, for instance, failed to break even every other year, or every fourth year, or every tenth year, etc. If this approach were taken, it would also seem useful to determine how often growers have program failures at the present time. Some studies in France (Boussard and Petit, 1967) seem to indicate that a 10% failure rate may be tolerable. In other situations a much higher failure rate might be acceptable. An example might be where a severe pest is devastating crops year after year. Growers might well view even a 50% failure rate as an eminently acceptable improvement.

VI. DESIGN THE MONITORING SYSTEM

The objective of this section is to combine the previous results so as to determine the number of points to be monitored and where best to locate them. Figure 5 illustrates one method for determining the number of points. The contents of the square brackets tell the source of various quantities; Roman numerals refer to previous sections of this paper.

In the graph at the upper left, the maximum acceptable risk is drawn as a constraint. The horizontal axis is time during the research and development (R & D) process. Early in the program models may well be quite crude and the risks of using them unacceptably high even with very accurate monitoring. During this phase no solution to the IPM problem is possible. After continued research, however, models will improve and the risks associated with their use (as assessed by the methods of Welch et al. (1981)) will fall. As this happens the models can be used with lower and lower quality data and yet still meet the maximum risk constraint. The rising curve in the upper left is determined by subtraction and represents the increment of acceptable risk due to lowered intensities of monitoring.

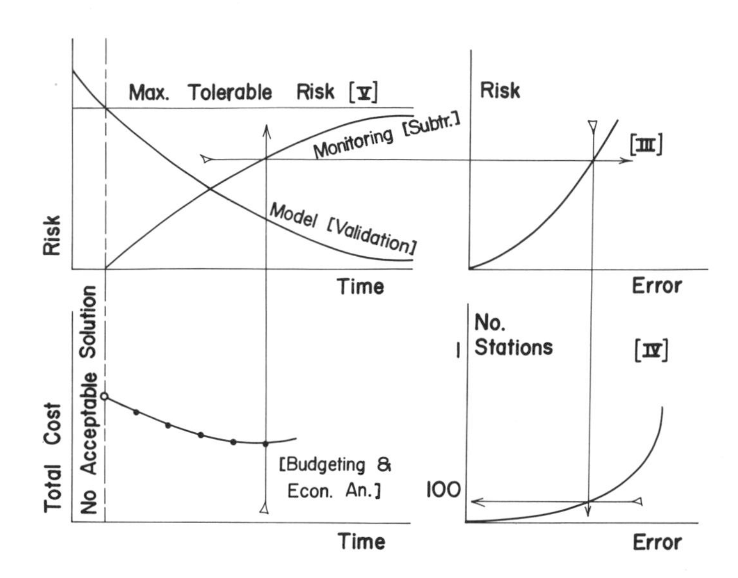

Fig. 5. Assembling the results of the analysis.

Concomitant with the passage of time are shifts in the economics of the research/implementation program as shown in the lower left of Fig. 5. Early implementation of crude models may be successful but require expensive, very accurate, and intensive monitoring systems. On the other hand, delaying implementation may permit the use of simpler monitoring systems supporting more complex models. However, total costs may still be high because of (1) greater research costs, (2) the accumulation of grower losses that even crude models could have moderated, and (3) inflation.

Thus there is likely to be some R & D program length with lowest total cost. Ideally, we could determine this duration by the appropriate budgetary and economic analysis and plan to have our program implemented in this period of time. A more practical approach, however, is to consider iterative methods (indicated by the solid circles) which evaluate recent research gains in terms of total costs (grower losses included). In this way, project managers could sense when research phases were approaching the point of diminished returns and begin to initiate implementation.[3]

At this point the determination of the number of stations required becomes relatively straightforward. For a given level of risk due to monitoring, the graph at the upper right in Fig. 5 (derived in Section III) gives the required root mean square error in the monitoring system. The graph at the lower right (the transpose of Fig. 4) yields the corresponding number of stations.

This leaves the question of where should the stations be located. This author is not certain how to approach this problem in a completely de nova fashion. However, the problem of selecting an optimal subset of stations in an existing network seems to be tractable. This paper will only give a brief summary of the method; details of the analysis will be published elsewhere.

The basic desideratum of any network is to be dense enough to make possible adequate predictions of conditions throughout the patch surrounding each monitoring point. In this context, of course, "adequate predictability" means sufficient accuracy to make the desired management decisions within the constraints of maximum acceptable risk.

These considerations suggest an approach via clustering procedures. Beginning with an existing network, one can grade the dissimilarity of each pair of stations based on the inability of the measurements at one station to predict the conditions at the other station. This unpredictability, of course, can have many indices. For the purposes of this study, the root mean square

---

[3] While it is admittedly only conjecture, this author feels that because of the huge monetary sums involved in agriculture, an analysis of this type would put a premium on accelerated implementation of IPM programs.

errors of the previously discussed univariate regression models were used. Through the use of a specially designed hierarchical algorithm the 103 Kansas weather stations were clustered. This yielded a series of successively smaller weather networks ranging from 103 stations to a single station. These networks are optimal in the sense that, as the number of stations is successively reduced, the minimum amount of unpredictability is introduced at each step. Associated with each optimal network is a figure of merit relating to the overall error rate.

Given that a certain number of stations is known to be required, the optimal network of that size shows where they would best be located. For illustrative purposes, Fig. 6 shows the optimal eight-station network for the state of Kansas.

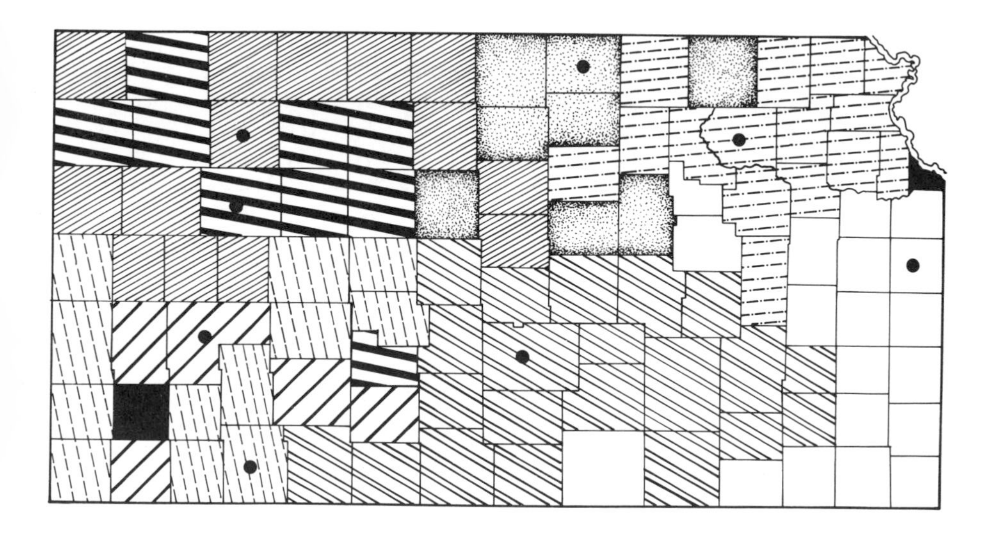

Fig. 6. An optimally-sited, eight-station network for Kansas. Cross-hatching show the areas served by each station. The two counties colored solid black were not included in the analysis.

VII. SUMMARY

This paper has laid out a program to determine the type of variables and degrees of resolution necessary for meteorological inputs to IPM decision making. This program is, of course, not perfect but at least appears to incorporate the major elements and steps necessary to achieve this goal. Over the next five years, it is hoped that this program can be refined and specific examples worked out. If nothing else, this should do much to create a closer working relationship between meteorologists and agricultural biologists.

REFERENCES

Boussard, J. and Petit, M. (1967). Am. J. of Agric. Econ. 49:869-880.

Fulton, W. C., and Haynes, D. L. (1977). Environ. Entomol. 6:393-399.

Haynes, D. L., Brandenburg, R. K., and Fisher P. D. (1973). Environ. Entomol. 2:889-899.

Rabbinge, R. (1976). "Biological control of fruit-tree red spider mite." Center for Agricultural Publishing and Documentation, Wageningen, The Netherlands. 228 p.

Reamur, R. A. D. de. (1735). Acad. des Sci., Paris, 1735:545.

Ruesink, W. G. (1976). An. Rev. of Entomol. 21:27-44.

Welch, S. M., Croft, B. A., and Michels, M. F. (1981). Environ. Entomol. (in press).

THE MANAGEMENT OF PLANT PATHOGENS

S. P. Pennypacker and R. E. Stevenson
Department of Plant Pathology
Pennsylvania State University
University Park, Pennsylvania

I. INTRODUCTION

Plant disease is "a dynamic interaction between an organism and its environment which results in abnormal physiological, and often morphological changes in the plant" (Merrill, 1980). These alterations may be caused by many biotic and abiotic agents. Biotic pathogens include among others certain fungi, bacteria, viruses, viroids, nematodes, mycoplasmas, spiroplasmas, and parasitic seed plants, while abiotic causes of disease encompass inanimate agents such as air-, water-, and soil-pollutants, cultural practices, products of plant metabolism, and chemicals used as herbicides, insecticides, and fungicides.

The presence of a pathogenic agent in a host population seldom results in the occurrence of diseased plants without a favorable environment. This "triad concept of disease," i.e., the need for a host, pathogen, and favorable environment for disease development, is the basis for several control tactics in an integrated management scheme for plant pathogens. For example, successful control of most biotic agents requires detailed knowledge of the organism's life cycle, behavior of the organism in, or on the plant, and the effect of environmental factors on the interaction of the pathogen and host.

ISBN 0-12-332850-0

Numerous articles have been published on relationships existing between plant pathogens and atmospheric phenomena. This voluminous assemblage of biometeorological literature describes the influences of weather factors on various segments of the life cycles of many individual organisms and/or relates the resultant disease to the meteorological data. Reviews published on the effect of weather and climate on plant disease include: Bourke, 1970; Heptig, 1963; Reichert and Palti, 1967; Rotem, 1978; Stevens, 1934; Waggoner, 1965; and Waggoner, 1968. Rather than resummarize the microclimate/plant disease literature, we will briefly describe the tactics currently available for the integrated management of plant pathogens and describe how weather data is used for both simulating plant disease epidemics and scheduling chemical applications to manage selected plant pathogens.

II. TACTICS FOR MANAGEMENT OF PATHOGENS

Maintaining the incidence of disease below a level that causes significant economic loss is a very complex problem. The diversity and variability among the pathogenic fungi do not allow the use of a common control technique. Some fungi, for example, have spore stages specialized for dispersal by air movement, some by water movement, and others spread in the soil. Some fungi move great distances from plant to plant by air-borne spores, while viruses and mycoplasmas may be spread hundreds of miles by insect vectors, and man moves pathogens varying distances in plants, plant parts, and soil. Other fungi may be moved only short distances by splashing rain. Control is further complicated by the adaptation of some pathogens to survival under adverse growing conditions. For example, the eggs of some nematodes and the spores of certain species of Fusaria remain viable in soil several years in the absence of suitable host plants. Seed of the parasitic plant *Orobanche ramosa* may also remain viable up to 13 years in soil in the absence of host plants (Garnan, 1903).

Although a specific pathogen may at times by controlled be a single procedure, the complexity of many of the pathogens requires the use of more than one method to achieve satisfactory control. Chemical, biological, cultural, and physical techniques may be used to manage plant pathogens.

A. Cultural and Physical Control

Cultural practices of various kinds aid in protecting plants from disease through the promotion of plant vigor and creation of unfavorable conditions for infection. The cultural practices affecting these conditions include crop rotation, controlling soil pH, maintenance of optimum soil moisture, climatic modification, and sanitation.

The causal agents of plant disease may be avoided to some degree by selecting the best sites and dates for planting. Planting sites may be selected to avoid fields containing high populations of root infecting organisms and to avoid fields containing crop residues that may serve as a source of the disease causing organisms. The freedom to choose an appropriate planting date may allow for selection of a time when environmental conditions are favorable for the growth and development of the crop and unfavorable for the pathogen. A natural reduction in the pathogen population may also be obtained by crop rotation. Take-all of wheat caused by *Gaeumannomyces graminis* var. *tritici* is an example where cultural control can be obtained by rotating wheat with a nonhost crop (Cook, 1981).

Greenhouse, nursery, and field crops may be protected by modifying their aerial and soil environment. Plant spacing, weed control, and cutting practices in young tree plantations, pruning and row orientation in vineyards, and soil mulches in high value crops are excellent examples where microclimate modification is quite successful in controlling pathogen activity. Environmental modification may also be the only feasible means to protect low value crops, such as forest trees.

The modification of the environment in storage areas to maintain plant quality and to promote rapid curing and wound healing had greatly reduced the incidence of storage rots in root crops such as potatoes (Cunningham, 1953). Environmental control of pathogens is also of value in greenhouse operations. Correct heating and venting techniques help reduce the incidence of free water on plant surfaces and thus removes a necessary environmental factor for infection by many pathogens. Black spot on rose, for example, was completely eliminated in greenhouse culture by simply eliminating overhead watering systems (Pirone et al., 1960).

Disease may also be avoided by using pathogen-free planting stock. This is one of the most effective means of controlling seed-, tuber-, and plant-borne pathogens. Removing diseased plants and plant parts and infested plant debris are other useful approaches. Destroying alternate hosts, e.g., cedars, currants, and barberries, interrupts the life cycle of rust fungi for the control of cedar-apple rust, white pine blister rust, and the leaf and stem rusts of wheat, respectively.

Additional management practices used to provide protection include the flooding of fields (Stoner and Moore, 1953; Stover, 1962) irrigating, and the removal of crop debris by burning (Hardison, 1980; Parameter and Uhrenholt, 1975; Papendick and Cook, 1974). The low thermal death point of many pathogens also allows for their eradication from seed through the use of heat treatment. Steam treatment of soil before planting is a physical method of control that is especially useful in greenhouses (Aldrich and Nelson, 1969; Baker et al., 1957). The heat and radiation treatment of seed and other plant parts are additional practices used to control several pathogens (Raychadhuri and Verma, 1977).

Certification and quarantine programs are additional physical control techniques sometimes used to exclude plant pathogens. These inspection procedures help reduce the potential dissemination of pathogens via intra- and interstate transportation of plant materials.

B. Biological Control

Several biological techniques are currently used to manage plant pathogen activity. Each method uses an organism, either directly or indirectly, to control the disease-causing agent.

At present the use of resistant plant varieties is a major method of biological control and is based on an incompatability between the plant and the pathogen. Plants may possess a physiological and/or mechanical form of resistance. While physiological resistance is usually based on some form of chemical incompatibility, mechanical and functional resistance is usually due to a physical characteristic that prevents the organism from penetrating and infecting the plant. This inherent ability of the plant to overcome the effects of the pathogen may be obtained in certain plants through plant nutrition. In some cases, the plant produces a "phytoalexin" compound that offers plant resistance to the pathogen (Paxton, 1975). Plant varieties displaying these various types of resistance to specific organisms are normally selected or developed through breeding programs. Varieties of this type have been used for many years and are often the most effective and economical means of controlling disease. Biological strategies currently playing only a minor role in pathogen control include multiline varieties, klendusic varieties, cross-protection, hyperparasites, and predators.

The management of diverse pathogen populations has received some degree of success through the physical mixing of different varieties of a crop. While the varieties used for these multilines all have similar agronomic characteristics, they differ in resistance to particular races of the pathogen. The extensive breeding programs required for this type of control have, however, limited its commercial use to only two known multilines; one for the control of oat crown rust in Iowa, and the other for control of wheat stem rust in Mexico (Browning and Frey, 1969).

Klendusic plant varieties possess resistance to factors other than the pathogen. These varieties may, for example, escape infection by possessing resistance to an insect that acts as a carrier of the disease-causing organism.

The natural or controlled infection of seed, or a crop, with an avirulent or a weakly virulent strain of virus or virus-like agent is used against some bacteria and viruses. This cross-protection technique either prevents crops from becoming infected later by a more virulent strain, or reduces the severity of later infections. Although not commercially used in the U.S., cross-protection is used in the Netherlands for controlling tomato mosaic virus (Rast, 1979) and in Brazil for controlling tristeza of citrus (Costa and Muller, 1980).

Another potential form of biological control is the use of hyperparasites or predators to control pathogens (Baker and Cook, 1974; Papavizas and Lumsden, 1980). In this method of control nonpathogenic microorganisms protect plant surfaces against infection by directly exploiting pathogenic agents. Presently, those hyperparasites that attack sclerotia of pathogenic fungi appear to be the most likely capable of reducing inoculum in commercial field crops (Ayers and Adams, 1981). Although only limited success has been reported under commercial field conditions, this technique of reducing inoculum may hold promise for the future.

The treatment of cotton seed with *Pseudomonas fluorescens* to control damping-off (Howell and Stipanovic, 1980), the control of crown gall on fruits and ornamentals by applying a strain of *Agrobacterium radiobacter* (Kerr, 1980), and the use of ectomycorrhize to protect pine from root infection by *Phytophthora cinnamomi* (Marx, 1975) are examples of another type of biological control. Here, microorganisms protect the plant surface by creating a biological barrier against the pathogen. Although numerous reports of controlled studies display the potential of these minor types of biological control, the determination of their utility in the control of pathogens within practical ecosystems will require additional research.

C. Chemical Control

Natural and synthetic chemicals have been used many years for the control of plant pathogens. Sulfur, for example, was used as a fumigant over 2000 years ago. This led to the use of mercury as a wood preservative in 1705, mercuric chloride as a means of controlling stinking smut of wheat in 1755, copper compounds as seed treatments to prevent wheat smut in the early 1800's, Bordeaux mixture in the late 1800's for controlling downy mildew of grape, and organo-metallics in the early 1900's when ethyl mercuric chloride was used as a seed treatment to control diseases of cereals (Merrill, 1980). Since the 1930's, many organic fungicides have been developed and used to control pathogens that cause various diseases.

Producing acceptable yields of good quality crops still often depends on the use of fungicides since some pathogens cannot be controlled satisfactorily by other means. These chemical formulations are commonly applied to protect plants or plant parts from infection by the pathogen, or to prevent colonization and often reproduction of the pathogen. Applying fungicides, nematicides, and bactericides offers direct control of some pathogenic fungi, nematodes, and bacteria. Additional disease control may be obtained indirectly with nematicides, insecticides, and herbicides. Nematicides may control nematode carriers of disease-causing organisms as well as provide direct control of pathogenic nematodes. Nematicides may also eliminate nematodes that wound plant roots and create infection sites for fungi and bacteria. Insecticides have an analogous effect on insects.

Eliminating alternate hosts, for example, weeds existing in field borders by the use of herbicides, prevents a buildup of inoculum on the noncrop host. This treatment subsequently prevents the transfer of pathogens from alternate hosts to the crops.

To a minor extent, fungicides are applied to the soil to suppress pathogens. These soil treatments protect roots from infection and also reduce seedling mortality caused by damping-off organisms. Applying fungicides directly to seeds also protects emerging plants from damping-off.

There are, however, only a few chemicals that cure or treat both new and established infections. In these limited cases, chemotherapy is achieved by "systemic" fungicides that enter the plant and produce an internal change. These chemicals probably either inactivate toxins produced by the pathogens, or have a direct toxic or inhibitory effect on the pathogen.

III. FORECASTING PLANT DISEASE EPIDEMICS FROM WEATHER DATA

The timing of fungicide application is very important for successful crop protection. The sprays are often applied, however, on a time-interval basis rather than on the basis of pathogen activity. Our objective in forecasting plant disease is to indicate the ideal time to apply chemicals for effective pathogen control and to minimize the number of sprays required to do so.

Fungicides are most commonly applied to the foliage, stems, and fruit of plants. When treating these above ground parts, it is often necessary to apply fungicides repeatedly to insure protection of newly emerging plant tissue. The scheduling of fungicide applications based on an analysis of the local microclimate has, however, allowed for a substantial reduction in the number of sprays required to control selected pathogens (Beaumont, 1947; Bourke, 1970; Hyre, 1954; Hyre et al., 1962; Kerr and Rodrigo, 1967; Krause et al., 1975; Madden et al., 1978; Miller and O'Brien, 1957; and Stevens, 1934).

The triad concept of disease is the basis for several control tactics that may be used in an integrated pest management scheme. This "disease triangle" model implies that disease may occur only when the three points of the triangle favor the disease interaction, i.e., if a susceptable host, a virulent pathogen, and a favorable environment coexist. The environment segment of the model is the sum of all external biological, chemical, and physical conditions that affect the organisms in the host-pathogen system under consideration. Critical environmental factors include: the air and soil temperature; atmospheric moisture in the form of humidity, dew, and rainfall; the moisture content, aeration, type, pH, and fertility of the soil; and vectors of the plant pathogen.

The development and implementation of disease management systems often utilize models for the environment and the pathogen, and a few schemes also consider the growth and development of the host. Presently, pathogen management schemes are commonly derived from system science and mathematical concepts. The independent variables of the quantitative models normally consist of biological data and local meteorological parameters.

The local climate, i.e., the average pattern of atmospheric conditions of a specific area, is frequently used in developing pathogen management schemes. Although the altitude to which these conditions apply is not often clearly defined, phytopathologists usually limit the microclimatic conditions to those within the plant canopy. The sensitivity of many pathogens to fluctuating weather variables, however, leaves little room for the use of macroclimatic data. A majority of the pathogen management systems currently in existance, therefore, actually use weather data rather than climatic data.

"Weather-disease" models used in the management of plant pathogens are commonly referred to as forecasters of plant disease. These models express a phase of pathogen response as a function of the present and/or past weather conditions. Based on an analysis of the weather data, fungicides which kill or prevent spore germination are scheduled for application when conditions favor the pathogen, e.g., sporophore or spore formation. Disease control models of this type have been used since the late 1800's for scheduling tactics for pathogen management. Although these forecasting models are driven primarily by weather data, there are a few systems which require the input of biological information.

The types of data required to drive ten of the many forecasting systems reported for the control of plant pathogens are presented in Table 1. The systems included for the control of Stewart's wilt (Stevens, 1934) and tobacco blue mold (Miller and

Table 1. Types of data used in ten forecasting or warning systems for the management of plant pathogens.

System	Input data						
	Weather (a)						Biological
	T	RH	RAIN	DEW	Td	LIGHT	
Stewart's Wilt	X						
Tobacco Blue Mold	X						
Northern Leaf Blight of Sweet Corn							X
Tea Blister Blight				X		X	X
Downy Mildew of Grape	X	X	X	X	X		X
Potato Late Blight							
-Dutch Rules	X		X	X		X	
-Blitecast	X	X	X				
Apple Scab							
-Mill's	X		X				
-MSO	X	X		X			
Tomato Early Blight	X	X	X	X			

(a) T = temperature
RH = relative humidity
RAIN = rainfall
DEW = dew period or leaf wetness
Td = dewpoint temperature
LIGHT = photoperiod or cloudiness

O'Brien, 1957) are based solely on temperature data while the fungicide applications for control of northern leaf blight of sweet corn (Berger, 1972) are scheduled on the basis of biological data, i.e., crop phenology. On the other hand, tea blister blight (Kerr and Rodrigo, 1967) are downy mildew of grape (Miller and O'Brien, 1957) forecasts are based on an analysis of both biological and weather data.

The "Blitecast" system is used primarily in the Northeastern United States for forecasting outbreaks of potato late blight from weather data (Krause et al., 1973). The disease caused by the fungus *Phytophthora infestans* (Mont.) de Bary, is considered to be the most important disease affecting potato crops. This particular blight ocurs in almost all areas of the world where potatoes are grown and was the cause of the potato crop failure in Ireland which led to the great famine of 1846. The same fungus also causes considerable damage to tomato crops. The control of late blight is normally obtained by applying fungicides which kill or prevent germination of the fungus spores.

Systems based on the analysis of weather data are not new concepts in pathogen management and several such warning systems are available for late blight control. In 1926 the environmental criteria for late blight onset were outlined and became known as the "Dutch Rules" (Everdingen, 1926). In this early system, the meteorological variables associated with an outbreak of late blight were identified as temperature, rainfall, dew, and cloudiness.

Systems developed in the United States which accurately schedule late blight control sprays include a method used in the northeast United States which is based on air temperature and relative humidity (Hyre, 1954) and a rainfall-temperature method used in the north-central region (Wallin, 1951). These two methods have now been integrated to form the "Blitecast" computerized forecasting system which analyzes the temperature, relative humidity, and rainfall data and advises spray applications when necessary (Krause et al., 1975). After several years of using the computerized version in a batch-processing environment, a battery operated micro-computerized system was assembled and programmed to monitor and analyze the weather data (MacKenzie and Schimmelpfennig, 1978). This farmer-owned automatic late blight warning system displays upon request the current temperature and relative humidity, the rainfall for the past 12 hrs, and coded messages for a no-spray, a five day and a seven day interval spray schedule.

Mills (1944) and Mills and LaPlante (1951) outlined the system for the control of apple scab caused by Venturia inaequalis and it serves as an example of a management system used for many years in apple orchards. This system identifies apple scab infection periods on the basis of the number of hours of wetness required for infection to occur at various temperatures. Based on these relationships and information on temperature and when rainfall began, growers are advised when infection is likely to occur and if a protectant or eradicant fungicide should be applied. This system has since been modified to consider the influence relative humidity has on infection (Jones et al., 1980). The modified system has been incorporated into a portable unit similar to the Blitecast system for potato late blight. The electronic instrument is designed to be located in individual apple orchards where it monitors the required temperature, leaf wetness, and relative humidity data and predicts apple scab infection periods.

A forecaster for early blight on tomato (FAST) is another example of a recently developed disease forecasting system (Madden et al., 1978). Early blight, caused by the fungus Alternia solani (Ell. and G. Martin) Sor., is the major foliar disease of tomato in the northeastern United States. Currently fungicides are the most effective control measure and they are commonly applied at seven- to ten- day intervals regardless of the environmental conditions. Relatively infrequent "critical" periods, when environmental conditions favor early blight development may, however, be identified by the analysis of temperature, relative humidity, rainfall, and dew period data. The likelihood of disease is greatest when these temperature and moisture conditions favor spore formation and infection of the tomato crop.

Structured similarily to the late blight system, FAST thereby provides an efficient spray schedule by integrating two empirical models driven by temperature and moisture data. One model uses measurements of temperature and leaf wetness, while the second model expresses disease severity as a function of temperature, relative humidity, and rainfall.

The weather data required to drive this computerized system may be obtained easily and simply. The temperature and humidity data may be recorded from a standard hygrothermograph in an instrument shelter placed 35-40 cm above ground and located beside a tomato row. The leaf wetness data may be estimated with a Taylor dew meter (Taylor, 1956) and rainfall measured with a simple rain gauge.

The FAST system was initially developed and field tested in 1976. Plant disease progress curves for the first year illustrate the equal effectiveness of three and seven sprays applied on the basis of weather data and disease control obtained through a rigid schedule of 10 or 13 sprays (Fig. 1). Four years of testing under field conditions reveal that this system provides an efficient fungicide application schedule for early blight of tomato. The system identified the time for initial spray applications, scheduled additional applications when the environment favored the pathogen, and resulted in a 44 to 78 percent reduction in the number of sprays commonly applied to control early blight with concurrent savings in fuel, fungicides, labor, equipment, and possible environmental damage.

The ability to relate disease to environmental data, therefore, provides a valuable tool for the management of plant pathogens. To successfully minimize the number of spray applications, forecast systems of this type must, however, accurately identify a critical period of pathogen activity. If applied too early, new plant growth and fungicide degradation may reduce the degree of protection below that required to control the disease. Likewise, fungicides applied after penetration and infection have taken

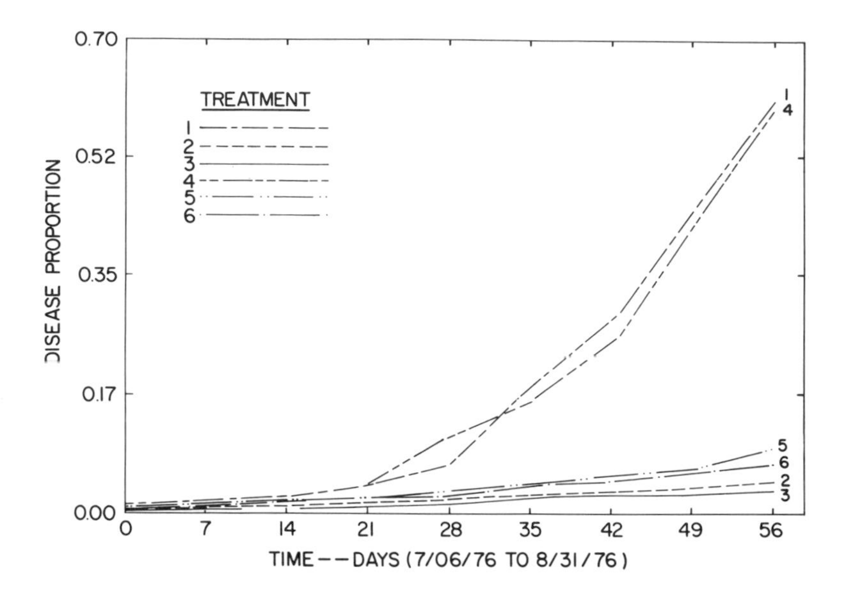

Fig. 1. Disease progress curves for six tomato early blight treatment plots. Treatments: 1 = no spray, 2 = ten sprays, 3 = 13 sprays, 4 = one spray, 5 = three sprays, and 6 = seven sprays. Reprinted by special permission from Madden et al. (1978), American Phytopathological Society.

place provide no protection. Therefore, the accuracy of timing fungicide applications is crucial for the successful implementation of forecasting systems within pathogen management schemes.

IV. SIMULATING PLANT DISEASE EPIDEMICS FROM WEATHER DATA

Modeling each stage of the life cycle of pathogens enables the simulation of the development of the resulting disease (Waggoner, 1978). The modeling of Alternaria solani as a function of the environment resulted in the development of EPIDEM in 1969 (Waggoner and Horstall, 1969). This was the first computer simulator of a plant disease and it was designed to simulate early blight epidemics on potato and tomato. This initial computer simulator of a plant disease provided the initiative for the development of: the MYCOS system in 1971 for Ascochyta blight of chrysanthemum caused by Mycosphaerella ligulicola (McMoy, 1971); EPIMAY in 1972 (Waggoner et al. 1972); and EPICORN in 1973 (Massie, 1973) for the southern corn leaf blight caused by Helminthosporium maydis and EPIVEN in 1973 for apple scab caused by Venturia inequalis (Krantz et al. 1973). The rigid framework of these simulators inspired the structuring of a system which could more easily be modified for other host/pathogen systems. To meet this objective, a flexible plant disease simulator (EPIDEMIC) was assembled and tested for simulating stripe rust of wheat epidemics caused by Puccinia striiformis (Shrum, 1975). The commonality among these simulators is such that each mimic a disease by generating disease progress curves for a specific host and set of environmental data.

The biological responses, e.g., expressed plant disease and plant growth, are commonly estimated as a function of time and weather data. Simulators of plant disease may, however, also require initialization of specific variables. Estimates of the initial amount of inoculum, plant density, proportion of diseased tissue at a given time, and atmospheric spore concentration are examples of biological data that may be required to initialize computer simulated epidemics. The physical data required to initialize a simulator may include greenhouse volume, a parameter relatively easily obtained.

A review of simulators of plant diseases discloses, however, the need to usually monitor many more variables than are required as input to the forecasting systems mentioned in the previous section. Temporal measurements of the temperature and movement of air, light conditions, and moisture comprise the weather data which drive the simulators of plant diseases (Table 2). Moisture is a very important variable in relation to disease epidemics and may be expressed in terms of relative humidity, dew point temperature, dew period, rainfall, and soil moisture. Pathogen take-off and dispersal models commonly use measurements of direction and speed of air movement along with estimated diffusion coefficients. Photoperiod and the quality and intensity of solar radiation complete the listing of variables used to drive most simulators.

Although the forementioned simulators were not originally designed for the management of plant pathogens, they, or modifications thereof, may play a future role in pathogen management.

Table 2. Types of data used in six computer simulators of plant disease.

	Input Data						
	Weather(a)						Biological
System/Reference	T	RH	WIND	RAIN	DEW	LIGHT	
EPIDEM, Waggoner and Horsfall, 1969	X	X	X	X	X	X	
MYCOS, McMoy, 1971	X	X	X	X	X	X	X
EPIMAY, Waggoner, 1972	X	X	X	X	X	X	
EPICORN, Massie, 1973	X	X	X	X	X		X
EPIVEN, Krantz et al., 1973	X	X	X	X	X	X	
EPIDEMIC, Shrum, 1975	?	?	?	?	?	?	?

(a) T = Temperature
RH = Relative humidity
WIND = Wind speed
RAIN = Rainfall
DEW = Dew period or leaf wetness
LIGHT = Photoperiod or cloudiness
? = Requirement is dependent upon the host/pathogen system being simulated.

V. MONITORING METEOROLOGICAL DATA

The development, vertification, and implementation of certain components of pathogen management systems require the acquisition and analysis of quantitative data characterizing biological and physical variables. Computer technology has advanced to the stage that it is now relatively easy to select appropriate models to describe data sets. Although statistically sound procedures are used in fitting and selecting the model, the chosen models commonly only describe the test data set for the conditions under which the experiment was conducted. Therefore, the models may not necessarily consider the actual fluctuating environment nor variation among host plants, isolates, etc. The failure to consider such factors could cause discrepancies among field, controlled laboratory, and _in vitro_ investigations. The differences among results from such studies may not, however, necessarily be due to organisms in culture and controlled environments not responding the same as organisms respond on host tissue in field conditions, but rather that they were subjected to different environments (Pennypacker and Stevenson, 1980).

It is relatively easy to measure the magnitude of many of the physical variables in field and controlled environment studies. The types of instrumentation used to acquire weather data associated with plant disease epidemics has recently been summarized (Pennypacker, 1978). In controlled studies, accuracy and representativeness of the data can be very good if care is taken in the study design, selection of the sensor, and the placement of the sensor. These three items must be considered very carefully when conducting studies to obtain data required to derive algorithms for constructing forecasters and simulators of plant disease. We must also be very cautious of using published data from studies that were not designed specifically to answer the objectives of our task, i.e., modeling biological response as a function of the ambient environment. Reports often do not provide the true conditions under which the biological response actually occurred.

It is possible to obtain accurate and representative data in the laboratory for modeling many biological responses as a function of the environment. However, the acquisition of data in actual field conditions is seldom of the same degree of accuracy as that for which the models may have been developed for disease management systems. Measurements from the sensors may be of sufficient accuracy but their limited number and placement may fail to reflect spatial variation within plant canopies or variation between or among fields. This should be a major area of concern, especially in implementing disease management systems on a state, regional, or national scale. To implement disease management systems on a large scale, the availability of weather data which characterize large areas must be considered.

The basic data for formulating empirical models which describe plant and pathgen responses to environmental factors may be obtained manually from controlled studies. Management systems developed by integrating several of these models which describe associated responses are, however, commonly driven by field data. Such data characterize the biological and weather variables which regulate plant growth and reflect the microclimate to which the plant-pest system is subjected. This type of environmental data may be obtained from several sources, e.g., Weather Bureau Station records, computerized data acquisition units, and miniature weather shelters near or within individual production areas. When working in relatively small growing areas, the data required to drive forecasting systems may be obtained quite easily through the use of hygrothermographs, rain gauges, and dew meters which may be monitored manually on a daily basis.

The environmental parameters we associate with plant pathogens may also be monitored through the use of appropriate sensors and automated data loggers. Recently developed, battery-operated microcomputer and microprocessors may be user programmed to monitor, store, and transmit data. These electronic systems are especially well suited for monitoring the large volume of temporal weather data that is required to drive simulators of plant disease. In addition to monitoring, these units may also be programmed to process the weather data and display the forecast warning for the pathogen being managed (Jones et al., 1980; and MacKenzie and Schimmelpfennig, 1978).

Some pathogen management systems require the processing of real-time information and the dissemination of recommended management tactics. Computerized data systems will play an important role in the acquisition of weather data for developing and driving such pest management models. These systems may also allow us to account for the spatial and temporal changes of the local microclimatic conditions.

The acquisition of the weather data is important so that we may implement empirical models and adapt theoretical models to real-life situations. To do so, we need to know how sensitive each model is to the variable by which it is driven, a subject on which little has been reported (Pennypacker et al., 1978). Items of equal importance include the representativeness and accuracy of the field data which will be used to drive the models.

The sensor-based, computerized system outlined in Fig. 2 was implemented to support our data acquisition, control, and data management requirements. The overall system utilizes two computers which may function independently of each other. The larger of the two units, an 8 megabyte, IBM S/370 Model 3033, is used for data management. The sensor-based system, an IBM S/7, is a small unit that operates in an unattended mode and is designed primarily as a centralized data acquisition and reporting unit. The monitoring system utilizes a central telemetry unit (CIU) that is capable of supporting from 1 to 32 remote terminal units (RTU). Each RTU may monitor and/or control from 1 to 56 parameters at any distance from the central system. The flexibility of the system, therefore, allows for expansion to support a wide range of application. The acquired data may be displayed on a teletypewriter (TTY) in the form of a hard copy, punched paper tape (PPT), or stored on a magnetic disk and later routed to the host computer (S/370) for data reduction and analysis. The remote job entry (RJE) system provides a means to enter associated biological data and computer programs into the system. We may also gain access to the data and receive pathogen management decisions via the RJE

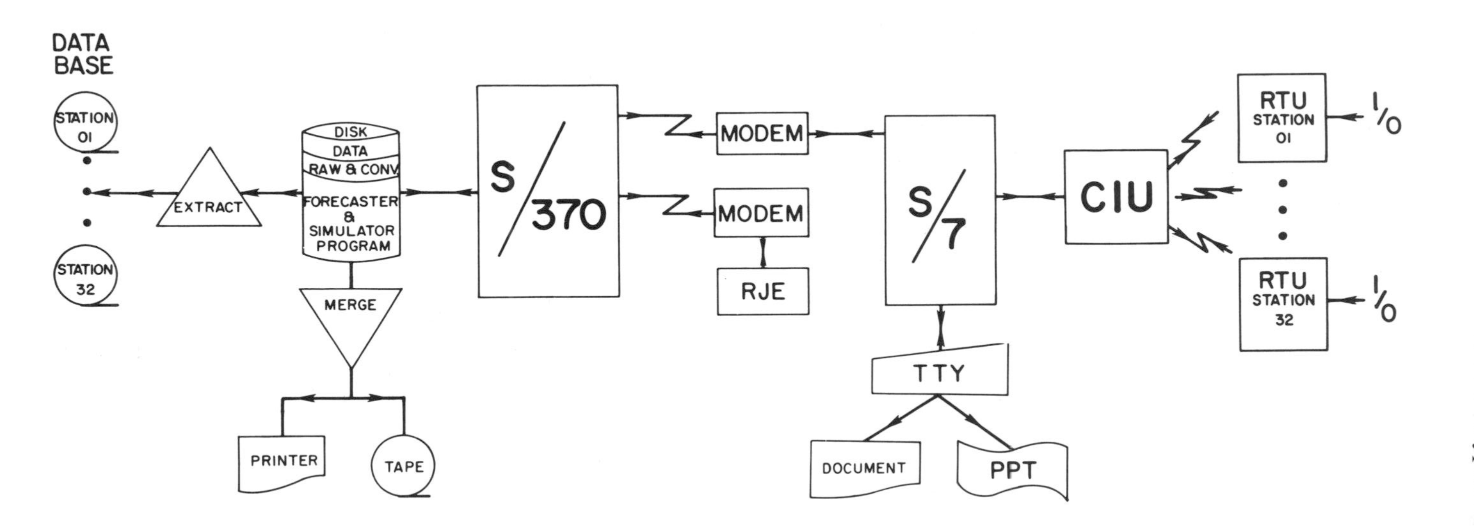

Fig. 2. A computerized network for monitoring meteorological variables, managing biological and physical data, and delivering pest management decisions. (S/370 is a host computer; S/7 is a sensor based computer; CIU is a central telemetry unit; RTU's are remote terminal units; TTY is a teletypewriter; PPT is punched paper tape; and RJE is a typewriter terminal providing remote job entry and output from the host computer).

units. Capable of continuously monitoring microclimatic data from several study sites, the system is a very powerful research tool for use in the development and verification of computerized forecasters and simulators of plant diseases.

In addition to managing the weather data, the same host computer is used for the storage, retrieval, and dissemination of state-wide plant disease information recorded in Pennsylvania (Pennypacker et al., 1976). This portion of the management system includes data verification algorithms to insure data integrity and translation algorithms for retrieving data which conforms to variable but specific criteria. The described system is, therefore, capable of monitoring weather data for developing, testing, and implementing pathogen management models, making pathogen management decisions, and managing records of plant disease occurrence.

REFERENCES

Aldrich, R. A., and Nelson, P. E. (1969). Plant Dis. Reptr. 53:784-788.
Ayers, W. A., and Adams, P. B. (1981). In "Biological control in crop protection," (G. C. Papavizas, ed.), Beltsville Symp. Agric. Res. Vol. 5, Allanheld, Osmun & Co., Montclair, New York.
Baker, K. F., and Cook, R. J. (1974). "Biological control of plant pathogens." W. H. Freeman, San Fraicisco. 433 pp.
Baker, Kenneth F., et al. (1957). Calif. Agric. Exp. Sta., Man. 23:332.
Beumont, A. (1947). Trans. Brit. Mycol. Soc. 31:45-53.
Berger, R. D. (1972). Proc. Fla. State Hort. Soc. 85:142-144.
Bourke, P. M. A. (1970). Ann. Rev. Phytopathol. 8:345-370.
Browning, J. Artie, and Frey, K. J. (1969). Ann. Rev. of Phytopathol. 7:355-382.
Cook, R. J. (1981). Phytopathol. 71:189-192.
Costa, A. S., and Muller, G. W. (1980). Plant Disease 64: 538-541.
Cunningham, H. S. (1953). Phytopathol. 43:95-98.
Everdingen, E. van. (1926). Tijdschrift Plantenziekten. 32: 129-140.
Garnan, H. (1903). Ky. Agric. Expt. Sta. Bull. 105. 32 pp.
Heptig, G. H. (1963). Ann. Rev. Phytopathol. 1:31-50.
Hardison, J. R. (1980). Plant Disease 64:641-645.

Howell, C. R., and Stipanovic, R. D. (1980). Phytopathol. 70: 712-715.
Hyre, R. A. (1954). Plant Dis. Rept. 38:245-253.
Hyre. R. A., MacLeod, John, and Davis, Spencer, H. (1962). Plant Dis. Rept. 46:393-395.
Jones, A. L., Lillevik, S. L., Fisher, P. D., and Stebbins, T. C. (1980). Plant Disease 64:69-72.
Kerr, A., and Rodringo, W. R. P. (1967). Trans. Br. Mycol. Soc. 50(4):609-614.
Kerr, D. (1980). Plant Disease 64:24-30.
Krause, R. A., and Massie, L. B. (1975). Ann. Rev. Phytopathol. 13:31-47.
Krause, R. A., Massie, L. B., and Hyre, R. A. (1975). Plant Dis. Rept. 59:95-98.
Kranz, J., Mogk, M., and Stumpf, A. (1973). Z. Pflanzenkrankh. 80:181-187.
MacKenzie, D. R., and Schimmelpfennig, H. G. (1978). Blitecast System. Am. Potato J. 55:384-385.
Madden, L., Pennypacker, S. P., and MacNab, A. A. (1978). Phytopathol. 68:1354-1358.
Marx, D. H. (1975). In "Biology and control of soil-borne plant pathogens," (G. W. Bruehl), pp. 112-115. Am. Phytopathol. Soc., St. Paul, Minnesota.
Massie, L. B. (1973). Modeling and simulation of southern corn leaf blight disease caused by race T of Helminthosporium maydis Nisik. and Miyake. Ph.D. Thesis. Pennsylvania State University, University Park, Pennsylvania. 84 pp.
McMoy, R. E. (1971). Epidemiology of chrysanthemum Ascochyta blight. Ph.D. Thesis. Cornell University, Ithaca, New York.
Merrill, W. (1980). "Theory and concepts of plant pathology." 2nd Ed., The Department of Plant Pathology, Pennsylvania State Unversity, University Park, Pennsylvania.
Miller, P. R., and O'Brien, M. J. ((1957). Annu. Rev. Microbiol. 11:77-110.
Mills, W. D. (1944). N. Y. Agric. Exp. Stn. (Ithaca) Ext. Bull. 630:4.
Mills, W. D. and LaPlante, A. A. (1951). New York Agric. Exp. Stn. (Ithaca) Ext. Bull. 711:18-22.
Papavizas, G. C. and Lumsden, R. D. (1980). Ann. Rev. Phytopathol. 18:389-413.
Papendick, R. I., and Cook, R. J. (1974). Phytopathol. 64: 358-363.
Paxton, J. D. (1975). In "Biology and control of soil-borne plant pathogens," (G. W. Bruehl, ed), pp. 185-192. Am. Phytopathol. Soc., St. Paul, Minnesota.
Parameter, J. R. Jr., and Uhrenhold, B. (1975). Phytopathol. 65:28-31.
Pennypacker, S. P., Knoble, H. D., and Longenecker, J. L. 1976. Proc. Am. Phytopathol. Soc. 3:246.

Pennypacker, S. P. (1978). In "Plant Disease: An Advanced Treatise," (J. G. Horsfall and E. B. Cowling, eds.), II: 97-118. Acad. Press.
Pennypacker, S. P., Stevenson, R. E., Madden, L., and Knoble, H. D. (1978). Proc. 3rd ICPP., Munich. p. 327
Pennypacker, S. P., and Stevenson, R. E. (1980). Protection Ecol. 2:189-198.
Pirone, P. P., Dodge, B. O., and Rickett, H. W. (1960). "Diseases and pests of ornamental plants." 3rd Ed. Ronald Press Co., New York. 776 p.
Rast, A. B. (1979). Neth. J. Plant Pathol. 85:223-233.
Raychaudhuri, S. P., and Verma, J. P. (1977). In "Plant Disease: An Advanced Treatise," (J. G. Horsfall and E. B. Cowling, eds.), I:177-189. Acad. Press.
Reichert, I. and Palti, J. (1967). Mycopathol. Mycol. Appl. 32:337-355.
Rotem, Joseph. (1978). In "Plant Disease: An Advanced Treatise," (J. G. Horsfall and E. B. Cowling, eds.), II:317-337. Acad. Press.
Shrum, R. D. (1975). Progress Rept. 347. Agric. Expt. Sta., Pennsylvania State University. 81 p.
Stevens, N. E. (1934). Plant Dis. Rept. 12:141-149.
Stoner, W. N., and Moore, W. D. (1953). Plant Dis. Rept. 37: 181-186.
Stover, R. H. (1962). Commonw. Mycol. Inst. Phytopathol. Pap. 4:1-117.
Taylor, C. F. (1956). Plant Dis. Rept. 40:1025-1028.
Wallin, J. R. (1951). Phytopathol. 41:37-38.
Waggoner, P. E. (1960). In "Plant Pathology: An Advanced Treatise," (J. G. Horsfall, and A. E. Dimond, eds.), 3: 291-312. Acad. Press, New York.
Waggoner, P. E. (1965). Annu. Rev. Phytopathol. 3:103-126.
Waggoner, P. E. (1968). In "Biometeorology," (W. P. Lowry, ed.), pp. 45-66. Oregon State University Press.
Waggoner, P. E., and Horsfall, J. G. (1969). Conn. Agric. Exp. Sta., New Haven. Bull. 698. 80 p.
Waggoner, P. E., Horsfall, P. E., and Lukens, R. J. (1972). Conn. Agric. Exp. Sta., New Haven. Bull. 729. 84 p.
Waggoner, P. E. (1978). In "Plant Disease: An Advanced Treatise," (J. G. Horsfall and F. B. Cowling, eds.), II: 203-222. Acad. Press.

RADIATION QUALITY AND PLANT DISEASES

C. M. Leach
Department of Botany and Plant Pathology
Oregon State University
Corvallis, Oregon

and

Anne J. Anderson
Department of Biology[1]
Utah State University
Logan, Utah

I. INTRODUCTION

Plant diseases in the broadest sense are caused by parasitic organisms such as fungi, bacteria, nematodes, and mistletoes, but also result from abiotic causes attributable to phytotoxic chemicals, nutrient problems, etc. (Agrios, 1978). The severity of these diseases is dependent on the interaction of host, pathogen and environment. In this chapter, we will be concerned with the influence of a single environmental factor, light, on plant diseases. The influences of solar radiation in the visible and near visible spectrum on plant diseases is a large and diverse topic which has many ramifications. In most instances, effects of radiation are poorly understood both qualitatively and quantitatively; even less is known about the fundamental biochemical and biophysical nature of these phenomena.

---

[1] Technical Paper No. 5707 of the Oregon Agricultural Experiment Station.

ISBN 0-12-332850-0

In this chapter, we have restricted our discussion to the relation of radiation quality to plant diseases caused by fungi; we have also narrowed our coverage to radiation at the earth's surface within the visible and near visible spectrum (Fig. 1). The effects of ionizing radiation such as vacuum-UV, gamma rays, X-rays and cosmic rays, and nonionizing radio waves, will not be discussed because there is no evidence that they influence plant diseases under natural conditions.

Much has been written on the influence of light on fungi (Smith, 1936; Marsh et al., 1959; Carlile, 1965, 1970; Leach, 1971b) but far less has been published on the influence of light on the plant's response to disease. We will not attempt a comprehensive review of this large body of knowledge but we will be concerned with an overview of the subject as is summarized in Fig. 2 and suggest possible directions for future research.

For light to influence either the host or the pathogen, it must be absorbed by a specific photoreceptor (Clayton, 1970, 1971) which then initiates various biophysical and biochemical events. Absorption of light by photoreceptor is wavelength specific and thus will vary considerably from one phenomena to another dependent on the molecular nature of the photoreceptor. Blue light, for example, is very effective in inhibiting sporulation in *Botrytis cinerea* (Tan, 1974) but red-far red light is needed to trigger ascospore discharge by the apple scab pathogen, *Venturia inaequalis* (Brook, 1969). Our emphasis in this chapter will be the quality of radiation involved in various phenomena, though readers should recognize that with few exceptions, precise action spectra are woefully lacking.

II. GENERAL CONSIDERATIONS

A. The Plant Disease Cycle

The events of the typical "disease cycle" of a foliar disease (Fig. 2) are directly or indirectly regulated by physical and biotic factors of the environment such as atmospheric humidity,

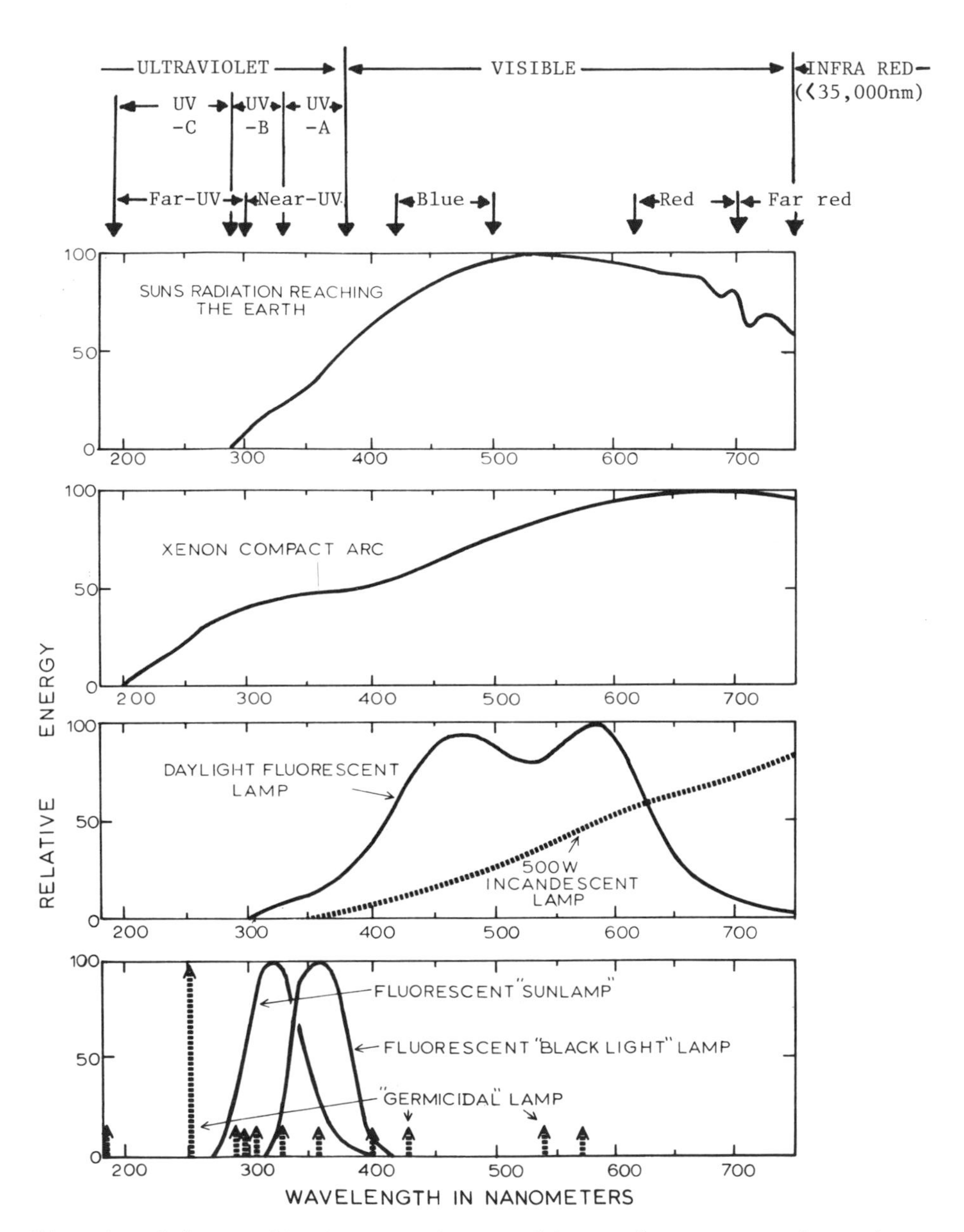

Fig. 1. Solar radiation at the earth's surface compared to that emitted by various lamps. Wavebands of radiation that influence plant diseases are indicated in the upper portion of the figure and include two different systems of UV nomenclature. (Modified after Leach, 1971b.) (nanometers = 1000 μm)

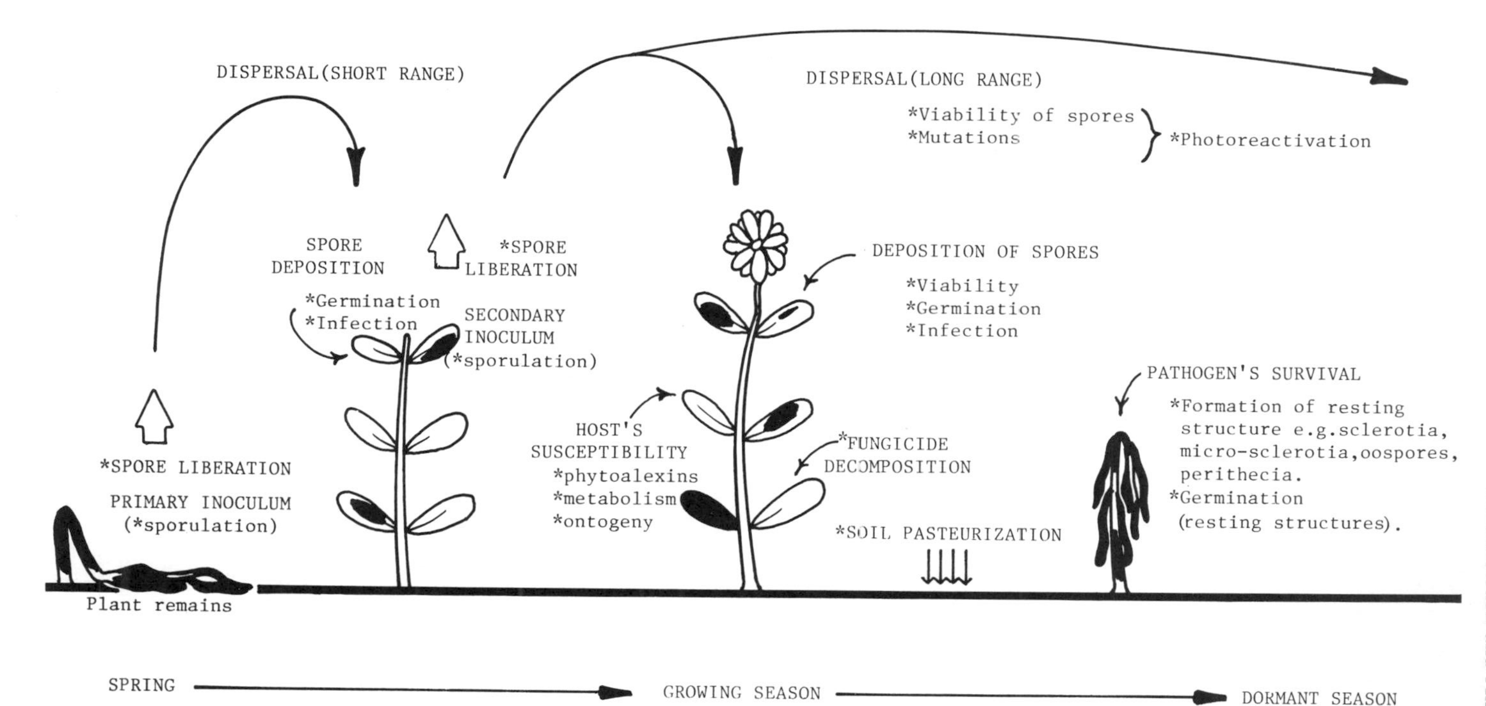

Fig. 2. Light influenced phenomena (*) associated with a typical foliar disease cycle.

surface moisture, temperature, air movement, hosts physiology, light, etc. (Zadoks and Schein, 1979). For a disease to successfully develop within a crop, a pathogen must find conditions favorable for its rapid increase and dispersal and then have conditions favorable for spore germination and infection on its new host. If certain critical factors are unfavorable during any of this series of events, the disease will fail to develop to its maximum potential. Though light influences each of the events shown in Fig. 2, currently, we do not know how general these responses are for different diseases. Most light studies have concentrated on a single phenomenon such as spore germination or sporulation rather than examine all possible light-influenced events for a single disease. In addition, most light studies have been conducted within the laboratory or glasshouse and their relevance to disease in field grown crops must remain largely conjecture.

B. Interaction of Light with Other Factors

A plant disease involves a dynamic interaction between host, pathogen and environment, an old concept conveyed by the plant disease pyramid (Fig. 3). Light is but one of several important physical factors that may influence plant diseases (Calhoun, 1973; Rotem, 1978) and its effects can never truely be divorced from these other interacting factors. For example, light induced sporulation of most fungi will proceed only when the atmospheric humidity is near saturation, and the temperature is within a favorable range (Hawker, 1957). Other factors will also be limiting in all light-triggered phenomena. These interrelationships are poorly understood in most of the phenomena that we shall consider.

A farmer's management of a cultivated crop can effect light quality and intensity. Merely by varying the density of planting, altering the date of sowing, planting at high altitude versus a low altitude, growing a crop at different latitudes, cultivating plants in a greenhouse versus in the field, a farmer can alter the quality and intensity of light. Man, therefore, under certain circumstances can modify the influence of light on plant diseases (Fig. 3).

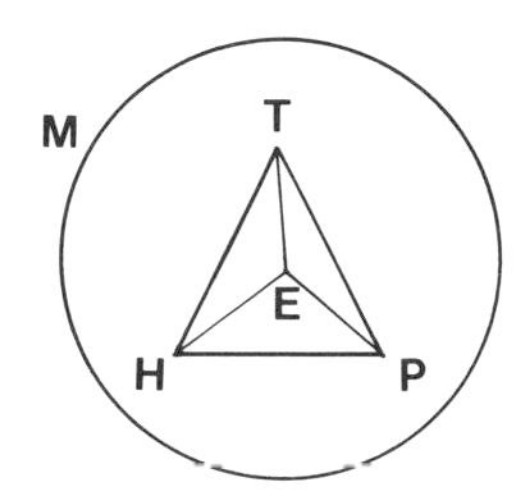

Fig. 3. A schematic portrayal of man's (M) ability to influence plant diseases by modifying one or more interrelated components of the plant disease pyramid (Leach, 1979). Components of the pyramid are host (H), pathogen (P), environment (E), and time (T).

C. Light Quality Under Natural Conditions

Crops are normally exposed to solar radiation extending from the far-ultraviolet (approximately > 0.28 μm (280 nm)) through the visible and infrared spectra (Fig. 1). The quality, intensity and periodicity of solar radiation are extremely variable and are dependent on such factors as time of day, time of year, latitude, altitude, as well as atmospheric conditions such as the presence of water vapor, air pollutants, etc. (Robinson, 1966; Henderson, 1970). The quality and intensity of solar radiation can also be modified by the absorption and reflection by foliage (Vezina and Boulter, 1966; Young, 1974) and by the absorption and reflective properties of soil. Man's activities, such as erecting greenhouses and using cloches over field crops, can significantly modify the quality of solar radiation received by plants.

Ultraviolet radiation is important in a number of disease phenomena. Very little far-UV (< 0.20 μm) reaches the earth's surface (Fig. 1) mainly because it is strongly absorbed by the stratosphere's ozone layer. Before the earth acquired a mantle of oxygen, much shorter wavelengths of UV reached the surface (Ponnamperuma, 1968; Berkner and Marshall, 1964). The minimum wavelength of far-UV now reaching the earth's surface is approximately 0.28 μm (Koller, 1965; Barker, 1968; Thorington, 1980). Though the percentage of far-UV in solar radiation is small, it is very effective photochemically and when absorbed can induce various changes in living organisms including higher plants and fungi (Klein, 1978).

Near-ultraviolet radiation, NUV, (0.30-.38 μm), which is also designated by some biologists as UV-B (0.28-.32 μm) and UV-A (0.32-.38 μm), is a common component of solar radiation at the earth's surface (Koller, 1965). It can influence both pathogen and host and also cause some indirect effects on plant diseases.

Visible light, i.e., that light normally seen by the human eye (0.39-.70 μm, approximately) is an important component of solar radiation. With some exceptions, its main influences on plant diseases are through changes induced in the host. The visible spectrum extend into the longer wavelength infrared radiation. In general, the infrared spectrum has little photobiological activity. However, there are possible exceptions that will be discussed in later sections. Infrared radiation is strongly absorbed by water vapor and much is reflected by plant leaves. Such factors are important in considering its influence on plant diseases caused by fungi. In addition, absorbed infrared radiation can cause significant heating of fungal and plant tissue.

D. Measurement of Radiation Quality

In recent years, spectral radiometers have been commercially available for measurement of light quality but their use under field conditions is complex and fraught with difficulties (Edwards and Evans, 1975; Young, 1974); not the least of which is accounting for biological effects of other interacting factors. We question the rationale for using spectral measuring devices under field conditions before knowing what wavelengths are effective in inducing or inhibiting a particular phenomenon under study. Carefully planned field experiments in which meteorological measurements are compiled with biological data will provide clues to possible light-disease relationships. The next step should be to take the problem into the laboratory and there determine whether light is truly implicated by experimentally defining the light relationship and its interactions with other critical factors. With this knowledge available, it is then a fairly straightforward matter to return to the field and measure only the spectral region known to influence the particular phenomenon under investigation. For example, it is well known from laboratory experiments that UV (< 0.34 μm) will induce sporulation in many fungi. With this information as a guide, it is a simple matter to adapt a radiometer, or use a photometer with filters (Leach, 1971b) to record these effective wavelengths at various locations within or above a plant canopy.

III. RADIATION QUALITY AND THE PATHOGEN

A. Production of Primary and Secondary Inoculum

The occurrence of a serious epidemic among foliar diseases is dependent on the rapid increase of the pathogen's population, usually in the form of spores. Spore production is profoundly influenced by various components of the physical environment, including such factors as light, temperature, and atmospheric and surface moisture (Hawker, 1957; Calhoun, 1973; Rotem, 1978; Rotem et al., 1978). Sexual and asexual reproduction, or spore production, can be stimulated in many plant pathogenic fungi by exposing them to certain wavelengths of radiation. Most investigations of the effects of light have been in vitro on facultative parasites and saprophytes (Marsh et al., 1959), far fewer have involved in vivo studies, and fewer still have been concerned with obligate parasites. Rotem (1978) is of the opinion that most in vitro studies are irrelevant to sporulation behavior in nature.

This is of a controversial viewpoint that has still to be supported experimentally.

The sporulation response of fungi exposed to light is not the same for all species of plant pathogens, and there are now known to be several distinct categories of responses. These range from fungi that sporulate profusely only when exposed to light to those that sporulate equally well in light or dark, to those in which light strongly inhibits sporulation. Near-ultraviolet radiation (< 0.34 μm) is highly effective in stimulating both sexual and asexual sporulation in many plant pathogenic fungi and precise action spectra (Fig. 4) have been determined for a small number of fungi including Leptosphaerulina trifolii (Leach, 1972) and Botrytis cinerea (Honda and Yunoki, 1978). In another smaller group of fungi, the effective wavelengths extend from the NUV into the blue region of the spectrum; e.g., Nectria haematococca (Curtis, 1971). While some fungi will sporulate under continuous exposure to NUV, others require a definite period of darkness following NUV. These latter fungi thrive under a diurnal cycling of NUV and darkness (Leach, 1967). In Helminthosporium oryzae and Stemphylium botryosum, exposure to NUV stimulates formation of conidiophores, but once conidiophores have formed, normal development of conidia requires a period of darkness (Honda, et al., 1968; Leach, 1968). During this dark phase, Aragaki (1962) discovered that blue light is inhibitory and even a brief exposure stops development (Lukens, 1966; Leach, 1968). An action spectrum determined for photo-inhibition of conidium formation in Stemphylium botryosum, revealed that blue light (0.48 μm, approximately) and far-UV (0.28 μm, approximately) were most inhibitory (Leach, 1968). Lukens (1966) reported for Alternaria solani that the inhibitory effect of blue light was negated by a subsequent exposure to red light; and similar results were reported for Botrytis cinerea by Tan (1975). More commonly, these same inhibitory effects of blue light are negated by a subsequent exposure to NUV (Honda et al., 1968; Kumagai and Oda, 1969, 1973;

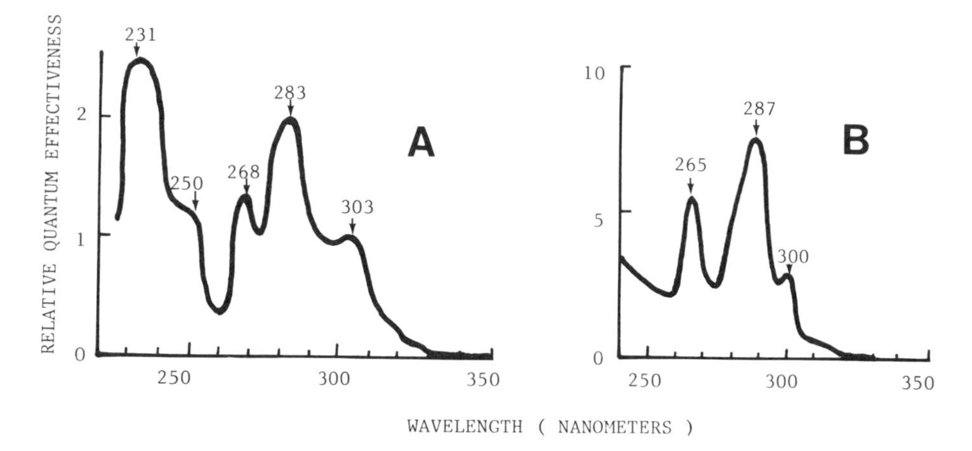

Fig. 4. Action spectra for UV-induced sporulation by two plant pathogenic fungi. A. Formation of conidia by Botrytis cinerea (Honda, 1977). B. Perithecium production by Leptosphaerulina trifolii (Leach, 1972) (nanometers = 1000 μm).

Tan, 1975), which involves a photoreversible pigment mycochrome (Kumagai, 1978; Tan, 1978). Most blue light inhibition studies have been on species of Fungi imperfecti. However, the same phenomenon is known for the obligate parasite Peronospora tabacina (Cruickshank, 1963) and the related Phytophthora infestans (Cohen, 1976). Even among plant pathogenic fungi that do not require light to sporulate, blue light can be inhibitory. This was demonstrated for selected isolates of Helminthosporium oryzae (Yamamura et al., 1978) and for other fungi (Cohen, 1979). Again, mycochrome appears to be involved.

One of the difficulties of generalizing about light requirements for plant pathogens is that all isolates of a single species may not react identically. For example, some isolates H. oryzae (Yamamura et al., 1978) and Ascochyta pisi (Leach, 1962) require an exposure to NUV to sporulate, others sporulate quite well in darkness. Chang (1980) has demonstrated for H. oryzae that these differences are genetically controlled. Even among those species of fungi that sporulate profusely in darkness, sporulation is generally enhanced by an exposure to NUV.

Any attempt to explain the role of light in crop diseases cannot ignore other environmental factors. Temperature, for example, is particularly important in the sporulation process and each fungus has its own unique optimum range. In a study of the interaction of NUV and temperature on sporulation by several *Fungi imperfecti* using temperature gradient plates, Leach (1962) demonstrated some very different responses dependent on specific light and temperature regimes. Among "diurnal sporulators" such as *Stemphylium botryosum*, the temperature requirements for the NUV induction of conidiophores was different from that of the dark phase needed for development of conidia. Aragaki (1961) first discovered that blue light inhibition of conidium development by *Alternaria tomato* could be nullified by low temperatures, and similar results were obtained for *Alternaria solani* (Lukens, 1966) and *Peronospora* sp. (Cruickshank, 1963; Cohen, 1976).

Humidity and leaf wetting are also important interacting factors in the sporulation process (Calhoun, 1973; Rotem, 1978). Thus, although light may be optimum for sporulation, if other factors are limiting, the influence of light can be nullified. In addition, the influence of the host on light-induced sporulation must be considered. For example, a positive relationship between photosynthesis and sporulation has been reported for several obligate parasites by Cohen and Rotem (1970), Reuveni et al., (1971), and Rotem et al., (1978). Photosynthesis is dependent upon absorption of blue (0.44 μm) and red (0.66 μm) light (Leopold, 1964).

1. Future Research Needs. Most light studies on sporulation have been in vitro and there is a need to assess the relevance of these findings to the behavior of pathogens in vivo experiments and under field conditions. Rotem and his associates (1978) have demonstrated that abundance of sporulation is related to photosynthesis in some obligate parasites. Further studies are needed to determine whether the influence of light on obligate parasites

is entirely an effect on the host or whether it may also involve the direct effects of NUV on the pathogen. A direct effect of light on the pathogen is suggested in Cole's (1971) studies on the powdery mildew *Erysiphe cichoracearum*, also by the investigations on blue light inhibition of sporulation of *Peronospora tabacina* by Cruickshank (1963).

The basic nature of photosporgenesis is still not understood. Several precise action spectra have been published both for induction and inhibition of sporulation but so far no one has isolated the photoreceptor(s) involved with the possible exception of mycochrome. An understanding of the biochemistry and biophysics of light-induced sporulation, in addition to strengthening the foundations of Mycology, could lead to new and innovative ways of forecasting and controlling plant diseases.

In the field, it is generally impractical for growers to modify the quality of solar radiation absorbed by the plants, however, where plants are grown commerically in glass houses or in the field under cloches, it is possible to alter the quality of radiation received by plants. Honda et al., (1977) and Honda and Yunoki (1977) recognizing the practical application of their laboratory findings constructed experimental greenhouses that filtered out NUV wavelengths effective in inducing sporulation. This resulted in a significant reduction in the incidence of grey mold of tomatoes and cucumbers caused by *Botrytis cinerea*. In similar experiments, white mold of eggplant and cucumber caused by *Sclerotinia sclerotiorum* were also reduced (Honda, et al., 1977). These pioneering studies by Honda and his associates now need to be extended to diseases of other glasshouse and cloche-grown plants The use of NUV absorbing glass or plastics may prove to be a simple and economically feasible means of reducing disease losses in glasshouse crops.

The inhibitory effects of blue light on sporulation are widespread, yet it is still too early to assess the universality of this phenomenon. Nearly 20 years ago Lukens (1966) attempted to control early leaf blight of tomato, caused by *Alternaria solani* by exposure of plants to blue light. Lukens trials with blue light were unsuccessful and he attributed his failure to the negation of blue light inhibition at low temperatures, a phenomenon first reported by Aragaki (1962). This single failure should not deter further investigations on the practical possibilities of using blue light to reduce plant diseases in glasshouses and other semi-controlled environments.

B. Spore Liberation

Under optimum environmental conditions, it is not uncommon for foliar diseases to rapidly spread through healthy crops. This rapid dispersal usually results from the release of spores into the air following their dislodgement by wind, rain, or by various active discharged mechanisms (Ingold, 1971; Meredith, 1973). Spores may be liberated as primary inoculum (initial spread) from various overwintering structures or as secondary inoculum (secondary spread) usually associated with lesions on plants (Fig. 2). Release of spores into the air can be triggered in some pathogens by exposing them to light (Ingold, 1971). An example, the powdery mildew *Erysiphe cichoracearum*, is shown in Fig. 5. The wavelengths of radiation effective in stimulating active spore liberation varies considerably which suggests the existence of several different mechanisms. To date, no precise action spectra have been published for light triggered spore discharge.

Studies on *Pleospora herbarum*, the cause of a leaf spot disease of alfalfa (*Medicago sativa*) have shown that only NUV will trigger discharge of ascospores (Leach, unpublished data). Mature perithecia exposed to broadband, filtered NUV as well as monochromatic NUV (0.01 μm bandwidth), discharged when exposed to NUV, but not when exposed to longer wavelengths extending into the infrared spectrum. This response was quite different to that reported for other pyrenomycete fungi. Ascospore discharge by *Sordaria macrospora* is most effected by blue wavelengths (Walkey

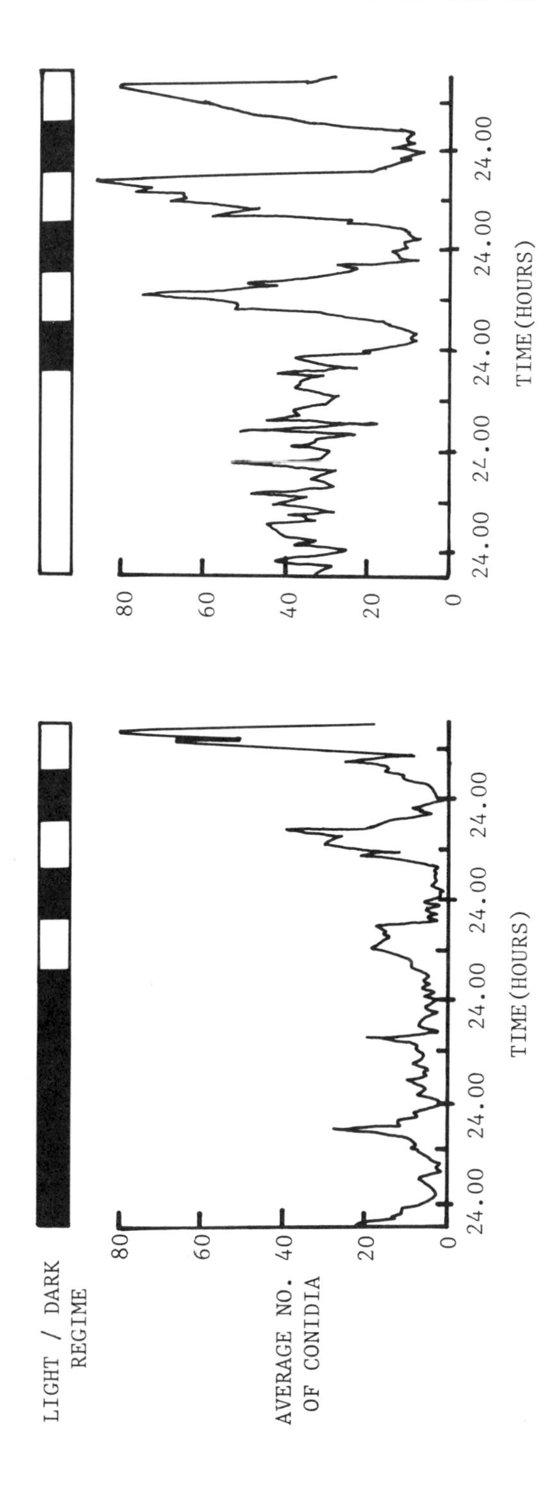

Fig. 5. Conidia of the powdery mildew _Erysiphe cichoracearum_ trapped under two experimental regimes of light and darkness (Cole, 1971).

and Harvey, 1967); *Venturia inaequalis* ascospores are released in response to red far-red radiation (Brook, 1969a, 1969b, 1975). Brook's studies on *V. inaequalis* demonstrated the existence of a red-far red photoreversible reaction not unlike the phytochrome response associated with higher plants. Among *Fungi imperfecti*, *Drechslera maydis* and *D. turcica* pathogens of maize, active discharge is most profuse following exposure to red-infrared radiation (Leach, 1980a, 1980b).

1. Future Research Needs. Many more pathogenic fungi must be examined before it is possible to assess the importance of light-stimulated spore discharge to the spread of plant diseases. The precise quantitative and qualitative relationship of light to spore discharge and how light triggered spore liberation is influenced by other environmental factors is generally not understood. In addition, the nature of the mechanisms of active discharge need to be unraveled.

B. Dispersal

Foliar plant diseases are generally spread by airborne spores. During dispersal and later when spores are deposited on foliage of new hosts, they are exposed to the full solar spectrum. The quality and dosage of absorbed radiation is dependent on such factors as the color, thickness and hydration of spore walls and whether they are dispersed over short distances of a few meters or long distances of thousands of kilometers. During dispersal over long distances we can assume that spores will be carried to higher altitudes where UV-B (0.28-.32 μm) intensities will be greater than for dispersal over short distances. Again, most studies on the influence of light on spores have been confined to laboratories, few field experiments have been conducted.

The adverse effects of far-UV (<0.30 μm) on fungi were reported many years ago by Fulton and Coblentz (1929). Dillon-Weston (1930) reported the death of spores and mycelium exposed to far-UV and Smith (1936) reviewed the mutagenic effects of far-UV. Currently there is large and widely scattered literature on the effects of far-UV on fungi, though much of it involves saprophytic fungi (Pomper, 1965). Far-UV is strongly absorbed by nucleic acids (Davies, 1980), proteins and a number of other molecular components of living cells (Jagger, 1967; Smith and Hanawalt, 1969). Absorption of UV by DNA may cause breaking of various chemical bonds and formation of thymine dimers, changes which are often lethal or result in mutations. Modern UV photobiology is a large discipline with fungi only one of many groups of organisms that are under investigation (Jagger, 1967; Smith and Hanawalt, 1969).

The importance of UV under natural conditions is largely conjecture. Solar UV at the earth's surface does include wavelengths that can be absorbed by DNA and proteins. At high enough dosages, these wavelengths could influence the inoculum density of a pathogen during dispersal by killing spores and also causing mutations that could result in new races of pathogens. In one of the few investigations conducted under natural conditions, Maddison and Manners (1972) determined the effect of sunlight on the viability of uredospores of several species of cereal rusts. They reported a marked reduction of germination among exposed spores; for example, a 20-hour exposure of *Puccinia graminis* uredospores to mid-summer sunlight in England, reduced germination to 10%. Using filters, they concluded that wavelengths close to 0.30 μm were partially responsible for the inactivation of spores, but longer wavelengths (0.315-.40 μm) were also involved.

Lethal and mutagenic changes induced in organisms exposed to far-UV, can be nullified by a subsequent exposure to longer wavelengths between 0.30 and 0.48 μm (Dulbecco, 1955). This phenomenon of "photoreactivation" (Kelner, 1949) has been reported for several plant pathogenic fungi (Buxton et al., 1957) and it could be widespread among plant pathogenic fungi naturally exposed to UV. Whether photoreactivation actually nullifies the lethal and mutational effects of UV under natural conditions is not known.

1. Future Research Needs. Future research should emphasize the effects of light on spore dispersal under natural conditions to answer such questions as: 1) Does UV-B significantly influence the viability of naturally dispersed spores over short and long distances? 2) Does UV-B affect the viability of spores deposited on exposed plant surfaces? 3) Does resistance of spores to UV-B damage depend on wall thickness and pigmentation as is indicated in the laboratory studies on *Cochliobolus sativus* by Tinline et al., (1960)? 4) Are UV-B dosages near the earth's surfaces sufficient to kill spores and induce mutations or must spores be carried to high altitudes to receive these exposures? 5) What is the frequency, if any, of UV-B induced mutations under natural conditions and does this frequency vary from species to species dependent on wall thickness and pigmentation of spores? 6) Are the lethal and mutagenic effects of UV-B under natural conditions generally nullified by photoreactivation as is suggested by the studies of Topolovskii et al. (1976).

Loss of viability of spores exposed to light is usually attributed to the effects of UV radiation but can we ignore the possible involvement of infrared radiation? The thermal death point for many fungi is within the range 40-60C (Hawker, 1957) though it is dependent on such variables as moisture level, length of exposure and other factors. In regions of high temperature and high-light intensity, it is possible that both airborne spores and those deposited on plant foliage are exposed to lethal temperatures resulting from absorbed IR radiation. This possibility should be investigated.

C. Spore Germination

Spores may be produced in profusion and effectively dispersed but unless they germinate and infect a new host, they will be of no consequence pathologically. The topic of spore germination is large and is covered by several review articles (Sussman and Halvorson, 1966; Madelin, 1966; Allen, 1976). Among the numerous physical factors known to influence germination, light is important. It can stimulate germination in some fungi and inhibit it in others. Few studies have precisely defined the stimulatory or inhibitory wavelengths. Light stimulation of germination occurs in the rust Puccinia sorghi (Neuhaus, 1969), in certain smuts and bunts, particularly species of Tilletia (Sussman and Halvorson, 1966; Schauz, 1968) and also among some of the lower fungi, e.g., conidia of Peronospora manschurica (Pedersen, 1964), and sporangia of Physoderma maydis (Hebert and Kelman, 1958). Light stimulates germination of oospores of Phytophthora species (Romero and Gallegly et al., 1963; Leal and Gomez-Miranda, 1965; Berg and Gallegly, 1966). Some reports indicate that blue light is effective in stimulating germination (Sussman et al., 1966). In Physoderma maydis (Hebert and Kelman, 1958), sporangial germination was most stimulated by blue light and least by yellow and red wavelengths. Among Phytophthora species, oospore germination was triggered by two wavebands, one in the blue and the other in the red-far-red (Berg and Gallegly, 1966; Ribeiro et al., 1975; Ribeiro et al., 1976).

Inhibition of spore germination by exposure to light occurs among several rusts (Staples and Wynn, 1965). Light inhibits germination of uredospores of Puccinia graminis (Dillon-Weston, 1932; Sharp et al., 1958; Givan and Bromfield, 1964a, 1964b; Lucas et al., 1975; Knight and Lucas, 1980), P. recondita (Zadoks, 1967; Zadoks and Groenewegen, 1967; Chang et al., 1973, 1974), Phragmidium mucronatum (Cochrane, 1945) and Uromyces trifolii (Thrower, 1964). Light (0.46-.60 μm) inhibits teliospore germination in the smut Ustilago tritici (Krajny, 1974). Calpouzos and Chang (1971) demonstrated that uredospore germination of P. graminis was inhibited by both blue and red-far-red wavelengths. Lucas and Kendrick (1975) reported that the far-red (0.720 μm) inhibition of uredospore germination could be nullified by a subsequent exposure to red (0.653 μm) similar to the phytochrome response.

1. Future Research Needs. The effects of light on spore germination have been examined for very few plant pathogenic fungi and until many more fungi have been investigated it is not possible to assess the relevance of this phenomenon to the spread of plant diseases. In addition, more in vivo studies are needed to determine whether the host might be indirectly influencing the response of spores to light. Spore germination experiments can be readily adapted to monochromatic irradiation studies and these should be initiated to define precisely the wavelengths affecting germination.

D. Infection

Various stages of the infection process among foliar pathogens are known to be influenced by light. In the penetration and colonization phases, there is evidence that light may directly influence the pathogen or influence the susceptibility of the host (discussed later). Studies on the direct influence of light on the infectivity of the pathogen are few and none have attempted to precisely define the effective wavelengths. Emge (1958) reported that infection structures produced by *Puccinia graminis* f. sp. *tritici* on wheat leaves were more abundant when germinating spores were exposed to 6 hours of sunlight at 2000-5000 ft-c; however, they were inhibited by higher intensities > 9000 ft-c. Infection structures also formed when plants were exposed to artificial light (daylight fluorescent lamps). Politowski and Browning (1975) also reported that infection by *P. graminis* was enhanced by exposure to light, though this was not true for *P. coronata*. In other studies on *P. graminis*, Sharp et al. (1958) observed many more appressoria with vesicles associated with plants grown in light (49%) versus darkness (19%). They also reported that uredospores germinated as well in darkness as in light up to an intensity of 300 ft-c, but uredospore germination was inhibited by higher intensities with complete inhibition at 1000 ft-c. A similar inhibitory effect was reported by Zadoks (1967). Benedict

(1971) reported an effect of light on infection of wheat by *Septoria tritici* without making it clear whether this was an effect on the host or the pathogen. Barley plants inoculated with *Rhynchosporium secalis* and grown in light (16,000 lx) developed fewer lesions than plants grown in darkness (Ryan and Clare, 1975) and light also inhibited spore germination and germ tube development. Blakeman and Dickinson (1967) studied the infection of detached alfalfa (*Medicago sativa*) leaves by four species of *Ascochyta* under different lighting. White light or long wavelength UV had no effects but a short exposure to far-UV caused a significant increase in susceptibility of leaves to *A. chrysanthemi*.

1. Future Research Needs. With the exception of *Puccinia graminis*, little is known about the effects of light on the infection process for most fungi. Even for *P. graminis*, virtually nothing is known about the quality of light that influences infection. Although he studies on *P. graminis* suggest that light enhances the infection process, many more studies must be conducted on other plant pathogens before the general significance of light to the infection process can be evaluated.

E. Survival Structures

Plant pathogenic fungi often survive adverse conditions in the form of resistant structures such as chlamydospores, sclerotia, micro-sclerotia, oospores, teliospores, cleistothecia and perithecia. In some fungi, light influences the formation, germination, and discharge of spores from these structures. In general, the qualitative relationship of light to the formation and germination of these survival units is poorly understood.

Many species of *Phytophthora* survive adverse conditions embedded in plant tissue either as chlamydospores or as sexually formed oospores. Both these structures can be influenced by light. Englander and Turbitts (1979) demonstrated a marked increase in numbers of chlamydospores of *P. cinnamomi* formed in colonies grown on a V8 sterol agar medium exposed to light as compared to those grown in darkness. Stimulatory wavelengths were reported in the 0.31-0.42 μm range. They also observed that when NUV intensities were increased about 400 W cm^{-2}, light became increasingly inhibitory with almost complete inhibition at 1600 W cm^{-2}. Englander and Turbitt also noted a strong interaction of light and temperature. Maximum NUV effect occurred at 26C with little effect at 21C. The influence of light on chlamydospore by production by *P. lateralis* is quite different to that just mentioned for *P. cinnamomi* according to Englander and Roth (1980). They discovered that light (a combination of NUV and daylight fluorescent lamps) strongly inhibited formation of chlamydospores *P. lateralis* in colonies growing on a V8-sterol agar medium. Oospore formation by *Phytophthora* species is also influenced by light. Most reports indicate inhibitory effects in vitro (Merz, 1965; Merz et al., 1968; Harnish, 1965; Klisiewicz, 1970; Ribeiro et al., 1976) and in vivo (Brasier, 1972). Though the inhibitory wavelengths have not been precisely defined, Ribeiro et al., (1976) demonstrated that NUV was most effective for *P. megasperma*. Harnish (1965) found inhibition even when NUV was removed from daylight fluorescent lamp radiation. The inhibitory wavelengths for *P. palmivora* were < 0.45 μm (Huguenin and Boccas, 1971; Huguenin and Jacques, 1973). Even germination of oospores of some *Phytophthora* species require light (refer to Ribeiro et al., 1976, for a review). Oospores of *P. infestans* formed in leaves required at least 30 minutes exposure per day to daylight fluorescent lamp radiation (Romero and Gallegly, 1962, 1963) to germination. Blue and far-red radiation were most effective in stimulating germination in *P. megasperma*, *P. capsici* and *P. cinnamomi*.

Formation and germination of sclerotia are also influenced by light in some species. Development of sclerotia by *Phymatotrichum omnivorum* is inhibited by blue light and favored by red light and darkness (Chavez, 1967). This is true also for Botrytis cinerea, as well as for other species of Botrytis and related genera (Jarvis, 1977). Tan and Epton (1973) reported that NUV exposures longer than 30 minutes inhibited development of sclerotia in B. cinerea. Sclerotial production in some isolates of Rhizoctonia solani is favored by light, while other isolates form sclerotia without regard to lighting conditions (Durbin, 1959). Trevethick and Cooke (1973) reported that numbers of sclerotia produced by Sclerotinia sclerotiorum, Sclerotium delphinii, and S. rolfsii, were all greatly increased by exposure to light. They did not determine which wavelengths were effective. Light also has a role in development of apothecia by both Sclerotinia and Monilinia species. Formation of mature ascocarps by S. sclerotiorum, and S. trifoliorum requires exposure of initials to light (Henson and Valleau, 1940; Ikegami, 1959; Bedi, 1962; Williams and Western, 1965; Coley-Smith and Cooke, 1971; Letham, 1975; Willetts and Wong, 1980). In S. sclerotiorum the effective wavelengths are confined to the near-ultraviolet spectrum, (Honda and Yunoki, 1975). In Willetts' (1969) in vitro studies on Monilinia sp., light stimulated formation of stromata in M. fructigena and M. laxa, but not M. fructicola. Willetts did not determine which were the effective wavelengths. Harada's (1977) extensive studies on Monilinia conflict with those of Willetts in that light inhibited formation of stromata in M. fructigena and M. laxa but not M. fructicola. He also reported that formation of mature apothecia was dependent on exposure to light (sunlight and daylight fluorescents were effective); also formation of macroconidia was light dependent in most species tested. Others have demonstrated the stimulatory effect of light on macrospore formation by Monilinia sp. (Terui and Harada, 1968; Jerebzoff, 1960; Sagromsky, 1952). The effective wavelengths were not determined.

Formation of basidiocarps from sclerotia of Typhula species is often dependent on the presence of light. Remsberg (1940) reported that the effective wavelengths were confined to the near-ultraviolet spectrum and this was later confirmed by Lehmann (1965).

In *Verticillium dahliae*, microsclerotium formation is inhibited by continuous exposure to light, particularly NUV (Brandt, 1964; Heale et al., 1965). Gafoor and Heale (1971) demonstrated that only NUV (0.30-.320 μm) was inhibitory in cultures of *V. dahliae*. Unpublished evidence by Flota et al., (1980, personal communication) indicates that formation of microsclerotia of *Pyrenochaeta terrestris*, the cause of onion pink root, is also inhibited by NUV.

The overwintering structures of Pyrenomycetes and Loculoascomycetes in some instances are immature perithecia (protoperithecia) submerged in plant tissue. These function much like the sclerotia of other fungi. In *Pleospora herbarum*, the cause of a foliar disease of alfalfa, protoperithecia form in response to exposure to NUV, however, maturation requires a cold period (Leach, 1966, 1971b). How widespread this phenomenon is has still to be determined. *Venturia inaequalis*, the cause of apple scab, behaves similarly to *P. herbarum*, though Wilson (1928) did not report any involvement of light.

1. Future Research Needs. The influence of light on the development and germination of survival structures has been reported for only a few genera of plant pathogens. Whether other fungi respond similarly to light will require many more studies.

The profound influence of light on the formation, germination and sporulation of survival structures cannot be disputed yet the significance of these findings to the occurrence of diseases under field conditions must at this time remain conjecture. Honda and Yunoki's (1977) report that the incidence of several diseases caused by *Sclerotinia sclerotiorum*, was reduced by screening out UV wavelengths known to stimulate apothecium development, suggests that the influence of light may be significant in some diseases.

IV. RADIATION QUALITY AND HOST-PARASITE INTERACTIONS

A. UV Radiation Effects

Plants strongly absorb radiation in the far and near UV range (Gates, 1965). Exposure of plants to UV radiation may cause cell death, which shows as lesions (Caldwell, 1971). Far UV radiation exerts several types of changes in the plant's DNA structure and mutations may occur (Caldwell, 1979). Presumably, if a UV-induced mutation altered a normally resistant response by a host, susceptibility could result. Plants do possess effective mechanisms to repair far UV-induced DNA damage (Caldwell, 1979). One such mechanism, photoreactivation, is dependent upon exposure to far-UV being followed by irradiation with NUV and visible wavelengths (Caldwell, 1979). Plants also contain pigments, such as anthocyanins and flavonoids in their cuticular layers that filter out far-UV (Caldwell, 1979).

UV radiation is also known to alter enzyme levels in plant tissue. For example, two groups of enzymes involved in biosynthesis of phenolic compounds are induced by UV radiation (Hahlbrock and Grisebach, 1979). One group converts phenylalanine to coumaryl CoA, which is the substrate for the second group of enzymes which synthesizes flavones (Hahlbrock and Grisebach, 1979). The increase in these enzyme activities explain the higher level of flavonoids and anthocyanins in UV irradiated plants (Caldwell, 1979 and Lindoo and Caldwell, 1978). Induction of the enzymes for phenolic biosynthesis could be significant in the disease pyramid (Fig. 3). Many flavonoids have phytoalexin activity and phytoalexins have been implicated in resistance of plants to fungal pathogens (Bridge and Klarman, 1972). Induction of phytoalexins by far-UV has been reported for soybean (Hadwiger et al., 1971) and pea (Wood, 1972). The increased phytoalexin levels in soybean after ultraviolet irradiation, enhanced resistance to the pathogen _Phytophthora megasperma_ var. _sojae_ (Bridge and Klarman, 1972). The significance of the UV effects on phenolic levels under natural conditions is uncertain because UV induction of phytoalexin synthesis in soybeans is photoreversible, that is, it can be prevented by a subsequent exposure to visible light (Bridge and Klarman, 1972).

Currently, the mode by which UV radiation alters plant enzyme activities is uncertain. One receptor may involve phytochrome. Both forms of the phytochrome protein absorb in the near UV range with maxima at 0.28 μm (Shropshire, 1977). Radiation at 0.28 μm influences the photoconversion of the phytochrome forms. Other evidence suggests that additional receptors, such as carotenoids, are responsible for mediating the UV effects.

Although exposure to UV may stimulate a plant's resistance mechanisms there are other reports where an increase in susceptibility occurs in irradiated plants. Chrysanthemum leaf discs exposed to far-UV were more susceptible to *Ascochyta* (Blakeman and Dickinson, 1967), and UV treatment of broad bean leaves increased the number of lesions produced by *Botrytis fabae* (Buxton et al., 1957). Barley plants lost their resistance to *Helminthosporium teres* and *H. sativum* upon exposure to far UV (Chakrabarti, 1968). Mechanisms accounting for the increased susceptibility are at present only speculative. In barley, Chakrabarti (1968) observed UV inactivation of inhibitors to *Helminthosporium*, and UV inactivation of phytoalexins was reported by Uehara (1958). Disease susceptibility resulting from exposure to UV could also result from stress through a reduction in photosynthetic activity. A depression of photosynthesis in plants irradiated with far UV has been demonstrated for some species (Caldwell, 1979).

The sensitivity of a plant's metabolism to UV indicates that radiation in this region of the spectrum may be of significance in the plant disease pyramid. In looking to the future, it is quite possible that the balance of the host-pathogen interactions could be significantly effected by an increase in UV irradiances resulting from a diminished ozone layer (Caldwell, 1971).

Many investigators have attributed the results of exposing plant to UV radiation from germicidal lamps to far-UV effects, yet these lamps often also emit longer UV radiation and visible light, a fact that has often been ignored. The effects of interplay between UV radiation with longer visible wavelengths on the host physiology and susceptibility to fungal pathogens needs further experimentation. Research is needed to determine whether the controlled use of UV radiation might have practical application. The possibility exists that seedlings grown in enclosed environments might be exposed to far-UV to enhance their resistance.

B. Visible Light Effects

Plants, unlike fungi, are almost entirely dependent on a daily exposure to visible radiation. Absorbance of specific wavelengths by chlorophylls and other pigments activates the photosynthetic system which accounts for carbon assimilation and supplies energy to the plant. Photosynthesis provides plant cells with ATP and the reducing power of NADPH, both of which are utilized in biosynthetic pathways. The sugar phosphates that are products of CO_2 fixation can be converted into other derivatives such as citric acid cycle intermediates and amino acids which are themselves the corner stones for biosynthetic routes. Clearly, conditions that limit photosynthesis will effect all of these dependent pathways. The biosynthesis of phenolic materials, some of which are involved in resistance mechanisms, is one example. Reduced levels of phenolic compounds have been reported for plants grown under low irradiance or short-day periods (Orellana, 1975; Thomas and Allen, 1971). Low irradiances and short-day periods, or both, have been associated with reduced resistance in several plants to certain pathogens. In experiments on *Verticillium* and *Fusarium* infection of tomato (Foster and Walker, 1947), and *Sclerotinia* stalk rot of sunflower (Orellana, 1975), plants raised under a regime of short days (four to eight hours of light) were more susceptible than plants grown in a regime of longer (12-18 hours). A reduction in irradiance increased susceptibility of tomato to *Fusarium* wilt disease (Foster and Walker, 1947), potato to Phytophthora infestans (Schumann and Thurston, 1977; Victoria and Thurston, 1974), safflower to P. dahliae (Thomas and Allen, 1971) and watermelon seedlings to Verticillium dahlia (Ben-Yephet, 1979). Field studies with potato have demonstrated that both the apparent infection rate and the size of lesions increased as plants were grown under more shaded conditions (Schumann and Thurston, 1977). Similarly, the sizes of anthracnose lesions on susceptible and resistant maize varieties were decreased by increased light intensities (Hamerschmidt and Nicholson, 1977). This well documented relationship between low-light levels and reduced resistance may be due to decreased amounts of phenolic materials and antifungal compounds in the plants. Thomas and Allen (1971) demonstrated that safflower grown under low irradiances had lesser amounts of antifungal polyacetylene compounds than did normally grown plants. A reduced level of lignification and extracellular compounds with phytoalexin activity was reported for sunflower grown under regimes of short versus long days (Orellana, 1975). This decrease in lignin levels could favor increased penetration and also permit the greater degradation and utilization of the plant cell walls by the pathogen (Wood, 1972). The apparent reduced ability of the light-stressed plants to produce antifungal compounds might also promote fungal development through lowered resistance (Wood, 1972). More research is needed to determine how general these light responses are for other host parasite interactions. Because of the effects of reduced light on increasing susceptibility, Thurston and associates (Schumann and Thurston, 1977; Victoria and Thurston, 1974) have cautioned that otherwise useful varieties may be discarded if screened under poor lighting conditions, such as those that occur in glasshouse experiments conducted during winter. However, a brighter viewpoint is that varieties that do perform well under these same conditions, could have an even better resistance potential in the field or under more intensive lighting.

The effect of light on resistance may not be the same for all plant species. A tomato variety having a vertical resistance to Verticillium, did not become more susceptible when subjected to a regime of short days (Jones et al., 1975). Likewise, the potato variety Russet Rural did not show greater resistance when grown under a regime of increased light (Schumann and Thurston, 1977). Although an increase in irradiance reduced the size of anthracnose lesion on resistant and susceptible maize varieties (Hammerschmidt and Nicholson, 1977), the varietal response still remained resistant or susceptible.

Diseases that are enhanced by high irradiances are also known. The size of lesions caused by Cladosporium fulvum on tomato was increased by high irradiances intensities (Langford, 1948). Similarly disease symptoms in carnations produced by Fusarium stalk rot were enhanced by high irradiances(Stack et al., 1978). An increase in pustulation at high irradiances has been demonstrated for some rusts and powdery mildews (Aust et al., 1977; Politowski and Browning, 1975; Rotem et al., 1978). Yirgou and Caldwell (1963, 1967) have speculated that light stimulated pustule formation may reflect an increase in the ease of penetration by the pathogen. They proposed that light decreased the levels of intercellular CO_2 through its effect on photosynthesis, and also promoted opening of the stomata. Another explanation for this enhanced sporulation is that the improved level of photosynthesis supplies greater quantities of photosynthates to the pathogen. Indeed chemical inhibition of photosynthesis decreased pustule formation through normally adequate lighting was provided (Rotem et al., 1978).

Photosynthesis is dependent on the absoprtion of blue and red wavelengths of light, but red and far-red wavelengths are also very important in the plant's phytochrome system (Schopfer, 1977; Shropshire, 1977). This system involves the conversion of a biologically inactive form of phytochrome to an active form by absorbance of red light (0.66 μm). Conversion back to the inactive form can be enhanced by absorbance of far-red radiation (0.73 μm). The phytochrome system controls a wide range of physiological responses including stem elongation, seed germination, leaf expansion, flowering and pigment synthesis. Included among the many enzymes that are regulated by phytochrome are two groups of enzymes involved in phenol metabolism for the biosynthesis of coumaryl CoA and flavonoids (Shropshire, 1977). As phenolic compounds have been implicated in defense mechanisms of plants, again there is potential for red and far red light to influence certain resistance mechanisms.

C. IR Radiation Effects

The absorption of light of longer wavelengths, particularly infrared radiation, may result in the heating of plant tissues. Plant surfaces, however, have evolved so that a high percentage of long wavelength radiation is actually reflected. Gates (1965) reported that above 0.70 μm, the absorbance of a poplar leaf rapidly declines from about 80% to less than 10% for wavelengths up to about 1.3 μm. In contrast, the absorption of light in the UV and visible ranges is about 90%. Inspite of this reflection of IR, incident radiation can cause heating in plants and the temperature of tissues can be appreciably higher than the surrounding air (Gates, 1965). Elevated temperatures increase the rate of metabolic processes such as photosynthesis up to the point where thermal damage can occur. As plant pathogen relationships involve metabolic processes, these may also be quite temperature dependent and consequently they may be sensitive to the heating effects of long-wave radiation. Diseases of root as well as aerial tissues could be effected because soil also warms in response to long-wave radiation. Elevation in temperature in the range of 10 to 25 C has been demonstrated to increase symptom formation in some diseases (Helms, 1977; Pieczarka and Abawi, 1978). Pretreatment of plant tissues at high temperatures (40-50 C) by immersion in heated water induces susceptibility in normally resistant plants

(Chamberlain, 1972; Yarwood and Hooker, 1966; Yarwood, 1976). On the basis of these reports, it is possible that the heating effects by long wavelength radiation might predispose plant tissues to infection by pathogen. Such temperature effects could be a contributing factor to variation in susceptibility recorded in varietal trials conducted in different geographic areas or in greenhouse versus field experiments. In an interesting recent report on the use of remote sensing instruments to detect infrared radiation emitted by healthy and diseased plants, leaves of plants with fungal root diseases were 3-5°C warmer than healthy plants (Pinter ct al., 1979).

1. Future Research Needs. The metabolism of plants is influenced by radiation throughout the solar spectrum. The wavelengths of radiation influencing such metabolic processes as photosynthesis and the phytochrome system have been precisely defined and the processes are well studied. In contrast, investigations on the interaction of pathogens and plants as they are influenced by light quality are only in their infancy. Results of various experiments suggest that phenolic biosynthesis and photosynthate supply may be biochemical sites for the effect of light on plant diseases. However, in a single spectral band there are examples where radiation may both promote and hinder disease development. Each plant pathogen system seems to have idiosyncracies which need to be resolved. From a practical viewpoint, further studies should be initiated to determine whether exposure of container- and greenhouse-grown seedlings can be used to enhance the ability of a plant to defend itself against potential pathogens.

V. INDIRECT EFFECTS OF RADIATION ON PLANT DISEASES

Light may influence the establishment and spread of plant diseases caused by fungi without directly affecting either host or pathogen. In the laboratory, these indirect effects mainly influence the pathogen, e.g., the fungitoxic effect of ozone formed from far-UV emitted by germicidal lamps, the fungitoxic effects of hydrogen peroxide formed in media exposed to far-UV, the photodecomposition of essential nutrients or vitamins incorporated in media and other effects (Leach, 1971b). These laboratory effects of light will be ignored and instead we will concentrate on those effects that may be of consequence under natural conditions.

A. Photodecomposition of Fungicides

Light influences the activity of a number of fungicidal compounds. In the words of Kaspers (1974), "It is well known that a number of fungicidal compounds produce differing results according to local climatic conditions. For example, Captan gives good control of *Plasmopara viticola* in Germany whereas in the south of France it has a relative weak effect against this disease." The effectiveness of a number of fungicides is significantly reduced after prolonged exposure to light (Kaufman, 1977; Woodcock, 1977; Kaspers, 1974). Most studies on the photodecomposition of fungicides have used lamps that strongly emit near- and far-UV wavelengths, presumably on the assumption that these are the wavelengths most likely to cause photodecomposition of fungicides. Though little has been done to precisely define these relationships, it can be assumed that the effective wavelengths will differ from compound to compound dependent on the molecules involved. Kaspers (1974), using a Xenon-lamp, found that the effectiveness of several ethylene bisdithiocarbamates (Maneb, Zineb, and Propineb) in controlling blight (*Phytophthora infestans*) of tomatoes, was significantly reduced when the fungicides were irradiated during in vitro and in vivo tests. When Maneb was exposed for 96h to radiation from a Xenon lamp, its effectiveness was reduced to 20% of that of freshly sprayed unirradiated fungicide. Numerous other examples of photoinactivation of fungicides have been published (Kaufman; 1977; Woodcock, 1971; Kaspers, 1974, Wells et al., 1979).

1. Future Research Needs. In 1967 Torgeson stated that there was no standard method for studying the effect of UV radiation on fungicides deposited on foliage. Although some progress has been made (Cavell, 1979), we consider this to be of high priority in future research. Future research might also explore the possibility of using fungicidally effective compounds that are photolabile as protectants on fruit and other edible crops where a short residual life could be beneficial.

B. Ozone and Plant Diseases

Ozone is a common air pollutant resulting from the UV-catalyzed reaction between oxides of nitrogen and hydrocarbons in the atmosphere. Though ozone in itself is an important phytotoxic chemical which probably causes more injury to vegetation in the United States than any other air pollutant (Hill et al., 1970), it can also directly influence fungal pathogens as well as alter the susceptibility of plants to diseases. Heagle (1973) has extensively reviewed the literature on the effects of air pollutants on plant diseases. Most of the reports cited by Heagle indicate that ozone generally stimulates sporulation, though there are also a few examples of inhibition of sporulation and spore germination. He noted that there is growing evidence that ozone decreases infection, invasion and sporulation by fungi, and also that quite small doses inhibit obligate parasitism. He concludes from the evidence, that ozone alters parasitism primarily through its effect on the host plant.

C. Solar Pasteurization of Soils

Soil-borne pathogens would appear to be well shielded from the direct effects of solar radiation under normal field conditions; however, the findings of Israeli investigators have demonstrated that solar radiation (presumably infrared), can be utilized to control soil-borne diseases. When field soil is mulched with clear polyethylene plastic, the soil temperature is raised to lethal levels for pathogenic fungi. This innovative and promising advance in the control of soil-borne diseases utilizing solar radiation is discussed by Katan (1980).

VI. CONCLUSIONS

Since solar radiation has been both an important driving force and a catalyst in the evolution of life on our planet, it is not surprising that light influences numerous facets of the plant disease pyramid. In our discussion we have attempted to present a broad perspective of the many ways that light can influence host-parasite interactions and we have indicated the effective wavelengths where these are known. We have, with a few exceptions, purposely skirted the subject of the interactions of light with other environmental factors for the sake of brevity and because in so many instances these relationships are poorly understood. Under natural conditions, light-sensitive responses can be significantly modified by other environmental factors. The negation of blue light inhibition of sporulation at low temperatures (Aragaki, 1962; Lukens, 1966) is an example of the importance of the multifactorial approach.

In general, very few light-regulated plant disease phenomena have been investigated extensively or in depth; none have been resolved at the biochemical and biophysical level and there is still much significant research to be done. Research on the influence of light on plant diseases has been largely in the domain of plant pathologists and mycologists. We are now at the stage where persons trained in the biochemistry and biophysics are needed to clarify the fundamental relationships. We must recognize that it is no longer possible for a single individual to be a plant pathologist, mycologist, biometeorologist, instrument expert, aerobiologist, crop specialist, biochemist, and biophysicist, yet all these disciplines are needed to resolve the relationship of light-to-plant diseases.

Many of the findings discussed in this article have resulted from experiments conducted in the confines of the laboratory or glasshouse. The field-oriented plant pathologist, or his colleague in agronomy, must surely be provoked to ask, "Yes, but what does this mean in the real world"? In the real world of controlling plant diseases, our answer at this time must be "We do not know." It is certainly not difficult to speculate on the possible importance of these findings to forecasting diseases, to control practices, to outbreaks of diseases, to the adaptation of plants by mutations, to the variation of susceptibility of plants to pathogens, and so on, but actually demonstrating or proving these relationships is difficult, particularly under natural conditions. Not all the findings we have discussed are esoteric and divorced from practicality. Honda and associates (Honda et al., 1977; Honda and Yunoki, 1977) have successfully demonstrated that diseases caused by *Sclerotinia sclerotiorum* and *Botrytis cinerea* on various plants grown in glasshouses, can be significantly reduced by using UV-absorbing glass or plastic that removes wavelengths effective in inducing sporulation. This suggests a new and innovative approach for controlling glasshouse diseases. For the critics of basic research, it is important to recognize that the foundation of this research are to be found in laboratory experiments conducted by mycologists and plant pathologists dating back at least 60 years.

Although the door to our understanding of the influence of light on plant diseases has been pushed slightly ajar, we have yet to enter into a comprehensive understanding of the relationship of light to plant diseases. The topic offers many opportunities for exciting and innovative applied and basic research and we urge young scientists to rise to the challenge.

REFERENCES

Agrios, G. N. (1978). "Plant Pathology." Academic Press, N.Y.
Allen, P. J. (1976). In "Physiological Plant Pathology" (R. Heitefuss and P. H. Williams, eds.), 4:51-85. Encycl. of Plant Physiol., Springer-Verlag, New York.

Anderson, R. A. and Kaserbauer, M. J. (1970). Photochem., 10:1229-1232.
Aragaki, M. (1961). Phytopathol. 51:803-804.
Aragaki, M. (1962). Phytopathol. 52:1227-1228.
Aust, H. J., Domes, W., and Kranz, J. (1977). Phytopathol. 67:1469-1472.
Barker, R. E. (1968). Photochem. and Photobiol. 7:275-295.
Bedi, K. S. (1962). Proc. Indian Acad. Sci. 55:213-223, 244-250.
Benedict, W. G. (1971). Physiol. Plant Path. 1:55-66.
Ben-Yephet, Y. (1979). Phytopathol. 69:1069-1072.
Berg, L. A. and Gallegly, M. E. (1966). Phytopathol. 56:583.
Berkner, L. V. and Marshall, L. C. (1964). Discuss. Faraday Soc., 37:122-141.
Blakeman, J. P. and Dickinson, C. H. (1967). Trans. British Mycol. Soc. 50:385-396.
Brasier, C. M. (1972). Trans. British Mycol. Soc., 52:273-279.
Brandt, W. (1964). Canadian J. Bot. 42:1017-1023.
Bridge, M. A. and Klarman, W. L. (1972). Phytopathol. 63:606-609.
Brook, P. J. (1969a). Nature 22:390-392.
Brook, P. J. (1969b). New Zealand J. Ag. Res., 12:214-227.
Brook, P. J. (1975). New Phytologist 74:85-92.
Buxton, E. W., Last, F. T., and M. A. Nour. (1957). J. Gen. Microbiol. 16:764-773.
Caldwell, M. M. (1971). Photophysiol. 6:131-177.
Caldwell, M. M. (1979). BioSci. 29:520-524.
Calhoun, J. (1973). Ann. Rev. Phytopathol. 11:343-364.
Calpouzos, L. and Chang, H. (1971). Plant Physiol. 47:729-730.
Carlile, M. J. (1965). Ann. Rev. Plant. Physiol. 16:176-202.
Carlile, M. J. (1970). In "Photobiology of micro-organisms" (P. Halldal, ed.), pp. 309-344. Wiley, N.Y.
Cavell, B. D. (1979). Pesticide Sci. 10:177-180.
Chamberlain, D. W. (1972). Phytopathol. 62:645-646.
Chakrabarti, N. K. (1968). Phytopathol. 58:467-471.
Chang, H., Calpouzos, L. and Wilcoxson, R. D. (1973). Canadian J. Bot. 51:2459-2462.
Chang, H. S., Wilcoxson, R. D. and Calpouzos, L. (1974). Phytopathol. 64:158.
Chang, H. D. (1980). Trans. British Mycol. Soc. 74:642-643.
Chavez, H. B. (1967). Diss. Abstr., 28B:2217-2218.
Clayton, R. K. (1970). "Light and Living Matter," McGraw-Hill, N.Y. Vol. 1.
Clayton, R. K. (1971). "Light and Living Matter," McGraw-Hill, N.Y. Vol. II.
Cochrane, V. W. (1945). Phytopathol. 35:458-462.
Cohen, Y. and Rotem, J. (1970). Phytopathol. 60:1600-1604.
Cohen, Y., Eyal, H. and Sadon, T. (1975). Canadian J. Bot. 53:2680-2686.
Cohen, Y. (1976). Australian J. Biol. Sci. 29:281-289.
Cohen, Y. (1979). Phytoparasitica 7:44-45.

Cole, J. S. (1971). In "Ecology of Leaf Surface Micro-Organisms" (F. T. Preece and C. H. Dickinson, eds.), pp. 323-337. Academic Press, N.Y.
Coley-Smith, J. R. and Cooke, R. C. (1971). Ann. Rev. Phytopathol. 9:65-92.
Cruickshank, I. A. M. (1963). Australian J. Biol. Sci. 16:88-98.
Curtis, C. R. (1971). Plant Physiol. 49:235-239.
Davies, R. J. H. (1980). Photochem. and Photobiol. 31:623-626.
Dillon-Weston, W. A. R. and Halnan, E. T. (1930). Phytopathol. 20:959-965.
Dulbecco, R. (1955). In "Radiation Biology" (A. Hollaender, ed.), II:455-486. McGraw-Hill, N.Y.
Durbin, R. D. (1959). Phytopathol. 49:59-60.
Edwards, D. P. and Evans, G. C. (1975). In "Light as an Ecological Factor: II" (G. C. Evans, R. Bainbridge, and O. Rackham, eds.), pp. 161-187. Blackwell Scientific, London.
Emge, R. C. (1958). Phytopathol. 48:649-652.
Englander, L. and Turbitt, W. (1979). Phytopathol. 69:813-817.
Englander, L. and Roth, L. F. (1980). Phytopathol. 70:650-654.
Foster, R. E. and Walker J. C. (1947). J. Agric. Res., 74:165-185.
Fulton, H. R. and Coblentz, W. W. (1929). J. Agric. Res. 38:159-168.
Gafoor, A. and Heale, J. B. (1971). Microbios 3:131-141
Gates, D. M. (1965). Sci. Am. 218:76-84.
Givan, C. V. and Bromfield, K. R. (1964a). Phytopathol. 54:116-117.
Givan, C. V. and Bromfield, K. R. (1964b). Phytopathol. 54:382-384.
Hadwiger, L. A. and Schwochau, M. E. (1971). Plant Physiol. 47:588-590.
Hahlbrock, K. and Grisebach, H. (1979). Ann. Rev. Plant Physiol. 30:105-130.
Hammerschmidt, R. and Nicholson, R. L. (1977). Phytopathol. 67:247-250.
Harada, Y. (1977). Bull. Faculty Agric., Hirosaki University (Japan), 27:30-109.
Harnish, W. N. (1965). Mycol. 57:85-90.
Hawker, L. E. (1957). "The Physiology of Reproduction in Fungi." Cambridge University Press, N.Y.
Heale, J. B., and Isaac, I. (1965). Trans. British Mycol. Soc. 48:39-50.
Heagle, A. S. (1973). Ann. Rev. Phytopathol. 11:365-388.
Hebert, T. T. and Kelman, A. (1958). Phytopathol. 48:102-106.
Helms, K. (1977). Phytopathol. 67:230-233.
Henderson, S. T. (1970). "Daylight and Its Spectrum." Am. Elsevier Publ. Co., N.Y.
Henson, L. and Valleau, W. D. (1940). Phytopathol. 30:869-873.

Hill, A. C., Heggestad, H. E., and Linzon, S. N. (1970). In "Recognition of Air Pollution Injury to Vegetation: A Pictorial Atlas" (J. S. Jacobson, and A. C. Hill, eds.), pp. B1-B5. Air Pollut. Cont. Assoc., Pittsburgh.
Honda, Y., Sakamota, M. and Oda, Y. (1968). Plant and Cell Physiol. 9:603-697.
Honda, Y. and Yunoki, T. (1975). Ann. Phytopathol. Soc. Japan 41:383-389.
Honda, Y., Toki, T. and Yunoki, T. (1977). Plant Disease Reporter 61:1041-1044.
Honda, Y. and Yunoki, T. (1977). Plant Disease Reporter 61:1036-1040.
Honda, Y. and Yunoki, T. (1978). Plant Physiol. 61:711-713.
Huguenin, B., and Boccas, B. (1971). Annales Phytopath., Paris, 3:353-371.
Huguenin, B. and Jacques, R. (1973). Comp. Rend. Hebd. Seances Acad. Sci., 276:725-728.
Ikegami, H. (1959). Ann. Phytopath. Soc. Japan, 24:273-280.
Ingold, C. T. (1971). "Fungal Spores, Their Liberation and Dispersal." Oxford University Press, London.
Jagger, J. (1967). "Ultraviolet Photobiology." Prentice-Hall, N.Y.
Jarvis, W. R. (1977). "Botryotinia and Botrytis Species: Taxonomy, Physiology, and Pathogenicity." Monograph No. 15, Research Branch, Canada Department of Agriculture, Harrow, Ontario.
Jerebzoff, S. (1960). Comptes. Rendus Acad. Sci. 250:1549-1551.
Jones, J. P., Crill, P., and Volin, R. B. (1975). Phytopathol. 65:647-648.
Kaspers, H. (1974). Pflanzenschutz-Nachrichten, 27:62-74.
Katan, J. (1980). Plant Disease, 64:450-454.
Kaufman, D. D. (1977). In "Antifungal Compounds" (M. R. Siegel, and H. D. Sisler, eds.), 2:1-49. Marcel Dekker, N.Y.
Kelner, A. (1949). Proc. Nat. Acad. Sci., U.S.A., 35:73-79.
Koller, L. R. (1965). "Ultraviolet Radiation." J. Wiley, N.Y.
Klein, R. M. (1978). Botan. Rev. 44:1-127.
Klisiewicz, J. M. (1970). Phytopathol. 60:1738-1742.
Knights, I. K. and Lucas, J. A. (1980). Trans. British Mycol. Soc. 74:543-549.
Krajny, P. (1974). Abstract No. 44, Rev. Plant Path. 54(1):1975.
Kumagai, T. and Oda, Y. (1969). Development, Growth, and Differentiation, 11:130-142.
Kumagai, T. and Oda, Y. (1973). Plant and Cell Physiol. 14:1107-1112.
Kumagai, T. (1978). Photochem. and Photobiol. 27:371-379.
Langford, A. N. (1948). Canadian J. Res. (C), 26:35-64.
Leach, C. M. (1962). Canadian J. Bot., 40:1577-1602.
Leach, C. M. (1966). Photochem. and Photobiol. 5:621-630.
Leach, C. M. (1967). Canadian J. Bot., 59:1060-1083.
Leach, C. M. (1968). Mycol. 60:532-546.
Leach, C. M. (1971a). Trans. British Mycol. Soc. 57:295-315.

Leach, D. M. (1971b). In "Methods in Microbiology" (C. Booth, ed.), IV:609-664. Academic Press, N.Y.
Leach, C. M. (1972). Mycol. 64:475-490.
Leach, C. M. (1979). Australian Plant Pathol., 11.
Leach, C. M. (1980a). Phytopathol. 70:192-196.
Leach, C. M. (1980b). Phytopathol. 70:196-200.
Leal, J. A. and Gomez-Miranda, B. (1968). Trans. British Mycol. Soc. 48:491-494.
Lehmann, H. (1965). Phytopathologische Zeitscrift, 53:255-288.
Leopold, A. C. (1964). "Plant Growth and Development." McGraw-Hill, N.Y.
Letham, D. B. (1975). Trans. British Mycol. Soc., 65:333-335.
Lindoo, S. J. and Caldwell, M. M. (1978). Plant Physiol., 61:278-282.
Lucas, J. A., Kendrick, R. E. and Givan, C. V. (1975). Plant Physiol., 56:847-849.
Lukens, R. J. (1963). Am. J. Bot., 50:720-724.
Lukens, R. J. (1966). Phytopathol. 56:1430-1431.
Maddison, A. C. and Manners, J. G. (1972). Trans. British Mycol. Soc. 59:429-443.
Madelin, M. F. (1966). "The Fungus Spore." Butterworth, London.
Marsh, P. B., Taylor, E. E., and Bassler, L. M. (1959). Plant Disease Reporter, Supplmt. 261:251-312.
Meredith, D. S. (1973). Ann. Rev. Phytopathol., 11:312-341.
Merz, W. G. (1965). "The Effect of Light on Species of Phytophthora." MS Thesis, West Virginia University. Morgantown.
Merz, W. G. and Vickers, R. A. (1968). Proc. W. Virginia Acad. Sci., 40:135-137.
Neuhaus, W. (1969). Rev. of Plant Pathol. 49, No. 3781. 1970 (Abstract).
Orellana, R. G. (1975). Phytopathol. 64:1293-1298.
Pederson, V. D. (1964). Phytopathol. 54:903.
Pieczarka, D. J. and Abawi, G. S. (1978). Phytopathol., 68:766-772.
Pinter, P. J., Stangellini, M. E., Reginato, R. J., Idso, S. B., Jenkins, A. D., and Jackson, R. D. (1979). Sci. 205:585-586.
Politowski, K. and Browning, A. J. (1975). Phytopathol. 65:1400-1404.
Pomper, S. (1965). In "The Fungi" (G. C. Ainsworth, and A. S. Sussman, eds.), 1:575-597.
Ponnamperuma, C. (1968). In "Photophysiology" (A. C. Giese, ed.), III:253-267. Academic Press, N.Y.
Remsberg, R. E. (1940). Mycol. 32:52-96.
Reuveni, R., Cohen, Y. and Rotem, J. (1971). Israel J. Bot., 20:78-83.
Ribeiro, O. K., Zentmeyer, G. A., and Irwin, D. C. (1975). Phytopathol. 65:904-907.

Ribeiro, O. K., Zentmeyer, G. A., and Irwin, D. C. (1976). Mycol. 78:1162-1173.
Robinson, N. (1966). "Solar Radiation." Elsevier Publ. Co., N.Y.
Romero, S. and Gallegly, M. E. (1962). Phytopathol. 52:165.
Romero, S. and Gallegly, M. E. (1963). Phytopathol. 53:899-903.
Rotem, J., Cohen, Y. and Bashi, E. (1978). Ann. Rev. Phytopathol. 16:83-101.
Rotem, J. (1978). In "Plant Disease an Advanced Treatise" (J. G. Horsfall, and E. B. Cowling, eds.), II:317-337. Academic Press, N.Y.
Ryan, C. C. and Clare, B. G. (1975). Physiological Plant Pathol. 6:93-103.
Sagromsky, H. (1952). Flora 139:560-564.
Schauz, K. (1968). Arch. Mikrobiol. 60:111-123.
Schopfer, P. (1977). Ann. Rev. Plant Physiol. 28:223-252.
Schumann, G. L., and Thurston, H. D. (1977). Phytopathol. 67:1400-1402.
Sharp, E. L., Schmitt, C. G., Staley, J. M. and Kingslover, C. H. (1958). Phytopathol. 48:469-474.
Shropshire, W. (1977). In "The Science of Photobiology" (K. C. Smith, ed.), pp. 281-312. Plenum Press, N.Y.
Smith, E. C. (1936). In "Biological Effects of Radiation" (B. M. Duggar, ed.), II:889-918. McGraw-Hill, N.Y.
Smith K. C. and Hanawalt, P. C. (1969). "Molecular Photobiology." Academic Press, N.Y.
Stack, R. W., Horst, R. K., Nelson, P. E. and Langhans, R. W. (1978). Phytopathol. 68:423-428.
Staples, R. C. and Wynn, W. K. (1965). Bot. Rev. 31:537-564.
Sussman, A. S. and Halvorson, H. O. (1966). "Spores, Their Dormancy and Germination." Harper and Row, N.Y.
Tan, K. K. and Epton, H.A.S. (1973). Trans. British Mycol. Soc. 62:105-112.
Tan, K. K. (1974). J. Gen. Microbiol. 82:191-200.
Tan, K. K. (1975). Trans. British Mycol. Soc., 64:215-222.
Tan, K. K. (1978). In "The Filamentous Fungi" (J. E. Smith and D. R. Berry, eds.), 3:334-357. J. Wiley, N.Y.
Terui, M. and Harada, Y. (1968). Bull. Fac. Agric., Hirosaki University (Japan), 14:38-32.
Thomas, C. A. and Allen, E. H. (1971). Phytopathol. 61:1459-1461.
Thorington, L. (1980). Photochem. and Photobiol., 32:117-129.
Thrower, L. B. (1964). Phytopathol. Z., 51:280-284.
Tinline, R. D., Stauffer, J. F. and Dickson, J. B. (1960). Canadian J. Bot., 33:275-282.
Topolovskii, V. A. and Zaar, E. I. (1976). Rev. Plant Pathol. 55, No. 5132. (Abstract).
Torgeson, D. C. (1967). In "Fungicides. An Advanced Treatise." (D. C Torgeson, ed.), I:93-123.
Trevethick, J. and Cooke, R. C. (1973). Trans. British Bycol. Soc., 60:559-566.

Uehara, K. (1958). Ann. Phytopathol. Soc. Japan, 23:230-234.
Vezina, P. E. and Boulter, D. W. K. (1966). Canadian J. Bot., 44:1267-1284.
Victoria, J. I. and Thurston, H. D. (1974). Phytopathol. Notes, 64:753-754.
Walkey, D. G. A. and Harvey, R. (1967). Trans. British Mycol. Soc., 50:241-249.
Wells, C. H. J., Pollard, S. J., and Sen, D. (1979). Pesticide Sci. 10:171-176.
Willetts, H. J. (1969). Trans. British Mycol. Soc., 52:309-313.
Willetts, H. J. and Wong, J. A. L. (1980). Bot. Rev. 46:101-165.
Williams, G. H. and Western, J. H. (1965). Ann. Appl. Biol. 56:253-260.
Wilson, E. E. (1928). Phytopathol. 18:375-416.
Wood, R. K. S. (1972). Proc. Royal Soc. London, B. 181:213-232.
Woodcock, D. (1977). In "Antifungal Compounds" (M. R. Siegel, and H. D. Sisler, eds.), 2:209-249.
Yamamura, S., Kumagai, T. and Oda, Y. (1978). Canadian J. Bot. 56:206-208.
Yarwood, C. E. and Hooker, A. L. (1966). Phytopathol. 56:510-511.
Yarwood, C. E. (1976). In "Physiological Plant Pathology" (R. Heitfuss and P. H. Williams, eds.), pp. 703-718. Springer-Verlag, Berlin.
Yirgou, D. and Caldwell, R. M. (1963). Sci. 141:272-273.
Yirgou, D. and Caldwell, R. M. (1967). Phytopathol. 58:500-507.
Young, J. E. (1974). In "Light as an Ecological Factor: II" (G. C. Evans, R. Bainbridge, and O. Rackham, eds.), pp. 135-160. Blackwell Scientific, London.
Zadoks, J. C. (1967). Netherlands J. Plant Pathol., 73:52-54.
Zadoks, J. C. and Groenewegen, L. J. M. (1967). Netherlands J. Plant Pathol. pp. 83-102.
Zadoks, J. C. and Schein, R. D. (1979). "Epidemiology and Plant Disease Management," Oxford University Press, N.Y.

ROLE OF BIOMETEOROLOGY IN INTEGRATED PEST MANAGEMENT SOIL-PLANT-WATER RELATIONS AND DISEASE

J. M. Duniway
Department of Plant Pathology
University of California
Davis, California

I. INTRODUCTION

Biometeorology has always had an important place in plant pathology because the climate conditions the behavior of all plant diseases. In fact, the microenvironment around aerial plant parts routinely cycles between highly permissive and very restrictive conditions for infection by many pathogens and the occurrence of sufficiently humid or wet periods is one of the criteria used to predict the occurrence of several important diseases (e.g., Van der Wal, 1978). In comparison to the aerial environment, plant tissues and the soil have a relatively high and stable water status and the influences of substrate water potentials on disease development are less apparent than are the effects of ambient conditions. Furthermore, much of the methodology that is most useful in water relations research has recently been refined and plant pathologists have really only begun to apply the principles of water relations research to examine the effects of plant and soil water potentials on disease development. Nevertheless, there is good evidence that many of the changes in plant and soil water potential which occur routinely during crop growth have profound interactions with plant diseases. In fact, as the examples

ISBN 0-12-332850-0

discussed here will show, the effects of water potential on disease development can be much greater than the effects of water potential on the growth and yield of healthy plants.

Interactions between plant diseases and water potential take many forms. Some of the interactions between pathogens and water potential in soil are similar to the interactions between many aerial pathogens and their environment in that the behavior of the pathogen is determined by its own water requirements for growth, reproduction, or survival. In the use of wood, seed, or dry produce conditions are often manipulated to avoid the water potentials which permit the growth of decay organisms (Christensen, 1978; French and Christensen, 1978). Many important plant pathogenic fungi, however, are more tolerant of low-water potentials than are their living host plants and the effects of water potential on disease development are frequently mediated by a response of the host or other competitive microorganisms to changing water potential. Selected disease examples are reviewed here to illustrate some of the diverse mechanisms by which diseases interact with plant and soil water potentials. More detailed reviews on several facets of this topic have been published (Cook and Duniway, 1981; Cook and Papendick, 1972; Griffin, 1972, 1978, 1981; Schoeneweiss, 1978).

II. DIRECT EFFECTS OF WATER POTENTIAL ON PATHOGENS

Relatively wet conditions enhance the development of white mold epidemics on the aerial parts of bean plants in the field. The fungus which causes white mold, *Sclerotinia sclerotiorum*, persists in the soil as sclerotia which can germinate to produce several apothecia, each on the apex of a stipe that extends above ground. Ascospores are ejected from the apothecia and are carried by air currents to the surfaces of host blossoms and leaves where they act as the principal form of inoculum for white mold epidemics (Abawi and Grogan, 1979). Although all stages in the disease cycle require wetness, the production of apothecia is frequently limited by a lack of continuously suitable moisture in the top few centimeters of soil, a fact that helps to account for some of the associations between rainfall or irrigation and epidemics of white mold (Abawi and Grogan, 1979). The growth of apothecia, under constant laboratory conditions, requires more than ten days at soil matric potentials (i.e., the negative pressure component of soil water potential) greater than -0.5 bars, with matric potentials in the -0.1 to -0.2 bar range being optimal (Duniway et al., 1977). These effects of soil matric potential were determined with hydrated sclerotia and at nearly saturated ambient humidities. If significant water uptake is required either for hydration or to compensate for transpiration, the growth of apothecia in nature might be confined to even higher soil matric potentials, i.e., those where the hydraulic conductivity is sufficient. Indeed, abundant production of apothecia in bean fields of western Nebraska is sensitive to the combined effects of irrigation and density of the plant canopy, and it may be possible to manipulate both to reduce disease development (Schwartz and Steadman, 1978).

A number of diseases incited by fungi in the genus *Phytophthora* have historically been associated with very wet conditions. Disease relationships to wetness are, of course, most apparent for *Phytophthora* diseases of aerial plant parts, but the excellerated development of *Phytophthora* diseases on roots or other subterranean organs under relatively wet soil conditions can be equally striking. For example, the late blight organism, *Phytophthora infestans*, only infects large numbers of potato tubers during crop growth if soil moisture remains well above field capacity for at least 24 hrs (Lapwood, 1977). When combined with considerations of inoculum levels on the foliage and temperature (Sato, 1979), the moisture requirements for *P. infestans* to infect tubers can probably be used to predict the incidence of tuber infection (Lapwood, 1977), or perhaps tuber infection can even be avoided under some irrigated conditions. The hope that

Phytophthora diseases can be reduced through irrigation management is even more realistic for the large number of *Phytophthora* species which inhabit soil and which incite root rots, for the diseases they incite on a wide variety of plants are all aggravated by prolonged saturation of the soil (Duniway, 1979; Zentmyer, 1980). Although *Phytophthora* diseases are the examples chosen for discussion here, it should be noted that root rots and seedling blights caused by a variety of related, zoospore-producing fungi, such as *Pythium* and *Aphanomyces* species, are also strongly associated with very wet or even waterlogged soil conditions.

The association of *Phytophthora* root rot with saturated soil conditions has generally been attributed to enhanced activity of the pathogen under such conditions. Among those phases in the life cycles of *Phytophthora* species for which water requirements have been investigated, asexual reproduction by means of zoospores has the highest water potential requirements (Duniway, 1979). The predominate cycle in asexual reproduction begins with formation of sporangia, within which flagellated zoospores are later formed by cleavage of the cytoplasm. The zoospores are finally discharged from the sporangia and swim actively in the soil solution. Zoospores exhibit chemotaxis toward host tissues and they can be an abundant and very effective form of inoculum (Duniway, 1979).

The water requirements for soil-borne *Phytophthoras* to form sporangia and zoospores are reviewed elsewhere (Duniway, 1979; Gisi et al., 1980) and the minimum soil water potentials for sporangium formation range from about -4 bars for species such as *P. cryptogea* and *P. cinnamomi* up to about -10 millibars (mb) matric potential for *P. cambivora* and some isolates of *P. megasperma*. In all probability, enhanced sporangium formation by some but not all *Phytophthora* species is one of the reasons certain *Phytophthora* root rots are aggravated by saturated soil conditions. Evidence that zoospore discharge from sporangia requires the same levels of soil saturation with which *Phytophthora* root rots are associated is stronger. For example, when the matric component of soil water potential was increased to induce the final differentiation and discharge of zoospores by sporangia of P. cryptogea and P. megasperma, discharge of zoospores was rapid at zero (fully saturated) and -1 mb, was retarded at -5 mb, greatly restricted at -10 mb, and did not occur at -25 mb matric potential (MacDonald and Duniway, 1978).

There is no doubt that once zoospores are discharged, their swimming activities are confined to liquid water and because soil structure and water status determine the size distribution of water-filled pores in soil, these soil parameters largely govern the extent to which zoospores can move through soil. Some workers such as Griffin (1972, 1981) have noted the probable importance of these interactions for all motile microorganisms in soil, and work on Phytophthora cryptogea (Duniway, 1976a) shows that zoospores can readily swim 25-35 mm in standing surface water or through a course-textured potting mix at matric potentials higher than -1 mb. Active movement in the potting mix was reduced at -10 mb matric potential and was barely detected at -50 mb. Active movement through finer loam soils was limited to a distance of 5 mm at -1 mb and was not detected in two of three soils at -10 mb matric potential. Of course, zoospores can also be moved passively by water, but their passive dispersal to new infection sites is probably confined to the same saturated conditions as is active movement.

While the water requirements of the pathogen establish some of the climatic constraints on disease development, in the final analysis, pathogen behavior and disease epidemiology represent an integration of pathogen physiology with a number of other biological and physical parameters. Nevertheless, the more narrow the water requirements of a pathogen, the more likely they are to influence disease development; and the zoospore stage of genera such as Phytophthora has some of the most stringent water requirements known among plant pathogens. Very little is known about the physical requirements for infection by plant pathogenic bacteria in soil, but if their motility contributes to infection, their water requirements for infection may be nearly as exacting as are those for zoospores (Griffin, 1972, 1981). The motile stages of plant parasitic nematodes are known to be very much influenced by soil structure and water status and their movement to roots is optimum at soil matric potentials less than zero (saturation) but greater than about -300 mb (field capacity) (Wallace, 1964).

In contrast to motile forms, many of the plant pathogenic fungi in soil, especially members of the Fungi Imperfecti and Ascomycetes, can grow and even sporulate over a wide range of water potentials, and a large number can do so at soil water potentials much too low to support crop growth (Cook and Duniway, 1981; Griffin, 1972, 1978). It is with these more xerophytic pathogens that the effects of water status on disease development are primarily related to the responses of competitive or antagonistic microorganisms, the host, or to other physical considerations including soil strength and aeration.

III. WATER POTENTIAL EFFECTS ON HOST SUSCEPTIBILITY

In addition to causing physiological damage, the highest extreme in soil water status (i.e., flooded or waterlogged conditions) can predispose plants to infectious diseases. Some *Phytophthora* root rots are among the more interesting examples of flood predisposition for it is now becoming evident that the same very wet conditions that enhance zoospore formation and dispersal can predispose the host to infection. Predisposition of alfalfa was demonstrated by flooding plants in sterilized soil before inoculating them with zoospores of *P. megasperma*. When this was done, the percentage of alfalfa seedlings killed by *P. megasperma* progressively increased from 50 to 90 as the duration of the flooding treatment before inoculation was increased from zero to five days (Kuan and Erwin, 1980). The predisposing effect of flooding on the susceptibility of the rhododendron cultivar Caroline to *Phytophthora* root and crown rot is even more dramatic. Caroline, which is normally resistant to *Phytophthora* attack, became severely diseased if it was flooded for 48 hrs before inoculation with zoospores of *P. cinnamomi* (Blaker and MacDonald, 1981).

The mechanisms of flood predisposition to *Phytophthora* root and crown rot are not clearly known but low oxygen levels may make roots more leaky and thus more attractive to zoospores (Kuan and Erwin, 1980). In fact, increased root exudation under wet soil conditions can attract or stimulate the growth of other pathogens. Damping off of seedlings, for example, is aggravated by relatively wet soil conditions and such conditions increase the exudation and diffusion of nutrients from seedlings to attract zoospores or the growth of pathogenic fungi (e.g., Stanghellini and Hancock, 1971). Oxygen deficiency has many impacts on root physiology (Drew and Lynch, 1980) and it may also function in predisposition by impairing a plant's ability to resist pathogenic invasion by *Phytophthora* species and perhaps other microorganisms. Furthermore, a lack of regeneration or growth of new roots under chronically oxygen deficient conditions may seriously impair a plant's capacity to tolerate pathogenic attack by root pathogens such as *Phytophthora* spp. While only a few isolated examples of flood predisposition or aggravation of diseases have been researched to date, the detrimental effects of soil anaerobiosis on roots suggest predisposition of roots to infection by flooding may be widespread.

Moist ambient conditions generally hasten the postharvest decay of produce, largely because they satisfy the water requirements for growth by decay organisms. In addition, the internal water status of produce may influence pathogen ingress and growth directly, but a change in water status may also physiologically predispose produce to decay. For example, high water potentials have been found by Kelman et al. (1978) to increase the susceptibility of potato tubers to bacterial soft rot caused by *Erwinia* carotovora var. atroseptica, with even a slight increase in tuber water potential from -8.0 to -6.7 bars being sufficient to cause a several fold increase in susceptibility to soft rot. The results of Kelman et al. have major implications for the design of handling and storage procedures to minimize the threat of soft rot in potatoes and it would be most interesting to know if similar considerations apply to postharvest decays of other produce.

There is no doubt that the internal water status of growing plants affects their susceptibility to certain diseases. For example, plant water stress has been found to slow the development of some diseases (Paulson and Schoeneweiss, 1971; Rotem, 1969) while it is more frequently found to increase plant susceptibility to other diseases, i.e., to cause predisposition. In fact, the possible importance of water stress as a predisposing factor has been recognized by plant pathologists for a long time, but as the reading of recent reviews will show (Schoeneweiss, 1975, 1978), only a few of the studies on interactions between water stress and disease have actually isolated and quantified plant water status as a variable in disease development.

One relatively well researched example of water stress predisposition, which may seem somewhat incongruous after the preceeding discussions of inoculum behavior and predisposition by flooding, is Phytophthora root rot of safflower. In experiments under controlled conditions, the induction of water stress in safflower plants before they were inoculated with zoospores increased their susceptibility to root rot caused by P. cryptogea (Duniway, 1977). Significant predisposition of the susceptible cultivar Nebraska 10 occurred if water was withheld to depress leaf water potentials from the values higher than -6 bars which characterized well-watered plants, to values less than -13 bars. Comparable depressions in leaf water potential predisposed the normally resistant cultivar Biggs to a greater extent. In addition, the combined effects of a mild depression in leaf water potential from -4.0 to -8.8 bars and inoculation of the Biggs cultivar were significant, whereas neither treatment alone had a significant effect on the growth of Biggs plants. Recent field experiments have confirmed the importance of water stress predisposition in safflower by showing that the occurrence of water stress before irrigation increased the incidence of Phytophthora root rot in several safflower cultivars (Duniway, 1978). The interactive effects of irrigation and water stress on Phytophthora root rot in a susceptible safflower cultivar are shown in Fig. 1. Plants grown on available soil moisture without irrigation did not develop visible symptoms of root rot and the application of one or four irrigations increased root rot incidence significantly. The occurrence of stress before one irrigation also increased disease incidence significantly, and disease incidence after stress was relieved by irrigation finally became equivalent to the level of disease that developed following four irrigations in the absence of stress.

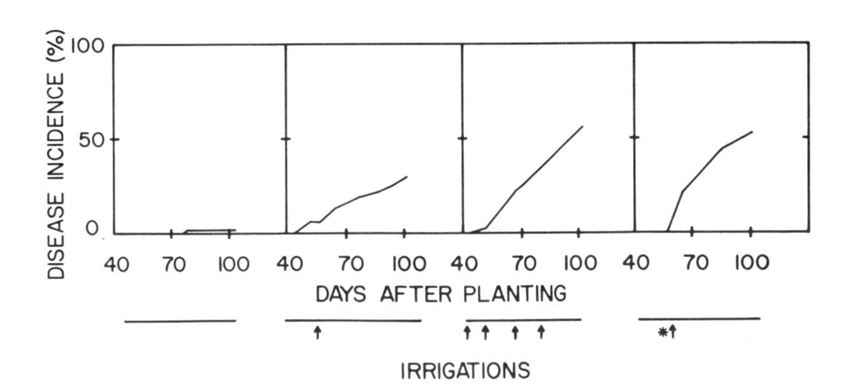

Fig. 1. The development of Phytophthora root rot in the susceptible safflower cultivar Nebraska 10 subjected to various irrigation schedules in the field. Surface irrigations were applied at the times indicated by arrows and the asterisk denotes the occurrence of stress before irrigations. There was no rainfall during crop growth. (Duniway, 1978, and unpublished data.)

Stresses due to drought or salinity have recently been found to predispose at least two kinds of ornamental plants to *Phytophthora* attack. Water stress treatments before inoculation rendered the normally resistant rhododendron cultivar Caroline susceptible to extensive attack by *P. cinnamomi* (Blaker and MacDonald, 1981); and the susceptibility of chrysanthemum roots to *P. cryptogea* was found to increase with the degree of salinity stress to which roots were exposed before inoculation (MacDonald, 1981). Furthermore, predisposition was demonstrated following exposures to NaCl levels, which in the absence of the pathogen, did not have lasting effects on the growth of chrysanthemum.

Phytophthora species cause root and crown rots in a very large and diverse number of plant species and predisposition to *Phytophthora* attack by flooding, drought, or salinity stresses has not been researched with hosts other than alfalfa, safflower, rhododendron, and chrysanthemum. Therefore, while the diversity of the few known examples suggests they are not unique, we do not really know how widespread stress predisposition is among the multitude of *Phytophthora* diseases. It should also be noted that while periods of drought or salinity stress are not conducive to inoculum formation or infection by soil-borne *Phytophthora* species, the inoculum of those few species which have been examined under stress can, nevertheless, effectively persist through the stress cycles which cause predisposition (Duniway, 1979).

There is considerable variation in the aggressiveness of facultative parasites and many of the fungi and bacteria which cause stem canker diseases of woody plants enter host plants but remain latent unless the host is weakened, usually by unfavorable environmental conditions. In the absence of irrigation, outbreaks of canker diseases among woody perennials are frequently associated with prolonged drought or several consecutive years of below normal rainfall. A recent review of these associations suggests that many of the trees which became diseased under drought conditions were predisposed by water stress to attack by relatively nonaggressive pathogens (Schoeneweiss, 1978). Unfortunately, most of the data or observations relating canker diseases to water stress are correlative and do not include useful measures of host water status. One notable exception for which causative relationships between disease and stress have been demonstrated is the stem canker disease incited by the fungus *Botryosphaeria dothidea*. Schoeneweiss (1975) maintained young trees in a relatively constant state of stress at the time of and for one week after they were inoculated with *B. dothidea*. He found that canker growth increased dramatically if sweetgum trees were stressed to less than -18 bars water potential or if red-osier dogwoods were stressed to between -24 and -30 bars. Canker expansion ceased in both tree species within one week after the time stress was relieved. Similar but more detailed experiments by Crist and Schoeneweiss (1975) with the same disease on European white birch showed that no canker growth occurred unless xylem water potential was depressed below -12 bars and that canker growth generally increased greatly with further depressions in water potential (Fig. 2). Exposure of white birch seedlings to defoliation or freezing stress also increased their susceptibility to attack by *B. dothidea*, but whatever the source of stress, Crist and Schoeneweiss (1975) found that the level of stress which first increased susceptibility to disease was less than the level of stress necessary to induce lasting damage in the absence of the pathogen. Furthermore, they found that if cankers had not girdled the stems, disease development ceased and trees returned to a healthy state within a short time after the stresses imposed on white birch seedlings were alleviated.

Some of the predisposing effects of water stress on canker diseases of woody plants can be long-lasting. For example, when white mulberry seedlings were stressed to the point of wilting before inoculation with *Fusarium solani*, 44% of the seedlings developed stem cankers in the three months after stress was relieved by regular watering, whereas no cankers developed following

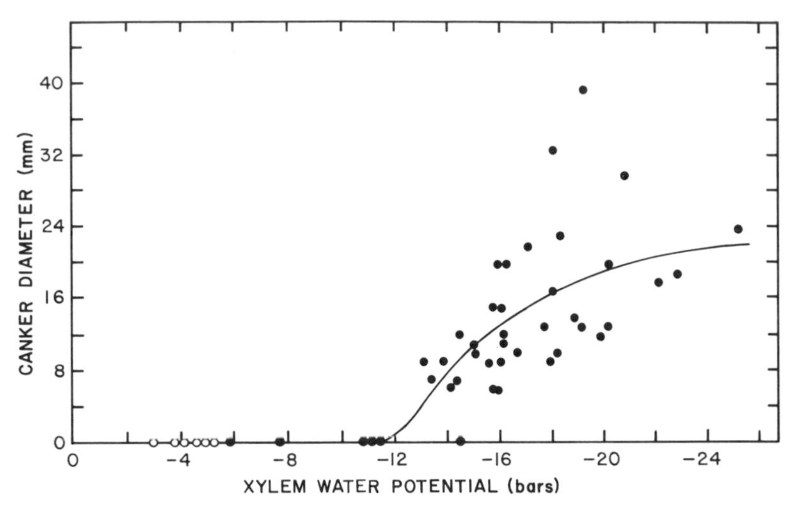

Fig. 2. Relationship of xylem water potential to canker diameter in stems of European white birch seedlings inoculated with *Botryosphaeria dothidea*. (From Crist and Schoeneweiss, 1975.)

inoculation of consistently well-watered seedlings (Schreiber and Dochinger, 1967). In California there is often a long delay before irrigation water is applied to prune orchards after harvest in August, and in some cases, growers hope for early fall rains and do not apply postharvest irrigations. A field experiment conducted by Bertrand et al. (1976) under such California conditions showed that the occurence of moisture stress in just one fall season in otherwise healthy French prune trees significantly increased *Cytospora* canker activity during the succeeding seven months. Their experiment was done by starting differential irrigation treatments in August and periodically measuring leaf water potentials at dawn through the time trees were inoculated in late October. Leaf fall and winter rains relieved any remaining water stress by December and longitudinal canker extension was evaluated monthly thereafter. Canker growth doubled in a linear fashion as the water potentials measured toward the end of the differential irrigation treatments decreased from about -7 to -22 bars. Bertrand et al. (1976) found leaf water potential to be a more sensitive measure of water status and predisposition than was bark moisture content. Various measures of bark moisture or turgidity have traditionally been used to evaluate proneness to canker diseases (Bertrand et al., 1976; Schoeneweiss, 1978).

The stalks and roots of field crops can be rotted by a number of fungi, some of which only attack a host under stressful conditions. For example, *Macrophomina phaseoli* is usually observed to cause charcoal rot of grain sorghum in the field when there is the proper combination of high temperature, low soil moisture, and fairly mature plants (Edmunds, 1964). In fact, the influences of environmental and host parameters on charcoal rot of sorghum are sufficiently complex that plant pathologists have only recently devised methods by which inoculations under controlled conditions will consistently reproduce the disease. Edmunds in 1964 found that inoculations of sorghum with *M. phaseoli* at 35 and 40C only yielded typical disease symptoms when they were done at between 15 and 30 days after anthesis and the plants were given just enough water to avoid permanent wilting. More recent experiments by Odvody and Dunkle (1979) have confirmed the importance of fertility and stress in charcoal root and stalk rot of sorghum. They grew fertilized (self-fertile) and nonfertilized (male-sterile) sorghum lines at 30-35C in soil infested with *M. phaseolina* and withheld water to impose stress at the time fertilized plants had grain in the soft dough stage. A large majority of the fertilized plants which were stressed to -25 bars leaf water potential developed charcoal rot while none of the fertilized plants at leaf water potentials between -13 and -18 bars and none of the sterile plants developed charcoal rot, even when the latter were stressed to -27 bars leaf water potential. *M. phaseolina* grows well in culture over a wide range of water potentials and it appears that water stress and the presence of grain increase the susceptibility of sorghum roots and stalks to pathogenic invasion by the fungus (Odvody and Dunkle, 1979). Although there are no data to show the level of stress which is required, an induction of water stress before inoculation with *M. phaseoli* has been clearly shown to predipose cotton to charcoal rot (Ghaffar and Erwin, 1969).

Hot dry weather also favors the development of charcoal rot in corn, but corn stalks and roots can be rotted by a number of pathogens and some are unlike M. phaseoli in that they cause stalk rot under relatively moist conditions (Dodd, 1980). Unfortunately, the situation in corn where the variety of pathogens which can be involved may broaden the range of conditions under which stalk rot can develop, is probably duplicated in other crops. Therefore, links between stalk rot and any specific set of environmental conditions are the most likely to occur if one pathogen (or perhaps a few similar pathogens) predominates, as is the case for charcoal rot caused by M. phaseoli. Research on the role of water stress in stalk rots is complicated further by the fact that stalk rot development in an individual host plant induces stress in that plant (Dodd, 1980; Schneider, 1981) and correlations between late season plant water stress and stalk rot are not likely to be valid. The role of water stress in one stalk rot other than charcoal rot is now becoming more clear. Schneider (1981) recently found that stalk rot of corn caused by Fusarium moniliforme in Californa was enhanced following the induction of a mild stress early in the growing season even though symptoms of stalk rot did not develop until late in the season.

IV. WATER POTENTIAL EFFECTS ON MICROORGANISMS ANTAGONISTIC TO PLANT PATHOGENS

Disease development may be confined to specific water potentials where a pathogen is able to escape the antagonistic or competitive activities of other microorganisms. Such interactions among microorganisms are most likely to influence soil borne diseases because the soil represents a relatively complex ecosystem on which changes in water potential can have very selective effects (Cook and Duniway, 1981; Griffin, 1981). However, while there is an intense interest in manipulating soil conditions to facilitate biological control (Baker and Cook, 1974), there are only a few disease examples for which soil moisture effects that are mediated by microorganisms other than the pathogen have been quantified from a water relations point of view.

Extensive field studies in the Pacific Northwest by Cook and his colleagues (e.g., Papendick and Cook, 1974; Cook, 1980) have revealed a distinct relationship between the development of plant water stress during the growing season and increased incidence of severe *Fusarium* foot rot in nonirrigated wheat. They found that water stress was earliest and greatest with those cultural practices that hastened early depletion of available soil water, i.e., high nitrogen fertility and high plant densities, and that the same cultural practices promoted the development of foot rot. In fact, foot rot was most prevalent in field plot treatments where midday leaf water potentials approached -40 bars, whereas plot treatments maintaining leaf water potentials higher than -30 to -35 bars had significantly less foot rot. Irrigated wheat and wheat in high rainfall areas would be expected to maintain much higher leaf water potentials and the disease is usually absent in such areas (Cook, 1980).

Predisposition of wheat to pathogenic invasion by *Fusarium roseum* 'Culmorum' probably hastens foot rot development at low water potentials (Papendick and Cook, 1974; Cook, 1980), but escape of the pathogen from the activities of antagonistic organisms may be equally important. Escape can occur because *F. roseum* 'Culmorum' has the ability to grow at soil water potentials below the lower limit for growth by most soil bacteria. For example, Cook and Papendick (1970) observed germ tubes from chlamydospores of the fungus to elongate and branch in natural soil for up to six or seven days at water potentials below about -15 bars. In contrast, the germ tubes lysed or the fungus formed new resistant chlamydospores within 24 to 48 hrs after germination if soil-water potentials were between 0 and -15 bars where bacteria are active.

The incidence of common scab on potato tubers varies with soil conditions, largely because the pathogen, *Streptomyces* scabies, is sensitive to the activities of antagonistic microorganisms (Baker and Cook, 1974). High levels of soil moisture (higher than about field capacity) during tuber formation and growth have been known to suppress the development of common scab for some time and recent experiments have shown that carefully scheduled irrigations can significantly reduce the incidence of scab. For example, Lapwood et al. (1970) found that irrigations applied during rain-free periods of tuberization in England significantly decreased scab on tubers of susceptible cultivars (Fig. 3). If English grading standards for potatoes are considered, the level of disease control obtained with a few well-timed irrigations was even more significant. Experiments in the United States (Davis et al., 1974) have found control of common scab to be facilitated by irrigation schedules that maintain high levels of soil moisture from four days before to about five weeks after tuber initiation. Furthermore, irrigation and chemical treatments had additive effects on the control of common scab.

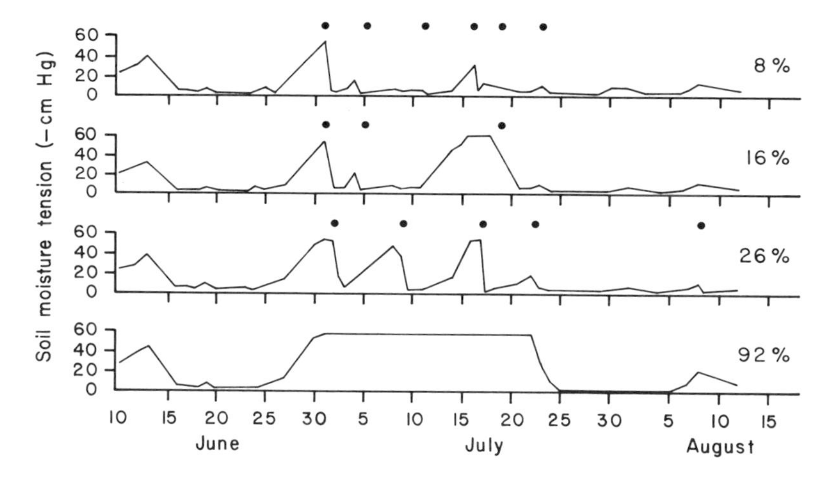

Fig. 3. Fluctuations in soil moisture tension and frequency of irrigation (indicated by large dots) during tuberization of potatoes in soil infested with Streptomyces scabies. The proportion of majestic tubers by weight (%) with one-eighth or more of their surface area affected by scab lesions is shown at the right for each irrigation treatment. 60 cm Hg = 0.8 bar. (From Lapwood et al., 1970.)

V. CONCLUSIONS

The few disease examples reviewed here illustrate some of the diversity that exists among disease responses to plant and soil water potentials. This diversity represents a significant research challenge if one wishes to progress beyond correlative data relating disease development and conditions in the field to obtain quantitive data on causative relationships. The usual approach toward isolating water potential as a functional variable in plant disease is to study its effects on pathogen behavior. This approach has served plant pathology well, but the examples reviewed here have clearly shown that studies on water potential effects on disease development should also consider the behavior of the host and antagonistic microorganisms.

This review has emphasized water potential effects on disease occurrence and development rather than water potential effects on the manifestation of disease as symptoms or a yield loss in infected plants. The space alloted does not permit review of the water relations of diseased plants, but other recent reviews (Ayres, 1978; Duniway, 1976b) show that diseases can affect almost every facet of plant water relations including root extraction of water from soil, internal resistances to liquid water movement, and transpirational behavior. In fact, the responses of plants to changing soil water potential and ambient conditions may be totally modified by diseases, and the impacts that disease development can have on host growth and yield are conditioned largely by the external environment. There is a critical need for more research on the ecophysiology of yield losses caused by plant diseases.

The few diseases which are reviewed here were selected because they represent some of the more studied examples among diseases which are known to be influenced by plant and soil water potentials. There is no doubt that many other diseases are influenced by water potential, but it is inherently difficult to recognize and experimentally cope with water potential as a variable in plant disease. Compare, for example, the volume of literature about temperature effects on disease epidemiology to the small volume of firm data showing the effects of water potential on disease development. Fortunately, the interest of researchers in soil and plant water potential interactions with plant diseases is growing rapidly.

REFERENCES

Abawi, G. S., and Grogan, R. G. (1979). Phytopathol. 69:899-904.

Ayres, P. G. (1978). In "Water Deficits and Plant Growth" (T. T. Kozlowski, ed.), V:1-60. Acad. Press, New York.

Baker, K. F., and Cook, R. J. (1974). Biological Control of Plant Pathogens. W. H. Freeman, San Francisco.

Bertrand, P. F., English, H., Uriu, K., and Schick, F. J. (1976). Phytopathol. 66:1318-1320.

Blaker, N. S., and MacDonald, J. D. (1981). Phytopathol. In press.

Christensen, C. M. (1978). In "Water Deficits and Plant Growth" (T. T. Kozlowski, ed.), V:199-219. Acad Pres, New York.

Cook, R. J. (1980). Plant Disease. 64:1061-1066.

Cook, R. J., and Duniway, J. M. (1981). In "Water Potential Relations in Soil Microbiology," pp. 119-139. Spec. Publ. No. 9. Soil Sci. Soc. of Am.

Cook, R. J., and Papendick, R. I. (1970). Plant and Soil 32: 131-145.

Cook, R. J., and Papendick, R. I. (1972). Annu. Rev. Phytopathol. 10:349-374.

Crist, C. R., and Schoeneweiss, D. F. (1975). Phytopathol. 65:369-373.

Davis, J. R., McMaster, G. M., Callihan, R. H., Garner, J. G., and McDole, R. E. (1974). Phytopathol. 64:1404-1410.

Dodd, J. L. (1980). Plant Disease 64:533-537.

Drew, M. C., and Lynch, J. M. (1980). Annu. Rev. Phytopathol. 18:37-66.

Duniway, J. M. (1976a). Phytopathol. 66:877-882.

Duniway, J. M. (1976b). In "Physiological Plant Pathology," (R. Heitefuss, and P. H. Williams, eds.) Ency. Plant Physiol., N.S. 4:430-449. Springer-Verlag, Berlin.

Duniway, J. M. (1977). Phytopathol. 67:884-889.

Duniway, J. M. (1978). Phytopathol. News 12:148.

Duniway, J. M. (1979). Annu. Rev. Phytopathol. 17:431-460.

Duniway, J. M., Abawi, G. S., and Steadman, J. R. (1977). Proc. Am. Phytopathol. Soc. 4:115.

Edmunds, L. K. (1964). Phytopathol. 54:514-517.

French, D. W., and Christensen, C. M. (1978). In "Water Deficits and Plant Growth," (T. T. Kozlowski, ed.) V:221-251. Acad. Press, New York.

Ghaffar, A., and Erwin, D. C. (1969). Phytopathol. 59:795-797
Gisi, U., Zentmyer, G. A., and Klure, L. J. (1980). Phytopathol. 70:301-306.
Griffin, D. M. (1972). Ecology of Soil Fungi. Syracuse Univ. Press.
Griffin, D. M. (1978). In "Water Deficits and Plant Growth," (T. T. Kozlowski, ed.), V:175-197. Acad. Press, New York.
Griffin, D. M. (1981). In "Water Potential Relations in Soil Microbiology," pp. 141-151. Spec. Publ. No. 9, Soil Sci. Soc. of Am.
Kelman, A., Baughn, J. W., and Maher, E. A. (1978). Phytopathol. News 12:178.
Kuan, T.-L., and Erwin, D. C. (1980). Phytopathol. 70:981-986.
Lapwood, D. H. (1977). Ann. Appl. Biol. 85:23-42.
Lapwood, D. H., Willings, L. W., and Rosser, W. R. (1970). Ann. Appl. Biol. 66:397-405.
MacDonald, J. D. (1981). Phytopathol. In press.
MacDonald, J. D., and Duniway, J. M. (1978). Phytopathol. 68: 751-757.
Odvody, G. N., and Dunkle, L. D. (1979). Phytopathol. 69: 250-254.
Papendick, R. I., and Cook, R. J. (1974). Phytopathol. 64: 358-363.
Paulson, G. A., and Schoeneweiss, D. F. (1971). Phytopathol. 61:959-963.
Rotem, J. (1969). Israel J. Agr. Res. 19:139-141.
Sato, N. (1979). Phytopathol. 69:989-993.
Schneider, R. W. (1981). Phytopathol. 71:253.
Schoeneweiss, D. F. (1975). Annu. Rev. Phytopathol. 13:193-211.
Schoeneweiss, D. F. (1978). In "Water Deficits and Plant Growth," (T. T. Kozlowski, ed.), V:61-99. Acad. Press, New York.
Schreiber, L. R., and Dochinger, L. S. (1967). Plant Dis. Reptr. 51:531-532.
Schwartz, H. F., and Steadman, J. R. (1978). Phytopathol. 68: 383-388.
Stanghellini, M. E., and Hancock, J. G. (1971). Phytopathol. 61:165-168.
Van der Wal, A. F. (1978). In "Water Deficits and Plant Growth," (T. T. Kozlowski, ed.), V:253-295. Acad. Press, New York.
Wallace, H. R. (1964). The Biology of Plant Parasitic Nematodes." St. Martin's Press, New York.
Zentmyer, G. A. (1980). Phytophthora cinnamomi and the diseases it causes," Monograph No. 10, Am. Phytopathol. Soc.

MODIFICATION OF PLANT CANOPY AND ITS IMPACT ON PLANT DISEASE

Joseph Rotem
Division of Plant Pathology
Agricultural Research Organization
The Volcani Center
Bet Dagan, Israel

and

Department of Life Sciences
Bar-Ilan University
Ramat-Gan, Israel

I. INTRODUCTION

The crop microclimate changes during the process of natural plant development. It becomes more humid, and the upper and lower limits of temperature become less extreme. Water deposited by rain or irrigation persists for a longer time; dew also persists longer, although its formation may be inhibited under a very dense canopy (Rotem, 1978). Modern agricultural practices have pushed us to dense plantings and high canopy densities, sometimes leading to an increase in foliage over and above its natural development.

Except for specific cases in which poorly developed plants are more predisposed to certain diseases (e.g., *Alternaria solani* on potatoes), and several other cases in which a greater incidence of disease occurs in plants seeded at a low rate, the too-lush foliage facilitates pathogen development. Modification of the plant canopy aims to reduce plant growth or plant density in order to make the habitat less favorable for the pathogen. This may be achieved by the proper rate of seeding, pruning, fertilization or

ISBN 0-12-332850-0

irrigation, by proper training or trellising, or by breeding cultivars with a sparser canopy. These practices are often feasible because in many cases the yield does not continue to increase with increase in the leaf area. Some of these practices are followed in order to increase the quality of yield, like pruning developed in the Biblical period, they become so rooted in agricultural tradition that their phytopathological effect tends to be forgotten.

In many cases the required inhibition of disease by canopy modification needs a drastic change in the plant canopy to be practical, and farmers, therefore, prefer to use fungicides. In other cases, diseases can be minimized by canopy modification alone or by the combined use of canopy modification and chemical methods. As discussed later, control by canopy modification should not be used in all habitats and against all diseases.

Control of plant disease through canopy modification usually requires a culturally induced change in the habitus of a single plant, or in the density of an entire crop, so as to make its microclimate less suitable for the pathogen. However, microclimatic change is but one of several ways by which canopy modification may affect a pathogen. The mechanism and the relative effects of these actions are discussed in this paper.

II. COMMON METHODS USED TO MODIFY THE PLANT CANOPY, AND THEIR EFFECT ON DISEASE

A. Pruning, Training and Trellising

In orchards, vineyards, and for some vegetables, pruning is the best example of a method which affects disease, although it is used mainly to improve fruit quality. In evergreen crops like citrus, openings are sometimes made in the crown in order to improve growth by allowing penetration of light into the shaded parts of the foliage. In all cases the thinned foliage provides for better ventilation and penetration of light, reduces humidity, allows quicker drying of rain or dew drops, and leads to improved coverage of the foliage by fungicides. These effects in turn inhibit disease development. Pruning, which is not necessary for improvement of yield and which is directed toward inhibiting diseases, is recommended in some tropical crops, e.g., cocoa suffering from Phytophtora palmivora (Asare-Nyako, 1977) and coffee attacked by Hemileia vastatrix (Schieber, 1975). Among vegetable crops, pruning is sometimes practiced with tomatoes, especially in glasshouse culture.

With all the beneficial effects on yield and disease in general, pruning is sometimes associated with increased incidence of diseases as in the case of Botrytis cinerea in tomatoes (Verhoeff, 1968). In trees these exceptions are easily avoided by disinfection of wounds.

Training and/or trellising is practiced with some fruit trees, vines and vegetables for reasons usually not related to disease control. The trained and/or trellised crops are better ventilated, more exposed to sunlight, and better covered by fungicides. Relevant to trellising, but not only to this, is the direction of rows: those parallel to the prevailing wind direction dry more quickly and exhibit less disease (Haas and Bolwyn, 1972; Palti et al., 1972).

B. Seeding Rate or Stand Density

The literature contains numerous observations on the relation between stand density and disease, and/or recommendations for inhibiting disease by decreasing the seeding or planting rate or the stand density (e.g., Asare-Nyako, 1977; Berger, 1975, 1977; Butler and Jones, 1949; Llano, 1977; Schieber, 1975; Scott, 1956; Wilcoxson, 1975).

Some exceptions have been found to the general rule that a dense stand promotes diseases. For instance, a dense stand means, in general, higher humidity. However, due to a reduction in dew points in shaded banana leaves in the Caribbean, and in chili peppers in India, the crops often escape damage from Cercospora musae and Colletotrichum capsici, respectively (Butler and Jones, 1949). It is indeed possible that under a very dense stand, dewfall is limited, but this also depends on the temperature and humidity at a given site (Rotem, 1978). These cases appear to be rather rare in properly cultivated crops. An exception to the general rule that a sparse stand inhibits disease by enabling better air movement within the plant canopy is the case of potato and tomato early blight in the Negev Desert of Israel. In this region, potatoes and tomatoes are predisposed to infection by sand storms, which wound the leaf tissue. A dense stand is, therefore, associated with less disease. Various cultivars of the same host species, for unknown reasons, exhibit different reactions to disease under different plant densities, as with Ustilago maydis on corn (Wilcoxson, 1975).

Variations in experimental techniques and in the prevailing weather conditions in a given area or season may explain, in part, some of the exceptional results. For instance, Berger (1975) reported that the spread of Cercospora apii on celery in Florida was partly inhibited by a sparse stand, whereas according to Strandberg and White (1978), it was not inhibited; however, the density of plants differed in the trials of these researchers. The reason for Kato's (1974) observation that rice blast in Japan is not affected by host density may possibly be attributed to the naturally humid habitat in the rice field. Finally, the rate of seeding did not affect Sclerotinia wilt of beans in Ontario because the increased distance of seeding resulted in denser individual plants with a microclimate favorable for spread of disease (Haas and Bolwyn, 1972).

In general, it is difficult to understand the reasons for diverging or exceptional effects of stand density on the spread of soil-borne pathogens, in which development is profoundly affected by biotic phenomena.

C. Fertilization and Irrigation

Fertilization in all climatic zones and irrigation in rain-deficient areas or seasons are the two major practices which influence the density of stand. Although deficiency or excess in most elements of nutrition may increase the severity of specific diseases, the action of excess nitrogen is the most common (Butler and Jones, 1949; Stevens, 1960). In most cases, nitrogen acts by changing the host's predisposition to disease, but the microclimatic effects created within the nitrogen-induced too-lush foliage cannot be ruled out.

Irrigation by sprinkling may affect disease directly; this can be minimized by the proper management of irrigation methods. More difficult is the management of the indirect effects of irrigation. Thus, whereas only sparse growth developed in the unirrigated crops, a dense stand – with all its microclimatic consequences – developed under irrigation (Rotem and Palti, 1969). Blad et al. (1978) tested the influence of furrow irrigation on canopy density and white mold development in two bean cultivars in the semi-arid conditions of Nebraska. The more frequently irrigated cultivar developed the denser canopy which was also cooler and wetter and had a higher disease severity.

D. Breeding

Breeding of ecologically appropriate cultivars is probably the most promising means of minimizing diseases by canopy modification. Numerous observations in the past have shown the value of cultivars with the desired growth habit. For instance, Gäumann pointed out in the 1940's that erect bean plants are generally less subject to anthracnose than squat, drooping cultivars. What he calls "standard" roses were less troubled by black spot disease than were bush roses; and potato cultivars with an open growth habit were less damaged by late blight than were dense cultivars (Gäumann, 1965). Foister (1946) mentioned the reaction of two potato cultivars to late blight in France. Biologically, the rather thinly growing Early Rose cultivar was as susceptible as the dense Saucisse cultivar. However, in the field, Early Rose was less subject to the disease because its foliage was not sufficiently luxurious to create a favorable microclimate (Foister, 1946).

Intentional breeding for ecologically suitable cultivars started much later. Coyne et al., (1974) described a cultivar of *Phaseolus vulgaris* with an architecture modified for improved air penetration within the canopy. This growth habit enables a quicker drying up of dew. As a result, the modified cultivar was less affected than the dense one by *Sclerotinia sclerotiorum*. This difference occurred in spite of an open space between the rows in plots with the compact plants (Coyne et al., 1974). The so-called "leafless peas" were bred in England; among other advantages, they are supposed to be less susceptible than the dense peas to attack by pea pathogens (Snoud, 1974).

Disease-related problems are associated with the dwarf cultivars generally attractive to breeders. In Israel, a dwarf tomato with short internodes and dense foliage was bred to protect the fruits from the hazards of radiation in the desert. This cultivar failed to serve its primary purpose because its compact foliage resulted in bad aeration, condensation of water within the foliage, and increased difficulty of penetration by fungicides within the foliage (N. Kedar, personal communication). The phytopathological effects of dwarf vs. tail cultivars and of dense vs. loose foliage in the case of *Septoria tritici* on wheat are described by Bahat et al. (1980) and discussed in Section IIIC. In some plant species, breeders can easily avoid ecologically unsuitable lines. For instance, in tomatoes the breeder can choose between parents with determinate and indeterminate growth habits, different leaf sizes, varying distances between internodes, etc. (R. Frankel, personal communication).

III. THE MECHANISM OF ACTION

As mentioned earlier, there are several mechanisms by which susceptibility to disease can be reduced in the dense stand. The first is analogous to the predator-prey principle well known in ecology (Odum, 1971). The second acts by improving the transfer of inoculum, and the third by improving the microclimate. There is also an indirect mechanism discussed in Section V. Often, the exerted overall effect results from the interaction of several mechanisms. However, each of them may operate under some conditions and not under others.

A. The Predator-Prey Principle

The only situations in which a deficiency in foliage leads to a decrease in the pathogen population are at the end of very severe epidemics, or when for biological reasons, the attacked organs have passed their stage of susceptibility.

A mechanism most analogous to the predator-prey mechanism and least associated with the microclimatic effects was described by Burdon and Chilvers (1975) for soil-borne *Pythium irregulare* in seedlings of *Lepidium sativum*. At high plant densities this damping-off disease was transmitted readily from one plant to another. At low densities, the greater distances between adjacent plants reduced the probability of successful transmission. It seems that this situation resulted from a coincidence of several factors, which would be rather rare in air-borne diseases. First, the buffered edaphic habitat of soil-borne disease is far less subject to environmental fluctuations than is the changing habitat of most air-borne diseases. Secondly, in contrast to the rapidly multiplying and dispersing populations of most air-borne pathogens, the soil-borne organisms multiply slowly and move slowly, step by step, from one site to another (Rotem, 1978). Thirdly, infection by *Pythium irregulare* is restricted to a short period at which the young seedlings remain susceptible (Burdon and Chilvers,

1975). Consequently, during the short period of host susceptibility, a crowded seed-bed with a steady soil climate created the rare situation in which disease development was governed by the density of stand, with little relation to microclimatic factors. Such an extreme situation is unlikely in the case of air-borne diseases.

B. Improvement in the Transfer of Air-Borne Inoculum

The following is a mechanism similar to that described in Section III A, but related to air-borne diseases and subject, in part, to micrometeorological influences.

In high foliage densities, the impact of wind on removal of air-borne spores is expected to decrease while the probability of their suitable deposition will increase. However, high winds will increase dispersal due to leaf flapping and rubbing of contiguous leaves. Pathogens which disperse easily (e.g., powdery mildews) are not expected to benefit from these mechanisms of dispersal in either high or low density stands. Pathogens which are not easily dispersed, such as *Rhynchosporium secalis* on barley, can be expected to improve their transfer in conditions of high wind and high foliage density.

C. The Microclimate Principle

The experiments of Berger (1975), Rotem and Ben-Joseph (1970) and Bahat et al. (1980) throw light on the microclimate principle involved in most air-borne diseases. Berger, in Florida, planted a standard number of celery seedlings in plots of different areas. Small amounts of *Cercospora apii* inoculum were already present in the seedlings and the disease started to spread immediately. At this early stage of plant development the very low epidemic rate was not affected by the density of stand. With the growth of the plants the crop microclimate changed according to stand density. The amount of inoculum increased and the difference in disease incidence between the dense and the sparse stands doubled within a period of several weeks. However, at a latest phase of the epidemic the progress of the disease in the dense plot decreased while at the same time it progressed slowly but markedly in the sparse plot. As a result, the final disease level was similar in all plots (Berger, 1975, and personal communication).

The equalization of the final level of disease in sparse and dense plots is common in heavy epidemics but may not happen during seasons less conducive to disease development. This was demonstrated by Rotem and Ben-Joseph (1970), in potato plots, maintained at different stand densities and subjected to late blight epidemics in the spring and autumn seasons in Israel. Spring, with its relatively short dewfalls, long days, and strong radiation, is conducive to epidemics which are light to moderate in severity. The prolonged dew periods, shorter days and relatively low radiation in the autumn, make this season most conducive to the heaviest epidemics possible. As shown in Fig. 1, the late blight epidemic was more severe in the autumn than in the spring. In both seasons, the dense plots were more heavily infected than the sparse ones, but the difference between disease levels in the dense and sparse plots was entirely different in the two seasons. Under very favorable conditions in the autumn, the final incidence of blight in the dense and sparse plots was 100 and 96%, respectively. In the less favorable spring season, the final incidence of blighted leaves was 74% in the dense plots but only 6% in the sparse ones. This means that canopy modification exerted by seeding density was more helpful in inhibiting late blight in the partly favorable meteorological conditions of the spring than in the very favorable conditions of the autumn.

In contrast to previous studies, Bahat et al., (1980) dealt with the effect of density on the spread of disease in individual plants. They studied the vertical progression of Septoria tritici in dwarf (70 to 80 cm) and somewhat taller (100 to 120 cm) wheat cultivars in Israel. All cultivars had the same genetic susceptibility. Infection started always in the lower leaves and moved

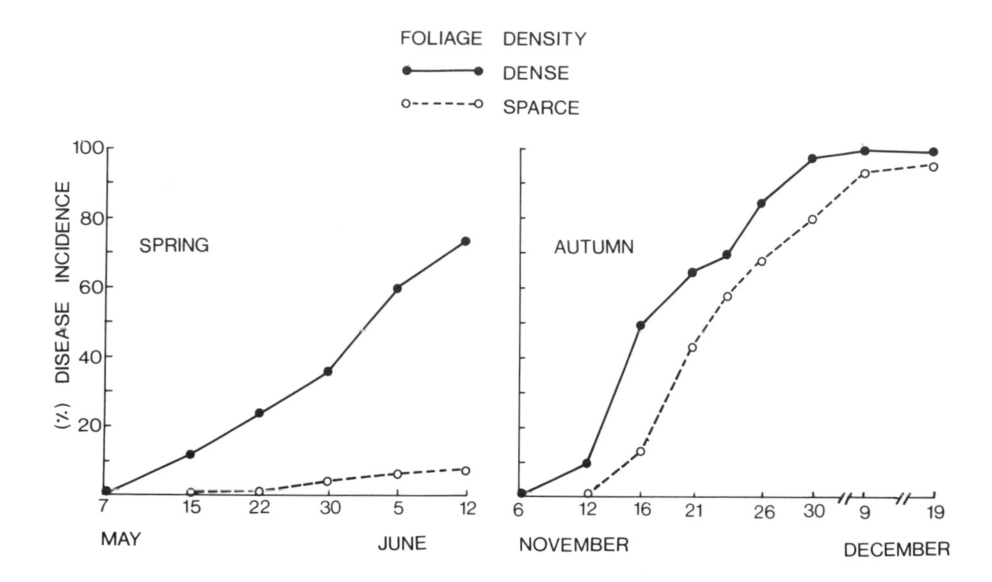

Fig. 1. The effect of foliage density on development of *Phytophthora infestans* in potatoes in Israel in the partly favorable spring and very favorable autumn season. (After Rotem and Ben-Joseph, 1970.)

upward, mainly by means of splash dispersal of spores. The vertical progress of the pathogen from lower to higher leaves was affected by the distance between consecutive leaves. Under conditions most suitable for *Septoria* epidemics, the rate of vertical progress was similar in the dwarf and taller cultivars, which means that under very favorable weather conditions the difference in foliage density is readily overcome by the pathogen. By contrast, in drier and hotter conditions, the pathogen progressed in the dense dwarf cultivars but not in the taller ones. The epidemiological disadvantages of the dwarf cultivars were thus more pronounced under unfavorable than under favorable weather conditions (Bahat et al., 1980). This is essentially similar to what has already been described in relation to the effects of plant density on the spread of potato late blight over an entire field.

IV. INTERACTION BETWEEN CANOPY MODIFICATION EFFECTS AND ENVIRONMENTAL AND BIOTIC FACTORS

It has already been mentioned that the desired effect of canopy modification on the crop climate is determined largely by the prevailing weather. Rotem and Palti (1980), who discussed this problem in relation to the effect of cultural practices in general, suggested a scheme which when adapted to canopy modification can be presented as follows: Modification of plant canopy as a means of minimizing a given disease is expected to be useless under conditions entirely unfavorable or extremely favorable for the pathogen, and has the best prospects for success where conditions are only partly favorable. In entirely unfavorable conditions, such as in a rainless season with high temperatures and low humidity, modification of the crop microclimate through canopy modification may be more pronounced than in less extreme conditions. However, under such conditions, pathogen development is restricted anyway, and the control effect of canopy modification is minimal. This occurs, for instance, with potatoes sown in the hot month of August in Israel. Until October, when the weather turns more favorable for late blight development, any modification of the plant canopy will not influence disease.

By contrast, the disease is always expected to develop under very favorable macroclimatic conditions. The effect of a microclimate somewhat modified by canopy modification in a constantly humid, rainy and mild area or season will then have little effect on the pathogen, as exemplified by late blight development in both the sparse and the dense plots in autumn in Israel (Rotem and Ben-Joseph, 1970) and by *Septoria* development in rather rainy and not too hot conditions in both the dwarf and somewhat taller wheat cultivars in the same country (Bahat et al., 1980).

However, when conditions for disease development worsen, the effect of canopy modification on control is expected to increase. This situation is illustrated by the effect of crop density on epidemics of potato late blight in the partly favorable spring season in Israel (Rotem and Ben-Joseph, 1970), and inhibition of Septoria in the taller wheat cultivar under hot conditions with little rain (Bahat et al., 1980). We may, thus, conclude that the efficiency of similar modification measures in the same host subjected to the same disease is conditioned by the specific local weather conditions and may differ widely under different environmental conditions. Considering these phenomena, a suggested schematic expression for canopy modification effects on disease development is depicted in Fig. 2.

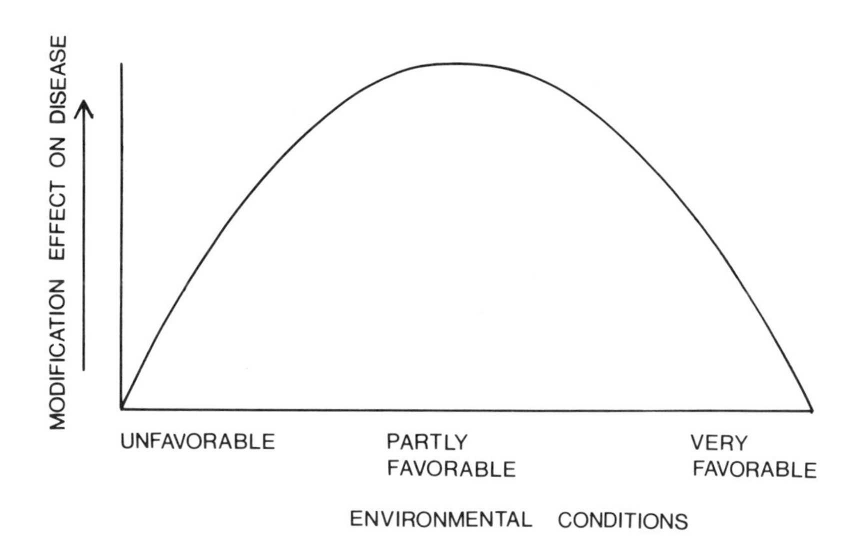

Fig. 2. Schematic illustration of the effect on disease by modification of plant canopy under different environmental conditions.

However, in addition to the meteorological characteristics of a given habitat, the effect of canopy modification on a disease is conditioned also by the specific characteristics of the pathogen involved. Rotem and Palti's (1980) principle for the effectiveness of cultural control measures will then work as follows when adapted to the action of canopy modification only: "Modification of plant canopy has better prospects of minimizing disease when the involved pathogen needs a long wetting period for infection, when its rate of inoculum build-up is slow, and when it is susceptible to environmental hazards." This is to suggest that the same canopy modification measured under identical environmental conditions may have different effects on diseases caused by pathogens with different characteristics. Probably the most important of these characteristics is the rate of inoculum build-up. With the slight changes induced in the crop microclimate, canopy modification is expected to be more effective against pathogens that multiply slowly, or early in the epidemic process, when the amount of inoculum is still small. With large amounts of inoculum, some spores may find favorable ecological niches in the plant canopy even under less favorable conditions. This may be deduced from Berger's trial which was directed specifically to the effects on canopy modification (Berger, 1975), and from many less specific experiments and observations (for pertinent literature, see Rotem, 1978; Rotem et al., 1978).

The rate of infection is largely dependent upon the duration of free-leaf moisture required by the spores of a given pathogen to complete the process of infection. Modification of the plant canopy is expected to act more effectively against pathogens which require long rather than short periods of wetness. Susceptibility of spores to environmental hazards is an important factor because of the many daytime hours which may pass between the early hours of the day when most of these spores disperse and the period when rain, irrigation or dew induce infection. Robust spores will survive that period well. Sensitive spores will then have a greater tendency to die when exposed, in canopy modification, to increased radiation and dryness. Wider spacing in the field, or the use of "thinner" cultivars which increase ventilation and shorten the wet period, especially in partly favorable environmental conditions, will enhance the success of control by canopy modification.

Pathogens like some powdery mildews able to develop under a wide range of temperature and humidity conditions are expected to be less effectively controlled by canopy modification than fungi with more specific requirements.

The theory behind ideas included in this section has been discussed by several authors among them Van der Plank (1963), Berger (1977), Rotem (1978), and Rotem and Palti (1980). However, in addition to the theory, we need better knowledge of the characteristics of different pathogens in order to make a more reliable prognosis for their expected reaction to canopy modification.

V. INTERACTION BETWEEN MODIFICATION OF THE PLANT CANOPY AND CHEMICAL CONTROL MEASURES

In addition to the mechanisms of action discussed in Section III, canopy modification may affect disease by increasing the efficiency of fungicidal action. This is achieved by the above-mentioned improvement in fungicide coverage of the sparser foliage and/or by widening the rows and thus enabling greater maneuverability of spraying equipment. Another possibility involves a combined action of those mechanisms discussed in Section III with those utilizing improved chemical control. To simplify, this means that although canopy modification substantially reduces the population of the pathogen, it is not in itself sufficient to bring about an economic reduction in pest development. Similarly, chemical control alone may not be effective under conditions of high inoculum pressure; moreover it does not achieve the desired level of control, or requires too dangerous or too expensive an amount of the chemical. Only a combination of canopy modification and chemicals brings about the desired results.

We lack specific experiments to assess the frequency of use and the degree of success of these possibilities. It is probable that canopy modification alone is the least frequently used and the least successful means of achieving economic control in the intensive kind of agriculture, but the best means in subsistence farming. Fungicidal action improved by canopy modification is apparently the most common and effective practice, especially in orchards and vineyards where pruning is needed to improve fruit quality. The combined effect of canopy modification and fungicidal treatments is often used with high-value crops such as flowers, where quality is a prerequisite for profit.

VI. CONCLUSION

This paper started with description of methods and ended with hypotheses. Due to differences in host and pathogen characteristics, variations in local climate, and unpredicted meteorological events, a particular method used to modify plant canopy may be more efficient in one place or crop than in another. In some cases, only a combination of several methods (e.g., using sparser cultivars and increasing the seeding rate) may bring about the desired effect. The number of possible combinations is large. In most cases, canopy modification alone is an insufficient or uneconomical means of preventing disease. However, through its direct action in reducing inoculum pressure, or through its indirect action in increasing fungicidal efficiency, it is expected to contribute to the success of disease control. This is a logical conclusion, but we lack experimental evidence of its correctness. We need more data to show the relative role of canopy modification in integrated pest control. It is difficult to obtain such data because of the varying effects of canopy modification under different microclimatic and biotic conditions. But it must be done.

REFERENCES

Asare-Nyako, A. (1977). In "Disease, Pests, and Weeds in Tropical Crops," (J. Kranz, H. Schmutterer and W. Koch, eds.), pp. 83-86. Verlag P. Parey, Berlin.

Bahat, A., Gelernter, I., Brown, M. B. and Eyal, Z. (1980). Phytopathol. 70:179-184.

Berger, R. D. (1975). Phytopathol. 65:485-487.

Berger, R. D. (1977). Ann. Rev. Phytopathol. 15:165-183.

Blad, B. L., Stedman, J. R., and Weiss, A. (1978). Phytopathol. 68:1431-1437.

Burdon, J. J. and Chilvers, G. A. (1975). Ann. Appl. Biol. 81:135-143.

Butler, E. J., and Jones, S. G. (1949). "Plant Pathology." MacMillian & Co., London. 979 p.

Coyne, D. P., Stedman, J. R., and Anderson, F. N. (1974). Plant Dis. Reptr. 58:379-382.

Foister, C. E. (1946). Botanical Rev. 12:548-591.

Gäumann, E. (1950). "Principles of Plant Infection." Hafner, N.Y. (Engl. ed.). 543 p.

Haas, J. H. and Bolwyn, B. (1972). Can. J. Plant Sci. 52: 525-533.

Kato, H. (1974). Rev. Plant Protect. Res. 7:1-20.

Llano, A. (1977). Plant Dis. Reptr. 61:999-1002.

Odum, E. P. (1971). "Fundamentals of Ecology." Saunders. 574 p.

Palti, J., Brosh, S., Stettiner, M., and Zilkha, M. (1972). Phytopathol. Med. 11:30-36.

Rotem, J. (1978). In "Plant Disease, Advanced Treatise" (J. G. Horsfall and E. B. Cowling, eds.), 2:317-337.

Rotem, J. and Ben-Joseph, J. (1970). Plant Dis. Reptr. 54: 768-771.

Rotem, J., Cohen, Y. and Bashi, E. (1978). Ann. Rev. Phytopathol. 16:83-101.

Rotem, J. and Palti, J. (1969). Ann. Rev. Phytopathol. 7: 267-288.

Rotem, J. and Palti, J. (1980). In "Comparative Epidemiology," (J. Palti and J. Kranz, eds.), pp. 104-116. PUDOC, The Netherlands.

Schieber, E. (1975). Ann. Rev. Phytopathol. 13:375-382.

Scott, M. R. (1956). Ann. Appl. Biol. 44:584-589.

Snoud, B. (1974). Euphytica 23:257-265.

Stevens, R. B. (1960). In "Plant Pathology," (J. G. Horsfall and A. E. Dimond, eds.), 3:357-429.

Strandberg, J. O., and White J. M. (1978). Phytopathol. 68: 223-226.

Van der Plank, J. E. (1963). "Plant Disease: Epidemics and Control." Academic Press, N.Y. 349 p.

Verhoeff, K. (1968). Neth. J. Plant Pathol. 74:184-192.

Wilcoxson, R. D. (1975). Plant Dis. Reptr. 59:678-680.

INTERACTIONS BETWEEN WEEDS AND OTHER PESTS IN THE AGRO-ECOSYSTEM

Robert F. Norris
Botany Department
University of California
Davis, California

I. INTRODUCTION

Weed control has been recognized as a required component of crop production systems since man first started to grow "crops" rather than gather wild plants. In the developing countries of the world, where modern weed science technology is not applied to any large extent, many people still spend much of their time weeding in order that they may eat (Holm, 1969, 1971). The reason for this is that removal of weeds is required because most crop plants will not survive without removal (by man) of the next level of successional species that invade land after man has cleared it to grow a crop. Wheat, for example, survived only three years when left untilled in a natural succession situation (see page 63, Lawes Agricultural Trust, 1974). Most annual row crops will produce little if any yield if weeds are not controlled, and will die out rapidly if natural succession is allowed to proceed; examples include cultivated varieties of sugarbeets, beans, tomatoes, potatoes, and most other annual crops. Tree crops would last longer but would not achieve the yield that man now demands of them. The weed science literature is replete with examples of the types of annual losses caused by weeds (see review by Zimdahl, 1979).

ISBN 0-12-332850-0

Hand pulling was the earliest method of control, coupled with various types of hand hoeing. Human energy was the main energy source for this type of weeding. Jethro Tull, in the late 16th century, developed the concept of cropping in rows with cultivation between the rows. Tull thus started the move to replace human energy with that of animals for controlling weeds. From the early horse-drawn "hoes" until the present day, cultivation has remained as a mainstay of most weed control programs. The sophistication of the cultivation equipment has increased in the last 50 years or so, and new types of equipment have permitted more rapid, more accurate, and more successful cultivation. During this time, we have almost completely substituted fossil fuel energy for that of animals.

The introduction of chemicals for weed control prior to the 1930's led to improved weed control in some crops, but it was not until the mid 1940's, when the phenoxy herbicides were discovered, that herbicidal weed control became widespread. In the intervening thirty years, the development of many herbicides has created a revolution in both how weed control can be achieved, and in the level of weed control that can routinely be attained (in relation to human effort expended). These developments have been one of the major factors that have freed the countries using the new technology from the requirement of having a large farm labor force that was employed primarily for weed control (Holm, 1971; Crafts, 1975). Herbicides have also permitted the development of new approaches to crop production, such as pasture renovation without cultivation, or the use of zero-till culture in several crops.

Weed control, up to the present, has remained an integration of techniques; herbicides have, however, become the mainstay of many weed management programs. It is now recognized for several of the relatively noncompetitive crops that they could not economically be grown commercially without herbicides (e.g., Hull, 1978; Young, personal communication). It is, however, still correct to state that cultivation remains a major component of most weed control systems; herbicides are used for weed control in the crop row coupled with cultivation for weed control between the rows. This is a common weed management practice in row crops. The situation is changing! Herbicides are making total reliance on chemical weed control more practical, thus reducing the need for cultivation. This, coupled with such factors as increasing costs of fuel for plowing and cultivation, reducing soil erosion, etc., may shift the economics in favor of this total reliance on herbicides for weed control. It, thus, seems likely that in some crops, complete dependence on herbicides may yet develop as we move from mechanical to chemical energy for vegetation management.

No attempt will be made here to further consider the role of various techniques for controlling weeds. The relationship of weed control to systems for the management of other pests will be addressed with emphasis on microhabitat alteration where such have been documented. It will be assumed that weed control is an essential component of the crop production system, and will, therefore, be practiced.

Weed control can interact with other components of pest management systems in three ways:

1. Habitat modification due to vegetation removal.
2. Direct and indirect responses of nontarget organisms to herbicides.
3. Alteration of weed growth due to nonweed pest control.

The most important aspect is that of indirect effects through habitat modification. Weed control necessarily removes some species of plants and leaves others; this alters the ecosystem (crop or otherwise). These changes in vegetation extend beyond the immediate response of decreasing the competition experienced by the crop plants. Changes in vegetation can alter the populations of insects, pathogens, nematodes, and the soil fauna and flora; these changes in population can be of no significance, they may be beneficial, or they may be detrimental. For pest management to be fully effective, these interactions must be known so that the best management decisions can be made. These types of indirect interactions occur regardless of the type of weed control method employed, be it cultural, mechanical, biological, or chemical.

The second type of interaction that can occur between weed control and other types of pest management is due to the use of herbicides; these chemicals can themselves have direct effects on nontarget organisms, such as insects, fungi, and nematodes, present in the ecosystem. Effects can also occur through herbicide modification of the physiology of the host plant, such that insect or disease organism growth is changed. Such effects can again be detrimental or beneficial in a pest management program.

The third interaction involves the reciprocal interactions that can also occur. The effects of crop damage by other pests can alter the severity of weed problem; mitigation measures applied for nonweed pests may have direct or indirect effects on weed growth and biological control, and pesticides may interact to modify the efficiency of one or all of the components of a mixed spray application.

II. WHAT IS PEST MANAGEMENT?

Integrated pest management has been defined as a "system that, in the context of the environment, and the population dynamics of the pest species, utilizes all suitable techniques and methods in as compatible manner as possible that maintain the pest populations at levels below those causing economic injury" (Glass, 1975). The Intersociety Consortium for Plant Protection (ICPP) has defined integrated pest management as "the use of multiple tactics in a compatible manner to maintain pest populations at levels below those causing economic injury while providing protection against hazards to humans, domestic animals, plants, and the environment."

None of the concepts are new; many have been employed intentionally or accidentally since man first started to control pests. This is especially true of weed control, where crop rotation and cultivation have been employed for several hundred years. In the 1950's and increasingly in the 1960's the concept of combining strategies, and attempting to use those that were least disruptive to the environment, began to take on increased importance. Due to the problems of insects becoming resistant to insecticides, due to resurgence problems, due to biomagnification of certain insecticides in food chains, and due to persistence of some insecticides, the adoption of IPM strategies has received the most attention for insect control (Stern, et al., 1959; Smith and Reynolds, 1966; Kilgore and Doutt, 1967; Huffaker and Croft, 1976; Smith and Pimentel, 1978; Thomason, 1978).

It is generally conceded (e.g., Glass, 1975; Huffaker and Croft, 1978), however, that if IPM is to achieve its ultimate goal, it must be utilized in the context outlined above. The ICPP has defined integrated as "a broad interdisciplinary approach taken using scientific principles of plant protection to fuse into a single system a variety of management strategies and tactics." They also define pest to "include all biotic agents (i.e., insects, mites, nematodes, weeds, bacteria, fungi, viruses, parasitic seed plants, and vertebrates) which adversely affect plant production." It is thus clear that IPM will require that weed, pathogen, nematode, and vertebrate control must be part of the system. It is stressed by several authors that the ultimate pest management system will consider all interacting factors in the agro-ecosystem before employing any particular pest control technique, as outlined by Steiner (1966). This would require a complete understanding of the ecosystem! It seems highly unlikely that this can ever be achieved.

The general types of interactions that should be considered as part of a pest management program are depicted in Fig. 1 (modified from Steiner, 1966). This shows the complexity of the

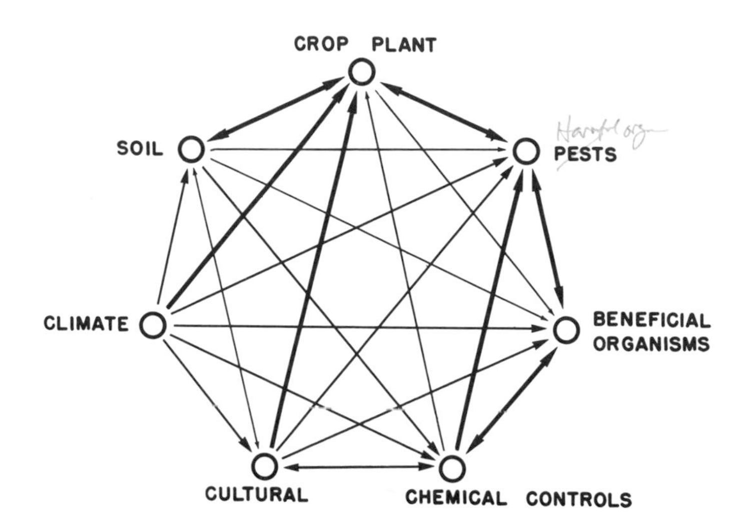

Fig. 1. Typical interacting factors that need to be considered as a part of a general pest management program (after Steiner, 1966).

interactions that can be occurring. Yet weeds did not enter in the discussions by Steiner (1966). If pest management is to approach the multidisciplinary level that is now generally recognized to be necessary then the "pest" box, as shown in Fig 1, will have to be expanded to allow for a consideration of the interactions that can occur between types of pests (i.e., weeds affect insects, insects transmit viruses, certain fungi parasitize nematodes, weeds interact with vertebrates, etc.). The "pest pentagon" approach is proposed (Fig. 2); this would allow for consideration of these types of interactions.

III. WEED SCIENCE IN PEST MANAGEMENT PROGRAMS (OVERVIEW)

Weed science suffers from an image problem! People accept weeds as part of life. Many weeds actually have beautiful flowers (e.g., larkspurs – yet these are very poisonous to cattle) and to

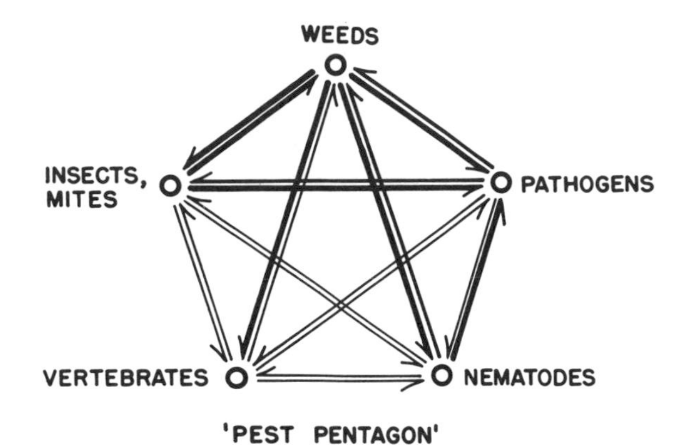

Fig. 2. The 'pest pentagon'; a diagrammatic presentation indicating the potentials for interactions between different types of pest organisms.

the layman are nothing more than wildflowers. And who wants to kill wildflowers?!

Insects, by contrast, are repulsive to most people and, therefore, create strong negative feelings; they must be controlled. This fact was emphasized recently by Olkowski and Olkowski (1976) who reported on the reaction of urban dwellers to insects; in their words many people show an irrational entomophobia. In discussing pest control in home vegetable gardens, this author has experienced considerable difficulty in convincing lay people that insects like lacewings, and syrphid flies, and spiders (most) etc., are in fact useful. Many diseases cause desirable plants to look sick, collapse, rot, or die; they too must be controlled. But weeds? Most people can "take them or leave them"; and leave them is probably the typical reaction.

Weed science suffers another problem; weeds do not cause cosmetic injury to crops. Insect damage, insects themselves, or disease blemishes mean reduced urban consumer acceptance of produce. Yield loss caused by weeds is of no significance to supermarket shoppers because they do not see the loss; but they do see the cosmetic insect or disease damage. An example might be leaf miner attack which can make celery unmarketable by current standards, but without actually reducing the yield to any great degree. The urban consumer can, therefore, relate to insect or disease control, but is totally unaware of the significance of weeds.

It can be argued that weed science has been too effective; when one travels in agricultural areas in the USA and Western Europe most of the fields are usually more or less free of weeds. So weeds are not really a problem! The casual observer does not see the management skills of the farmer, the time and effort that he and his workers had to use to achieve the clean field, or the chemicals that he employed. The farmer knows, however, that the economic losses from uncontrolled weeds can reach 100%; that is why so much effort is put into weed control programs.

The importance of weed control in practical pest management can also be judged as a reflection of the quantity of herbicides sold; recent figures for the USA indicate that over 57% of all pesticides sold were herbicides (Fowler and Mahan, 1978) and in the U.K. the herbicide proportion of pesticide use exceeds this figure (Mellanby, 1980). These figures did not reflect all the effort and energy used for weed control in the acres that were cultivated or the time and human effort used for hand hoeing.

The role of weeds in pest management is unique among organisms considered by man to be pests. This is most easily understood in the context of energy resource flow in the agro-ecosystem, similar to the original discussions of trophic dynamics by Lindeman (1942). Weeds are the only pest organisms that are primary producers; this contrasts with all other pest organisms which are primary, or even secondary, consumers (Fig. 3). The presence or absence of weeds in the agro-ecosystem, as they can serve as food sources for both pest and beneficial organisms, can thus have interactions with all phases of insect/mite, nematode, pathogen, or vertebrate pest control. The presence of weeds can, additionally, alter the microenvironment (see later section), and can serve as shelter for pest or beneficial organisms; the latter possibilities increase the complexity of potential interactions between weeds and other pest organisms. These are reasons why the development of a "pest pentagon" philosophy is essential.

From the viewpoint of overall crop production, weed control is an absolutely essential pest control component; if weeds are not controlled there will be, in many cases, no crop to protect from insects and diseases, as recently noted by Phillips et al. (1980). Because weeds are primary producers (Fig. 3) weed control becomes a component, intentionally or unintentionally, of IPM programs. It is, thus, vitally important that weed management be considered in such programs due to significance that weeds and their control have on other pests in the agro-ecosystem. The remainder of this chapter will emphasize and document the significance of these interactions.

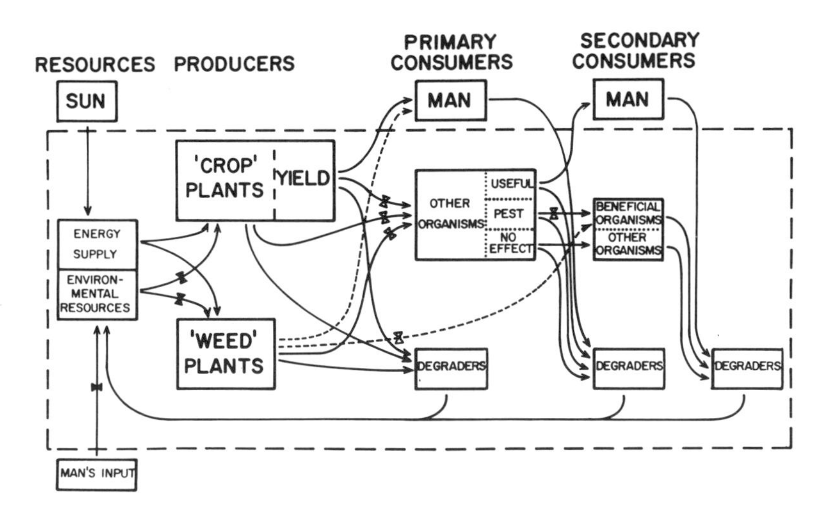

Fig. 3. Diagram depicting the position of weeds in the flow of energy/resources in the agro-ecosystem (solid 'valves' indicate direct effects on primary producers (weeds); open 'valves' indicate direct effects on consumers (nonweed pests).

IV. INTERACTIONS DUE TO PRESENCE OR ABSENCE OF WEEDS

A. Overview of Weed Modification of the Microhabitat

Plants modify the environment in which they live. They absorb and utilize certain fractions of solar radiation, they take up water and give it off into the atmosphere, they take up nutrients and redistribute them in the ecosystem, and they physically alter air movement. There are profound consequences of these microhabitat modifications on other organisms present in the ecosystem; this may be to the extent that the other organisms could not survive without the influence exerted by the presence of plants. Weeds are plants, and thus, their presence or their removal will alter the habitat for these other organisms.

Weeds frequently can occupy a relatively large component of the vegetation of the particular ecosystem in which they are growing. Seedling weeds in a widely spaced row-crop might contribute as much as 95% of the plant biomass present in the field, or weeds may represent the total transpirational component of a deciduous orchard ecosystem when the overstory trees are dormant or early in the spring before tree-growth resumes. In noncropped areas, such as ditchbanks, fences, and roadsides, the only vegetation present may be weeds. It is thus apparent that weeds and their control can modify the microclimate of the ecosystem, but the actual extent to which modification occurs will depend on the ratio of weed to nonweed vegetation present in that particular ecosystem. These modifications can include:

Light – quantity/quality: can have effects on all types of pests, but probably of greatest significance to other primary producers.

Water – in soil: of prime importance to other plants but can have direct effects on other soil-born organisms, especially pathogens and nematodes, as certain species require free water for specific stages of their life cycle.

Water - in air: affects organisms sensitive to relative humidity; probably most significant for certain pathogens.

Temperature - in soil/in canopy: environmental temperatures regulate the activity and rate of development of all poikilothermic organisms. This is the basis for the degree-day concept for predicting pest population development. Altered canopy and/or soil temperature can thus alter the rate of pest population change/development.

Shelter: this is a combination of the above parameters, can be critical for many vertebrate and invertebrate pests in that they will not survive in the open.

Nutrients - in soil: of greatest importance to other plants as they are the only direct users.

Nutrients - in plants: altered soil nutrient changes availability to plants, which can then modify the appearance/ food value of the plant to other pests and can alter the ability of the plant to tolerate the pest attack.

The magnitude of such microenvironment changes, due to weeds, have been poorly documented. In the ensuing sections of this chapter, many interactions between weeds and other pests are noted, but the mechanism(s) for the interactions have, in most cases, not been determined. The interactions could have been mediated through modification of the microhabitat to provide conditions suitable for the pest organisms; from increased relative humidity aiding in spore germination to provision of 'shelter' for rodents. The interactions could also be attributable to the provision of a food source by the weeds. The two possibilities may not always be separable, and may themselves be interacting.

B. Role of Weeds in Habitat Diversity

Species diversity probably has no advantage for weed control, and can actually make control more difficult. If weeds are left in a field for pest management reasons, even if they are present at a density that is sufficiently low they do not compete with the crop, it must be remembered that weeds set seed, and that a low density of a weed that causes no problem in this years crop could be a serious problem in next years crops. This fact seems to have been overlooked by those who advocate leaving weeds so that beneficial insect populations can be maintained (see below). The question could be asked "does monoculture of crops increase weed control problems?" Except in the case of problems caused by tolerant weeds, it seems unlikely. Monoculture may in many cases make weed control easier because of such factors as longer 'runs' for cultivation equipment and easier selectivity problems of herbicides to crop.

When habitat diversification is considered in the light of insect and disease management programs then the presence or absence of weeds can become significant (see reviews by van Emden and Williams, 1974; Alteiri et al., 1977; Zandstra and Motooka, 1978; Altieri and Whitcomb, 1979a and 1979b). Diversity of plant species in an agro-ecosystem has been demonstrated to lead to increased stability of insect populations for several crops (Pimentel, 1961; Root, 1973; Perrin, 1977; Mayse and Price, 1978). In these cases, the magnitude of attack by pest insects was reduced; this was attributed to increased plant diversity reducing the invasion or phytophagous insects and mites, or to increasing the numbers of beneficial insects present which then lowered the pest species populations. Weed species in the agro-ecosystem can act directly as a food source for beneficial insects, can act as a food source for phytophagous insects on which the beneficial insects can feed and maintain their population, and can modify the habitat such that beneficial insects can survive at times when environmental conditions would otherwise preclude their survival. There are numerous reports of weeds either decreasing pest insects, and/or increasing beneficial insects; the reasons why the weeds affect the population were frequently not documented.

Weeds on orchard floors have been shown to increase beneficial mites (Browne, 1975; Gerawan, 1975; Croft and McGroarty, 1977). Johnsongrass in grapes has likewise been shown to increase predatory mites (Flaherty, 1969). Two *Euphorbia* sp. in sugar cane helped maintain the Tachinid parasite *Lixophaga spenophori* when not in its parasitic stage (Leeper, 1974). Several insect problems have been reported to be reduced by weeds present in cole crops (Dempster, 1969; Smith, 1969, 1976a, and 1976b; Dempster and Coaker, 1974; Cromartie, 1975; Theunissen and Den Ouden, 1980). Leaf hoppers were reduced in beans by several weeds (Altieri et al., 1977) and by blackberries in grapes (Doutt and Nakata, 1973). Scale insects have been reported to be reduced in *Citrus* when weeds were present (Jordan et al., 1973), and *Poa trivialis* has been reported to increase aphid predators in cereals (Vickerman, 1974). These are but examples; it is likely that interactions such as those noted above occur quite frequently; and Zandstra and Motooka (1978), and Altieri and Whitcomb (1979a) reviewed additional examples of beneficial weed/insect interactions.

Removal of weeds from crops has been shown to increase the number of insects attacking the crop; this is the reciprocal to the preceding discussion. Aphids were greater in weed free brussel sprouts (Smith, 1969; Theunissen and Den Ouden, 1980), *Empoasca* leaf hoppers were higher in weed free beans (Altieri et al., 1977), and the larvae of Egyptian alfalfa weevil were higher in weed free alfalfa (Norris and Cothran, 1974). Root (1973) felt that "resource concentration" could explain the higher attack of insects in pure crop stands, in that herbivores are more likely to remain and build up in pure stands of the host. Other authors have felt that the background perceived by the insects will be different for a weed free versus a weedy crop stand; this is the plant apparency argument suggested by Feeny (1976). Basically this latter statement suggests that insects have greater difficulty finding their preferred host when there is a mixture of plant species present.

Increasing vegetative diversity can lead to increased insect attack, as weeds can also act as a food source or suitable habitat for many pest insects that later attack the crop. This aspect of habitat diversification is frequently ignored by those advocating its usefulness. Removal of weeds has thus been shown to decrease insects such as cutworms and aphids in Citrus (Jordan et al., 1973), to substantially lower cutworms in asparagus (Tamaki et al., 1975) and to decrease several insects in barley (Vickerman, 1974). London rocket has been shown to be a preferred host of the false chinchbug; the insect can build up on the weed then move to, and damage, grapes (Barnes, 1970). At the time that London rocket is actively growing, it may be the only food source and shelter in the vineyard and its presence thus encourages the false chinchbug; an irony of this is that the weed is often allowed to grow as a cover crop. Leigh (1961) had earlier noted that the false chinchbug built up on several weeds, including London rocket, shepherd's purse, and Lepidium nitidum; the insect completed at least one generation on the weeds prior to moving to, and causing severe damage in cotton. Green peach aphids will live and build up on many species of weeds; in orchards this insect has been recorded at over 5.3×10^8 aphids/ha on weeds. These could then move to the trees as the weeds died (Tamaki, 1975; Tamaki and Olsen, 1979; Tamaki et al., 1980). Cutworms (Agrotis sp.) in Florida prefer spiny amaranth on which to feed and oviposit; large numbers can build up if the weed is present in adequate numbers (Genung, 1959). Sherrod et al. (1979) also showed that cutworms, in Illinois, could build up on several weed species; the moths are actually attracted to weedy fields. These examples show clearly that weeds can act as a source of invasion of pest insect species. It is also recognized that when weeds are controlled, insects that have built up on the weeds in a crop can move to, and attack, that crop. Genung (1959) and Genung and Orsenigo (1970) note several examples of insects that attack either the following, or adjacent crop when weeds were removed. The latter authors suggested several tactics that could be used to minimize the problems caused when weeds are controlled, such as: discing early and baiting before sowing the crop; avoiding discing during critical bloom phases in the crop, if appropriate; elimination of wild hosts; and others. Genung (1959) has noted that control of spiny amaranth in beans resulted in almost complete crop loss due to cutworms being driven from the weeds; the weeds must never be allowed to grow or cutworms must be controlled prior to control of the weeds. Barnes (1970) noted that cultivation in late April for control of London rocket forced the false chinchbug to move to the grapes where it caused extensive damage; cultivation prior to bud break of the grapes would avoid the problem. Altieri and Whitcomb (1979b) also suggest that cultivation can be timed so that it provides maximum benefit in insect management. There are many unreported instances of this type of interaction between weed control and insect attack; further examples include: *Lygus* bug on pigweed in cotton and beans, and flea beetles on nightshade in tomatoes. It has been noted by Leigh (personal communication) that *Lygus hesperus* can build up on pigweed in maturing potato fields; when the potatoes are harvested the insects migrate to the adjacent cotton and other crops. It is thus clear that although increased plant diversity can decrease insect problems, there are also many situations when such diversity increases the insect problems.

The role of species diversity is less clear in the case of pathogens and nematodes. Lack of large areas of a single crop will reduce the spread of a disease over the whole area, but within a field the evidence is predominantly that weeds lead to greater disease and nematode problems. There is little evidence in the literature indicating that increased diversity of plants results in decreased disease problems; work in cereals has indicated that weeds may lead to decreased incidence of rust. There are, however, numerous references to diseases spreading from weeds in a crop back onto the crop, or being carried over the next crop; examples include transmissions of ergot from weedy grasses back to cereals (Mantle et al., 1977), transmission of Rhizoctonia solani from potatoes to weeds and then back to potatoes (Griesbach and Eisbein, 1975), and infection of several fruit and vegetable crops by diseases carried by weeds (Kavanagh, 1969, 1974). Johnson (1977) showed that weeds in cotton were responsible for survival of Verticillium dahliae in cotton fields; and that microsclerotia could be produced in live and senescent weed tissue; the weeds did not show symptoms of the disease. Weeds have also been shown to result in increased internal seed-born fungi in soybeans (Dhingra and da Silva, 1978); changes in the microclimate of the crop as a result of the presence of weeds was considered as a possible cause for the increased fungus attack. The parasitic weed dodder has been shown to be capable of transmitting pathogenic organisms from diseased to healthy crop plants (Benhamou et al., 1978). There are numerous examples of weeds acting as hosts for plant parasitic nematodes (see bibliography by Bendixen et al., 1979); some specific examples include nutsedges hosting nematodes in cotton (Bird and Hogger, 1973) and in rice (Hollis, 1972) pigweed and crabgrass hosting several species of nematodes (Miller and Ahrens, 1969). The latter authors noted that the weeds maintained high populations of the nematodes. It would thus appear that species diversity probably leads to increased levels of inoculum. Once a fungus or nematode has attacked a monoculture crop it will, however, spread more rapidly if there are only host plants present.

An implication of weed management programs in relation to species diversity that generally receives little attention is that of altered wild game habitat. Removal of weed vegetation can remove both shelter and food sources of wild game (Way, 1968). The latter can take the form of less seeds available, but can also be due to lowered numbers of insects available, such as has been reported in England in the case of partridges (Potts, 1970; Vickerman, 1974). Whether pest management will concern itself with this problem remains to be seen.

There is a considerable dilemma for the pest manager in the types of interactions just noted; is it better to leave the weeds, or some of them, or is it better to control the weeds and then look for alternatives for the management of insect and other pests (e.g., breed resistant varieties)? As noted by Smith (1969, 1976a) for brussels sprouts, weeds must be controlled otherwise there would be no crop to protect from aphids; Dempster and Coaker (1974) likewise noted the problem of competition from the species used to gain diversity. Gregory and Musick (1976) emphasized that weeds must be controlled in a zero-till situation, and that insect attack will have to be controlled by techniques other than tillage for weed control. Theunissen and Den Ouden (1980) pointed out the dilemma that in order to grow weeds to help maintain beneficial insects some potential crop acreage must be sacrificed; the economics of reductions in insect management costs will have to be judged against reduced productivity per unit area. In his review article, van Emden (1970) stated that the small benefits derived from insect control are far outweighed by the losses due to weeds. Another factor that enters into these considerations is the fact that weed free crops can often tolerate higher insect attack than a crop that is having to compete with weeds (e.g., alfalfa tolerates higher incidence of Egyptian alfalfa weevil when weed free; Norris, unpublished data). Much more information is required concerning these types of interactions if we are to make the most rational choices in future pest management programs.

Various authors have suggested that either mixtures of crops can be grown (Perrin, 1977) or that weeds be grown and or managed in such a way that they maintain beneficial insect complexes (Perrin, 1975; Altieri and Whitcomb, 1979b; Theunissen and Den Ouden, 1980). These suggestions must be judged extremely carefully in relation to the type of crop being grown. Any increase in plant diversity will substantially complicate weed control in that crop. The above authors did not explain how weed control can be attained other than by the use of hand labor, as mixed cropping will in most cases preclude the use of herbicides. Another problem is that a weed species that might be deemed beneficial in one crop might be a serious problem in another rotational crop. If the "beneficial" weed produced seeds, or vegetative perennating structures, then it would be unwise to allow it to grow for insect management reasons. The problem with weed population dynamics is that is is explosive, due to the large numbers of seeds that a single weed specimen can produce (see Section VIII). This coupled with the fact that seeds can remain viable in the soil for years makes the decision to leave weeds for insect control purposes extremely difficult to make. It is also clear that the presence of weeds in a crop, aside from the losses directly attributable to the weeds (competition, space allowed for growing the weed), can have serious repercussions on the management of pests other than arthropods. It is essential that the pest pentagon (Fig. 2) philosophy be considered when discussing the merits of habitat diversity for the management of a particular pest if undesirable interactions with other pests are to be avoided. It would be ironic if pigweed left for insect control reasons were at the same time increasing the population of a parasitic nematode, or serving as a reservoir host for a virus!

As a final note for weed control in crops versus the role of habitat diversification, we have to recognize that:

a) Monoculture of crops will be needed for large scale food production if the work force on the farm is low.

b) Weed control will be required, and thus the diversity of habitat will be decreased, regardless of the means employed for the weed control.

c) In some situations, it may be possible to selectively remove competing vegetation yet leave sufficient plants to maintain desired insect populations; this may be particularly true in perennial crops such as orchards and vineyards.

C. Role of Weeds in Sanitation Programs

Sanitation is a major factor in any pest management system, in that it reduces or stops invasion, or reinvasion, of plant pests. Virtually all sanitation procedures require destruction of plant material that is either directly (as a weed) or indirectly (as a host or overwintering site for insects, pathogens, nematodes, or vertebrates) involved with pest problems in adjacent crop land. Destruction of undesirable plants reduces the seed supply for weed spread (e.g., Kelley and Bruns, 1975). Sanitation also reduces or eliminates the habitat for overwintering insects and diseases, or destroys plants on which pest populations can build up before moving onto crops (Barnes, 1970; Yarris and Wharton, 1978) and removes shelter for pest organisms.

There are numerous records of weeds acting as sources of virus diseases which are then spread back to the crop (see reviews by Heathcote, 1970, Duffus 1971). The literature contains many new references every year that report weeds as hosts of viruses that attack crops (Weinbaum and Milbrath, 1976; Magyarosy and Duffus, 1977; D'Arcy and de Zoeten, 1979). Weeds are frequently symptomless carriers of viruses, an example would be common groundsel, stinging nettle, shepherd's purse and sowthistle infected with cucumber mosaic virus (Tomlinson et al., 1970). If the weed is perennial then the virus source is perpetuated; in other cases the virus can be transmitted in the seeds of weeds (Tomlinson and Carter, 1970). Removal of weeds around crop areas is a well-recognized component of virus management programs; reductions of beet western yellows virus incidence in sugar beets when weeds were killed on adjacent ditches is a good example of this type of program (Powell and Wallis, 1974).

Weeds also act as sources of disease innoculum, which then can reinfect crops (Greisbach and Eisbein, 1975; Mantle et al., 1977; Bains et al., 1978; Hepperly et al., 1980). The role of weeds as alternate hosts for nematodes has also been well documented (Bendixen et al., 1979; Franklin, 1970). Weed removal can thus play a major role in a sanitation program aimed at control of diseases and nematodes.

Weeds can also act as a shelter and refuge for vertebrate pests; this is especially true of ditchbanks and fence lines where such pests as ground squirrels and meadow mice can build up.

The fact that weeds can act as an overwintering site, and as alternate hosts, for insects is also well documented (van Emden, 1965, 1970; Genung and Orsenigo, 1970; Khattat and Stewart, 1980). A particularly striking current example is the hosting of the sugarbeet leaf hopper by many winter annual weeds in California. This then permits the reintroduction of curly top virus to several crops in the San Joaquin Valley (Magyarosy and Duffus, 1977). These authors even indicate that the weeds may allow the virus to develop more virulent strains.

It is also well documented that weeds can act as a source of beneficial insects (van Emden, 1965a, 1965b, 1970; Perrin, 1975; Syme, 1975; Croft and McGroarty, 1977; Zandstra and Motooka, 1978; Altieri and Whitcomb, 1979a, 1979b). Weeds can provide shelter from environmental extremes which can drastically alter insect population dynamics (Smith, 1954; Messenger, 1971); the Tachinid parasite (*Lixophaga spenophori*) could not survive in the open and weeds provided shelter (Leeper, 1974). Weeds also act as a direct food source for certain beneficial insects, especially those that feed on nectar during part of their life cycle (Chumakova, 1960; Leeper, 1974; Syme, 1975). It is interesting to note Lindeman's (1942) suggestion that the further an organism is removed from the primary producer the less likely it will be fully dependent on the preceding trophic level. Beneficial insects often occupy two trophic levels and thus feed directly on weeds (primary producers) as a food source (e.g., nectar) at one stage of their life cycle, while feeding on phytophagous insects (primary consumers) at another phase of their life cycle (Fig. 3). Alternately weeds can act as an indirect food source in that they support a population of insects on which beneficial insects can feed, permitting both population carry-over or population build up when the crop is absent (van Emden, 1965b; Doutt and Nakata, 1973; Perrin, 1975; Altieri and Whitcomb, 1979b). The utility of leaving weeds on which the beneficial insects can build up is questionable if the weeds are either serious problems in adjacent crop land, or are considered noxious by regulatory agencies (Syme, 1975).

The preceding discussions show that weed removal in a general sanitation program can have both beneficial and detrimental effects within an overall pest management program even when judged solely on the basis of the interactions with insects. This dilemma is clearly demonstrated by the fact that it is desirable to kill Russian thistle in a beet leaf hopper/curly top control program, but that it is useful to leave blackberries in the vicinity of vineyards in a grape leaf hopper control program. When the role of weed removal in sanitation program is additionally judged in relation to pathogen, vertebrate pest and nematode management, due to the almost universally detrimental affects of weeds as alternate hosts, it is clear that no easy decision can be made regarding whether to leave weeds as part of an IPM program; the gain from increased beneficial insects may be offset by increases in disease, nematode or vertebrate pest problems. Some authors have discussed the significance of weeds in relation to insects only. It is obvious from the foregoing discussions that the overall situation will have to be judged by the pest pentagon type of considerations if we are not going to "lose on the merry-go-round what we gain on the swings." It seems that each situation will have to be judged on its own merits, using as broad-based inputs as possible.

Other areas of pest management that may relate to sanitation include such practices as crop desiccation. This can stop insect development on the crop, and facilitate plow-down of the trash. The use of defoliants in cotton can aid in the control of pink bollworm (Adkisson, 1972). The use of crop-free periods is a sanitation practice used to combat disease, such as viruses in sugarbeets and celery, nematodes, and in some instances insects. These practices require that wild hosts of the pest organisms be controlled if the sanitation programs are to be effective.

D. Role of Weeds in Rotations

Rotations are used for all types of pest management programs, including control of weeds, insects, diseases, and nematodes. There has been a trend away from rotations for pest management reasons in recent years; this has placed added reliance on pesticides. The major reason for this trend is economics. Rotation for pest control may require the growing of a less profitable crop in order that a particular pest problem can be controlled. In the cost/price squeeze many growers have opted to grow the more profitable crop, and to rely more heavily on pesticides to control the problems that used to be controlled by the rotation. It seems unlikely that this situation will change unless the economics of crop returns to growers changes and/or political decisions force changes in cultural practices. These statements are particularly applicable to the use of fallow in a rotational scheme. Rotations can be very effective for weed control as they permit the farmer to use different weed control measures or different herbicides in different crops, and thus stop any one, or group of, weed species becoming a serious problem.

The role of fallowing for pest management needs to be covered in greater detail. Provided weeds are controlled, this technique can a) reduce weed populations. This applies especially to those species that are hard to control in crops. The lack of the crop presents techniques for weed control not available when the crops are present, such as flooding, fire, general cultivation, dessication, and the use of nonselective herbicides. b) Reduce pathogens and nematodes (Zadoks and Schein, 1979), and insects (soil borne mainly). The technique is only effective if host plants (weeds) are not allowed to grow. Weeds growing in rotations for control of tobacco root knot nematode negated the value of the rotation completely (Clayton et al., 1944) and weeds in cotton were the reason that rotations failed to control *Verticillium* wilt (Johnson, 1977). c) Improve soil moisture status. In dryland farming fallowing can build up the soil moisture level. This is only possible if weeds are not allowed to grow.

The techniques used for weed control during fallowing include cultivation and/or herbicides. Cultivation has been widely used and is especially effective in a dry, hot climate. It is less effective, or even not practical, in a damp, cool climate. It has the serious disadvantages that it requires relatively large amounts of energy (due to quantity of soil moved), may contribute to soil compaction, and can lead to increased soil erosion. Herbicides reduce or eliminate these problems. Chemicals for this type of use must be nonselective and short lived in the soil; selective herbicides would rapidly build up tolerant weed species, and long residual herbicides might prove to be a problem in subsequent cropping.

To place fallowing in a pest management perspective, the farmer can only afford to put his land out of production if the problem that fallowing will solve is more costly than the loss of a crop. This situation is not likely to occur often at current prices, land taxes, rents, etc.

E. Role of Weeds in Biological Control

Biological control is considered to be a major approach for management of pests. It is probably more applicable to insect management due to the multitude of prey/predator relationships that exist. It does, however, also play a useful role in control of weeds, nematodes, and diseases. It is perhaps appropriate to point out here that biological control is frequently equated with IPM; as can be seen from the definitions this is wrong.

Weed management can have interactions with biological control programs particularly those aimed at insect control. This can be indirect through habitat modification, or through direct action of herbicides against insects. The role of habitat modification has already been noted to affect many predators and parasites of crop plant pests (see earlier section on habitat diversity). Additionally, weeds can be a source of nectar for many species of beneficial insects (Syme, 1975; Theunissen and Den Ouden, 1980). This has raised an interesting dilemma. Attempts are being made to establish biological control of yellow starthistle, yet bee keepers do not want this troublesome, poisonous, weed controlled as it is a good nectar source. If weeds are an important part of the life cycle of a particular beneficial insect, such as the blackberry which has a leaf hopper on which the grape leaf hopper parasitic wasp *Anagarus epos* can overwinter and build up (Doutt and Nakata, 1973), the management to maintain the weed so that it does not interfere with crops, yet will support the benefical insect, should be investigated as a management program. In the light of the established role of weeds for beneficial organisms (see section on habitat diversification), then each situation should be judged on its own merit. As pointed out previously (van Emden, 1965b), the balance between weeds acting as hosts for detrimental and beneficial organisms means this type of management decision is hard to make. Any weed removal practice can thus impinge on an insect biological control program.

Herbicides carry the additional possibility that they may be toxic to the pest or predator. This area will be discussed in a later section on herbicides in pest management programs.

F. Influence of Water and Fertilizer Management

The interaction between weeds and water or nutrients is primary, whereas that with insects and diseases is only secondary. Water and nutrient status of plants, crops, and weeds, can alter their interaction with insects; attack of the plants and the dynamics of the population may be changed (Leigh et al., 1969; see reviews by Hagen, 1974; Beck and Reese, 1976). The dilemma for the pest manager is that the same nutrient and water regimes that may be useful for insect management in a crop may also be conducive to better weed growth.

Water management in rice can be used to minimize grass problems, yet poor management of water can make barnyard grass problems worse. Overwatering can cause scald of the alfalfa which then allows greater weed invasion. Most annual row crops cannot tolerate excess water, yet many weeds thrive with such excess water. The basic difference between weeds and the other types of pests must again be emphasized here. Weeds are plants and thus actually compete directly for water and nutrients.

The relation between fertilizer use and weeds is hard to evaluate on a pest management basis. Weeds, as well as the crops, can use nutrients supplied by a fertilizer. The author has seen weeds stimulated in a band along the crop row by the application of fertilizer applied at planting. This certainly changes the microenvironment around the developing crop seedlings which would then alter attack by other pests. It has been shown that the banding of fertilizer for cereals along the crop row provided a better response than broadcast applications when wild oats were present; this was attributed to the cereal getting a greater percentage of the fertilizer when banded (Hill, personal communication). Changing the phosphorus fertilizer in corn has been shown to alter the competitive relationship between pigweed and corn (Vengris et al., 1955). Nitrogen fertilization has also been shown to alter the competition between rice and nutsedge (Okafor and DeDatta, 1976). Myers and Moore (1952) showed that differing fertilizer regimes used in citrus could substantially alter the relative abundance of several species of weeds. All the above noted crop/weed changes in response to fertilizer management have the potential for altering incidence of other pests, although such interactions were not investigated. The relationships are essentially the same as those outlined under the habitat diversity topic. There are other examples of fertilizer altering growth of weeds and how they compete with crops; how this information can be used in integrated management programs is uncertain.

V. INTERACTIONS DUE TO METHOD USED FOR WEED CONTROL

The two major techniques used for weed control are tillage and herbicides. Both techniques remove plants from the ecosystem and can thus indirectly interact with other pest management programs. Both techniques can also interact with other pests directly. This is because tillage causes physical changes in the soil microhabitat and because herbicides both alter the physiology of host plants and/or can be toxic to organisms other than plants. Other techniques like flaming and flooding can also have impacts on other pests but are not widely used; techniques, such as hand weeding and biological weed control only interact with other pests indirectly through habitat modification.

A. Tillage

The main reason for tillage is weed control, and it is still the single most important method of weed control for many crops. It has many advantages that are still of importance in establishing a programmatic approach to weed management, and due to the drastic changes that occur in the soil as a result of tillage it can have profound effects on all phases of pest management. The advantages to tillage include:

a) It is nonselective and therefore does not result in a tolerant weed problem (perennials might be considered an exception).

b) It does not leave any chemical residues in the product, and it does not leave any chemical in the soil to create rotational problems.

c) It does not require sophisticated equipment, can be conducted by relatively unskilled labor, and can be done by hand.

d) It is of prime importance in many pest management programs in that it alters insect and pathogenic fungus populations. This will be discussed later in greater detail.

The main reason for tillage, as noted above, is weed control (Yarwood, 1968; Periera, 1975; Scott Russell, 1977; Soil Conservation Society of America, 1977; Phillips et al., 1980), and as stated by Yarwood the response to tillage is infinite. Referring back to the introductory comments it can be seen that it has been tillage over the centuries that has permitted man to grow crops in the presence of weeds (see also Phillips et al., 1980). In the absence of weeds, there are strong suggestions that tillage is of no value (Triplett and Lytle, 1972) or is even detrimental (Scott Russell, 1977; Soil Conservation Society of America, 1977). The availability of zero-tillage techniques for certain crops now permits researchers to test whether modification of soil physical characteristics or weed control are the major benefits of tillage, as pointed out by Baeumer and Bakermans (1973). The move to zero-tillage culture of some crops (see later) such as corn, soybeans, and cereals is now clearly proving that tillage is not needed to grow crops provided weed control is available in another form (herbicides).

In terms of weed management, cultivation is only partially successful against perennial species; unless used with great dilligence it only offers a temporary control. In some situations cultivation may make perennial weed problems worse by spreading propagules (e.g., nutsedge tubers).

The greatest limitation of cultivation in crop weed management is the inability to remove weeds from within the crop row. This single fact is the major reason for the phenomenal success of selective herbicides, or the selective use of nonselective chemicals (e.g., paraquat in orchards). Cultivation also has limited capabilities in broadcast seeded crops such as cereals, safflower, and alfalfa.

There are several other reasons why tillage is not an ideal method of weed control. These include (not necessarily in order of importance):

a) Destruction of soil structure, and decrease in water penetration due to crusting of the soil surface (Goss et al., 1978; Aase and Siddoway, 1980). The use of tractors and heavy cultivation equipment can create a compacted zone in the soil just below the tillage level (plow-pan).

b) Can cause root pruning in crop plants. Particularly important in shallow rooted crops and in orchards (Haynes, 1980).

c) Can cause decreases in beneficial soil microorganisms and fauna due to habitat modification (Eijsackers and van der Drift, 1976); weed control with herbicides can be much less disruptive on the soil environment (Curry, 1970; Scott Russell, 1977; Haynes, 1980; Mellanby, 1980). Reduction of tillage in no-till cereal situations showed increased earthworm populations (Edwards, 1970; Edwards, 1975; Ehlers, 1975). Reduction in tillage has also been shown to lead to increases in population of several types of soil arthropods (Edwards, 1975; Edwards and Stafford, 1979).

d) Can spread pathogens and nematodes, and physical damage to crops can create wounds through which pathogens can enter the plant, such as sugarbeet roots, crown gall in trees (Gerawan, 1975), and several pathogens in various fruit crops (Kavanagh, 1969).

e) Can lead to more rapid drying of the soil (Baeumer and Bakermans, 1973; Unger and Phillips, 1973; Triplett, 1976a). Using herbicides in place of tillage in a fallow system improved soil moisture and resulted in higher yields (Wicks and Smika, 1974).

f) Necessitates moving large quantities of soil; this requires considerable amounts of energy.

g) Can lead to severe soil erosion problems by wind or water. Leaving plant cover, even if killed by herbicides, can dramatically reduce erosion problems.

Most of the preceding points were discussed by Baeumer and Bakermans (1973) in their review of zero-tillage. Due primarily to the problems of soil erosion, and to a lesser extent for some of the other reasons, the concepts of reduced, and zero-tillage, crop culture have been developed (Larson et al., 1970; Phillips and Young, 1973; Baeumer and Bakermans, 1973; Soil Conservation Society of America, 1977; Phillips et al., 1980). These approaches replace mechanical tillage for weed control with herbicides, thus reducing the tillage needed; in zero-till (no-till) all weed management is obtained with herbicides. The development of effective herbicides was the major key to the development of practical reduced tillage programs (Baeumer and Bakermans, 1973; Triplett, 1976b; Phillips et al., 1980).

The impact of zero-till cropping on soil erosion has been outstanding; 100-fold reductions in soil loss have been recorded (Soil Conservation Society of America, 1973; Triplett, 1976a; Phillips et al., 1980), and zero-till culture has allowed cropping on land normally considered too steep to cultivate due to erosion. It also seems that considerable savings in energy use can be achieved. Trips over the field may typically be reduced from seven or more to only three (Phillips and Young, 1973; Triplett, 1976a, Phillips et al., 1980), which can be translated to reducing diesel fuel use from about 5.0 gal/A for conventional cultivation culture to only 0.7 gal/A for zero-till (calculated from Alder, Klingman and Wright, 1976). These figures do not allow for the energy in herbicides used in zero-till, but neither do they allow for the energy used in hoeing or to produce cultivation machinery used in the conventional tillage. For other comparisons of cultivation and tillage versus herbicides in terms of energy use the reader is referred to the articles edited by Stevens (1975), prepared by CAST (1977) and the reviews by Greene (1976) and Nalewaja (1980). The general indication is that herbicides use less energy than cultivation for control of weeds.

Zero-till has ramifications in pest management that extend beyond the control of weeds, soil erosion, etc. This can be due to affects that the vegetation cover (alive or dead; weeds or crop) in a zero-till system has on the microhabitat near the soil surface, or in the soil. It is only through the utilization of zero-till systems that some of these changes due to cultivation have been documented. Soil temperatures are changed by several degrees C (1 to 6°C is typical but some authors report up to 10°C differences) as a result of the vegetation present; zero-till results in temperatures that are higher in winter and cooler in spring and summer (Bauemer and Bakermans, 1973; Griffith et al., 1977; Bennet, 1977; Aase and Siddoway, 1980). Soil moisture status, and relative humidity near the soil surface, are also altered by cultivation; tilled soils are usually drier at planting time, retain less water, and have a lower relative humidity near the surface (Bauemer and Bakermans, 1973; Griffith et al., 1977; Bennet, 1977; Aase and Siddoway, 1980). Microenvironment changes such as these will affect all pest organisms; temperature changes will alter the rate of development and can thus alter the degree/day predictions made from mesoclimate data. Altered moisture status could change the early development of pathogenic fungi. Lack of destruction of the old plant residues can lead to better overwintering sites for fungi and higher residual inoculum not destroyed by being buried. Some examples of the types of reported changes in pest populations due to alterations in tillage practices are outlined below.

Fungus diseases have been shown to increase in many cases when tillage was reduced (Yarham, 1975; Boursalis and Doupnik, 1976). Yarwood (1970) pointed out, however, that cultivation can produce diseases; he referred to them as "man-made plant diseases". Take-all disease of cereals appears to be due in large part to cultivation; using zero-till practices for growing wheat significantly reduced the disease (Boon, 1967; Brooks and Dawson, 1968). Alteration in the microenvironment at the soil surface was considered as a likely cause for the change in disease incidence. Eye-spot disease was also reduced in the zero-till system.

Insects can also be either increased or decreased by tillage, depending on the species involved and the plants attacked. Decreased tillage in cotton, and other crops, due to increased reliance on herbicides for weed control, has led to increased emergence of two *Heliothis* sp (Hopkins et al., 1972). Cutworms have been noted to increase under zero-till, and a four-fold increase in overwintering eggs of corn root worm was also recorded (Gregory and Musick, 1976). The incidence of corn root worm injury was, however, not greater under the zero-till regime than with conventional tillage, Griffith et al. (1977) suggested that the lack of increased attack may be due to increased numbers of predators and parasites also surviving under zero-till. Comparing conventional tillage with zero-till for cereals Liebee and Horn (1979) indicated that cereal leaf beetle and two of its larval parasites were about equally affected by the altered tillage practices. Tillage in orchards is recognized as a technique that reduces populations of navel orange worm in almonds and omnivorus leaf roller in grapes. The use of the "grape plow" has been shown to kill grape root borer (Sarai, 1969). This means that the best management approach in grapes includes a combination of cultivation and herbicides for best control of weeds and the grape root borer (Kennedy et al., 1979). Reductions in tillage do not always lead to increased insect problems; All and Gallaher (1977) and All et al. (1979) report that changing to zero-till decreased the severity of lesser corn stalk borer problems, and Edwards (1975) noted reduced numbers of wheat stem borers under zero-till.

Changes in cultivation practices, primarily adoption of zero-till, can alter rodent problems. Griffith et al. (1977) noted that under some zero-till situations mice problems may be worse; they attributed this to improved cover for the animals. Slug/snail problems have also been noted to be greatly increased by adoption of zero-till practices (Edwards, 1975; Kalmbacher et al., 1979).

The overall conclusion of most authors, expressed most recently by Phillips et al. (1980), is that adoption of zero-till is not likely to lead to substantially increased insect and disease problems. It is acknowledged that the benefits of no-till outweigh the altered insect and disease problems, and that the pests will have to be taken care of in another way, be it with chemicals or by such practices as of resistant varieties and rotations (Boursalis and Doupnik, 1976; Gregory and Musick, 1976; Phillips et al., 1980). It seems clear at this point that zero-till (and reduced till) agriculture should be approached from the standpoint of an integrated pest management system, as allowances must be made for the many interactions in the system. The reader who wishes to learn more about zero-till systems is directed to the following references: Baeumer and Bakermans, 1973; Phillips and Young, 1973; Soil Conservation Society of America, 1977; and Phillips et al., 1980.

It seems that cultivation will remain as a major tool in weed management, but that the development of effective herbicides has meant that there is much less reason to cultivate, and that in many situations the use of herbicides for weed control appears superior for a variety of reasons.

B. Herbicides

1. General

Pesticides are generally acknowledged to be a necessary component of integrated pest management programs (Stern et al., 1959; Smith and Reynolds, 1966; Pimentel and Goodman, 1979). Quoting from Glass (1975) "pesticides have been, are now, and will remain, in the foreseeable future, basic tools for management." There is growing concern amongst those authors who promulgate IPM that government regulation in the area of pesticides is going to (if it already hasn't) severely restrict the ability of pest management to utilize the best pest control technique, and will stop the development of new pesticides, see article by Tucker (1978).

The need for herbicides in economically viable weed management programs is now absolute in many instances. This is particularly true of those that are used selectively for the control of weeds in annual crops. There are several reasons for this:

a) Traditional methods of weed control using the hoe have become prohibitively expensive in relation to the value of the crop.
b) Herbicides are effective. In many instances they do a better job than any other method of weed control.
c) Herbicides can permit cultural practices that cannot be carried out without them; examples include zero-till, stale seed bed techniques, and allow weed control to be carried out when conditions preclude mechanical methods (e.g., aerial application to a field the day after a heavy rain storm).
d) Herbicides permit, or obtain, control of weeds in the first few weeks of a crop. This is very hard to do by hand, and competition data show it is a critical period in the life of most crops.
e) Herbicides may be less energy intensive than cultivation. It is difficult to determine the energy efficiency of hand labor versus herbicides, and it is quite probable that hand labor uses less energy than herbicides. The question is, however, academic because hand labor has become economically not feasible. Several studies show that an effective herbicide program uses less energy than repeat cultivations (Stevens, 1975; Alder et al., 1976; Greene, 1976; CAST, 1977; Nalewaja, 1980).
f) Selective herbicides control weeds in the crop row, and in broadcast crops (e.g., cereals, alfalfa) in a situation where mechanical control has serious drawbacks, and where we traditionally relied on the hoe or suffered the yield loss.

Herbicides are so effective that their use has been increasing (Fowler and Mahan, 1978). This trend will probably continue, unless the cost of herbicides (in terms of energy) outstrips the costs of cultivation (in terms of fuel for tractors). It seems unlikely that will happen.

As herbicides seem likely to remain as a major means of controlling weeds, it is then appropriate to evaluate how herbicides interact with other components of pest management programs.

At this point it is probably appropriate to ask, "what is an ideal pesticide for use in pest management?" It is unlikely that there is a chemical that fits all these ideals, and there may never be one. The following are not necessarily in order of importance; some topics will be covered in greater detail later in the section.

a. The pesticide should be effective against the target pests. These may be selective or nonselective for herbicides, depending on use.
b. Should be noninjurious to crops, even when used incorrectly (i.e., it should have an excellent safety margin). This consideration does not apply to noncrop herbicides, except when they are used selectively, e.g., paraquat in orchards.
c. The control should last through the duration of the pest problem, then the chemical should dissipate. Herbicides vary widely in this characteristic; many do not last long enough to give ideal weed control (e.g., cycloate), others last too long and thus are severely restricted in crop use (e.g., picloram).
d. The pesticide should not build up in the soil, or in the biota through magnification in the food chain. There is currently no evidence that this is occurring for herbicides (Hill and Wright, 1978; Audus, 1970; Grossbard, 1976; Eijsackers and van der Drift, 1976; Way and Chancellor, 1976; Fryer et al., 1980a).

e. Any breakdown products should be less toxic than the parent compound, unless such breakdown products enter into the activity of the pesticide.

f. The pesticide should have low mammalian toxicity, and be safe for workers; or it should be readily antidotable. Herbicide toxicities are generally relatively low (cf. insecticides) and typically do not present serious hazard to workers; some notable exceptions include paraquat, dinoseb based herbicides, and acrolein (WSSA, 1979).

g. Pesticides should have a minimum of detrimental effects on nontarget species; this topic will be expanded in a later section.

h. It should be easy to apply, and not exhibit any major drift problems. As herbicides affect plants drift is generally easily detected as symptons, or kill of, adjacent plants. Consequently, we are much more aware of drift problems with herbicides than with other pesticides. Drift can be controlled, or reduced by correct application; most problems can be attributed to poor application.

i. The pesticide should be compatible with other pesticides and fertilizers, so that application costs can be minimized, and trips over the field reduced. This area will also be covered later.

j. The chemical should not cause the pest species to become resistant.

k. The pesticide should not create resurgence problems or develop a tolerant weed flora (related to item #1).

l. The pesticide should be cost-effective; the benefits derived should outweigh the anticipated crop loss, in both the short- and long-term view. The benefits of herbicides judged against cost is typically very favorable, hence the widespread use.

m. It should not leave a residue in the crop at harvest, or an undesirable residue in the soil after harvest.

As in all other situations, the decision to use a pesticide in a pest management program must weigh the advantages against the problems; the importance of the points in the foregoing list, thus must be judged according to the situation and the alternatives available.

There are several attributes of pesticides that are of particular concern in pest management programs; to a considerable degree they are the reciprocal of some of the above points. They include:

a. Detrimental effects on nontarget organisms. These may be direct or indirect and may be a result of intentional spraying, or unintentional application (drift).
b. Persistence in the environment and biomagnification in the biota.
c. Selection of a resistant population, or build up of a tolerant population.

2. Herbicide Interactions with Pathogens and Nematodes

There are many records of effects of herbicides on disease organisms and severity of disease attack (Kavanagh, 1969, 1974; Heitefuss, 1970, 1972; Katan and Eshel, 1973; Altman and Campbell, 1977a, 1979). The effects can be positive or negative. Katan and Eshel (1973) tabulated many such responses of disease attack to herbicide use. They also enumerated four ways that a herbicide might interact with a disease; these were:

a) direct stimulation/inhibition of the pathogen
b) increased/decreased virulence
c) increased/decreased host susceptibility (altered host physiology)
d) suppression/enhancement of antagonistic microorganisms.

The following is a partial listing of some of the recent reports indicating that herbicides can decrease the severity of disease, or may actually inhibit growth of a fungus. Harvey, et al., (1975) reported that trifluralin reduced the incidence of root rot (Aphanomyces euteiches) in peas; this was not attributed to a direct fungidical effect. More recently Grau (1977) has shown that both trifluralin and dinitramine do suppress the growth of the fungus, and that less pea plants died due to fungal attack when the herbicides were present in the soil. The potential for control of pea root rot by dinitroaniline herbicides has been further demonstrated by Teasdale et al. (1979), and several of the environmental factors that can influence the level of control have been investigated. Protection of cabbages from attack by Plasmodiophora brassicae has also been reported when trifluralin was incorporated into the soil (Buckzaki, 1973); whether this was due to altered host susceptibility or to fungitoxicity of the herbicide was not determined. Grossbard and Harris (1976) tested glyphosate and paraquat against cereal diseases; paraquat showed more activity against the pathogens. It was felt that both herbicides might give some protection against glume blotch. Brooks and Dawson (1968) had earlier reported that take-all disease of cereals was reduced under zero-till culture; paraquat was used to attain preplanting weed control but was determined not to have any direct effect on the fungus. They attributed the altered disease severity to alteration in the microenvironmental conditions. Kavanagh (1974) has reported that in general, herbicides have reduced plant diseases in vegetables and root crops, and also in some fruit crops. Some of these effects may have been indirect through reductions in cultivation and thus less wounding of the crop. It is generally thought that the removal of weeds either reduces the inoculum or makes the environment less conducive to the attack of the pathogen. Early destruction of potato haulm by diquat has been shown to decrease the incidence of late blight in the tubers (Kavanagh, 1974); presumably rapid killing of the tops stopped the disease from spreading into the tubers. The use of herbicides as an alternative to cultivation in hops has been reported to decrease the incidence of wilt in this crop (Sewell and Wilson, 1974). Stedman (1976) has reported that the spores of *Rhyncosporium secalis* from paraquat sprayed stubble were less infective than those from untreated stubble. Paraquat has also been reported to alter the saprophytic attack of fungi on decaying plant tissue (Wilkinson, 1969; Wilkinson and Lucas, 1969), and it was also shown that paraquat itself may be fungistatic; *Trichoderma viride* is reported to be very sensitive. Paraquat was also moderately effective in inhibiting *Fusarium graminearum* in corn (Fischl, 1974). Suppression of root diseases of peas has also been reported by Sacher et al. (1977) when treated with dinitrophenol or dinitroaniline herbicides; the former gave more effect on the diseases, the latter gave more effective weed control. Dinitrophenol has also been reported to reduce the severity of *Sclerotinia* blight in peanuts (Porter and Rud, 1979); mycelial growth was inhibited and the disease was reduced by 17 to 28% in the field. Campbell and Altman (1977) showed that cycloate in a potato-dextrose agar medium inhibited the growth of *Rhizoctonia solani*, yet at low concentrations (below 8 μg/g) in the soil the herbicide increased fungus attack on bean hypocotyls, and on sugarbeets (Altman and Campbell, 1977b); higher rates in the soil inhibited fungus attack.

There are also many examples of herbicides increasing the incidence of diseases. 2,4-D has been shown to increase disease in corn (Oka and Pimentel, 1976) and to increase blight in wheat (Hsia and Christensen, 1951). In the latter case 2,4-D inhibited the fungus in culture, so that the increased attack must have been a reflection of altered host physiology. Oka and Pimentel (1976) attributed the increased disease in corn to altered host plant physiology. Wyse et al. (1976) also attributed increasd attack of *Fusarium solani* on beans following treatment with EPTC to altered host physiology. Other reports of increases in disease include soybean root rot by *Thielaviopsis basicola* after chloramben treatment (Lee and Lockwood, 1977) and damping off disease in tomatoes and peppers following diphenamid treatment (Katan and Eshel, 1974). Increased attack by Verticillium dahliae on oilseed rape occurred when alachlor, nitrofen or trifluralin were used (Nilsson, 1977). Pinckard and Standifer (1966) showed that trifluralin also predisposed cotton to damping off caused by Rhizoctonia; infection increased from 30% in the check to 69% when 0.75 lb/A of trifluralin was used. Miller et al. (1979) also reported recently that trifluralin increased Thielaviopsis basicola in cotton; the disease was successively increased by the herbicide over a three year period. Atrazine has been reported to cause maize to lose its resistance to maize dwarf mosaic (MacKenzie et al., 1968); infection was only 12% showing symptoms in the check, to 38% in 5.0 ppm atrazine in soil, to 100% showing symptoms in soil with 20 ppm atrazine. If this type of interaction were to occur frequently, the implications in virus disease control could be serious.

All these results indicate that there are both direct and indirect effects of herbicides and diseases. These interactions can be either positive or negative, depending on the crop species involved, the herbicide involved, and the species of pathogen. From the point of view of integrated pest management, it is necessary to know which interactions are detrimental and which are useful. This will probably have to be worked out on a situation-by-situation basis as there does not seem to be an easily predictable pattern. The fact that the same herbicide, trifluralin, can suppress fungus attack in one crop (peas), yet increase disease in other crops (cotton, soybeans), emphasizes the complexity of the interactions and also the lack of ability to predict the results of the interaction.

The influence of herbicides on soil microflora is more difficult to assess. Audus (1970) probably stated the situation about as well as possible; we cannot draw any overall conclusion that herbicides are disrupting or enhancing soil fungi, but that we should not be complacent. The whole area of effects of pesticides on soil microflora was recently reviewed by Grossbard (1976) Anderson (1978), and Wright (1978), and the reader is directed to these reviews for general coverage of the topic. The possibility exists that herbicides might affect mycorrhizal fungi; Kelley and South (1980) investigated affects of several herbicides on mycorrhizae of pine; with few exceptions they found that the concentrations of herbicides needed to affect the fungi were higher than would normally be present in the soil. They concluded that mycorrhizal fungi would not be affected by normal rates of the herbicides tested. Smith and Ferry (1979) had earlier shown that simazine had no effect on mycorrhizal development in _Pinus_ seedlings, and their data suggested that it might, in some situations, actually enhance mycorrhizal formation. This was not attributed to suppression of weed growth, but the mode of stimulation was not determined. In two recent papers Fryer and associates (1980a and 1980b) provide further evidence for several herbicides that there are no long-term adverse effects on soil, and that there is no herbicide build up after several years of use. It seems fair to state that at present there appears to be no cause for concern with the use of herbicides in relation to soil microflora and microfauna. The greatest effect is usually due to removal of vegetation rather than any direct effect of the herbicide (Fox, 1964; Anderson and Drew, 1976; Clay and Davison, 1976; Greaves, 1979).

It has been pointed out by Volz (1977) that reductions in soil microbial populations due to weed removal with herbicides may in fact be beneficial in some instances. When nutsedge was present there was less nitrogen in the total biomass than when nutsedge was removed; the suggestion was made that nutsedge maintained a higher level of denitrifying bacteria in the soil.

If herbicides can alter nematode attack this might be useful in a pest management program. Romney et al. (1974) showed that attack of beans and onions by _Meloidogyne hapla_ was reduced after application of DCPA. They indicate that this was not due to altered pathogenicity, but was due to changes in the plant. More recently Orum et al. (1979) reported that oryzalin reduced the development of Meloidogyne incognita in tomatoes; the herbicide was not nematicidal and the effect was attributed to alteration in the response of the tomato root to the nematode attack. These two reports (Romney et al., 1974; Orum et al., 1979) raise the interesting speculation that herbicides which alter root development might all interact with nematode attack. Yeates et al. (1976) reported that paraquat treatment in grass reduced the numbers and diversity of nematodes. Any vegetation control should help to reduce nemtaode populations, and could thus be beneficial to a management program. If the nematode population is, however, vectoring a virus it has been suggested (Cooper and Harrison, 1973) that removal of weeds could increase the incidence of the virus in the crop because there would be no other hosts for the nematodes to attack.

3. Herbicide Interactions with Insects and Mites

Interactions of herbicides and weed control programs with insects is probably of greatest significance to IPM. Probably of greatest concern is the immediate, direct effect of herbicides on beneficial insects (see Chapter 12, Brown, 1978). It should, however, be emphasized that in general herbicides are much less toxic to insects than most insecticides (e.g., Stam et al., 1978).

Several herbicides have been shown to be moderately toxic when sprayed on bees (Moffett et al., 1972; Morton et al., 1972; Morton and Moffett, 1972). Paraquat, MSMA, and cacodylic acid were the most toxic; compounds like 2,4-D, picloram, EPTC and dicamba had low toxicity or were essentially nontoxic. The fact that most herbicides are not normally sprayed on flowering plants but on seedlings or young weeds means that the chance of spraying honeybees in the field is low, and is much less of a problem than bee kills with insecticides. A problem with insecticidal bee kills is that the bees are often not foraging on the crop being treated with the insecticide but on the flowering weeds present in the field. Any weed control program then may be beneficial with respect to bee kills. The fact that a herbicide is used for orchard floor weed control could be considered as an IPM advantage in this situation.

2,4-D has been shown by several authors to be toxic to other insects. Adams (1960) reported that direct spraying of Coccinelid beetle larvae caused a four-fold increase in larval mortality; it was felt that this could have serious consequences on aphid populations, although no aphid outbreaks have been reported as a consequence of spraying 2,4-D. Beyer and Mey (1976) reported that 2,4-D decreased the hatch rate of dock beetles, but that it increased the egg laying of the females. An earlier report (Ingram et al., 1947) showed that 2,4-D treatment resulted in increased attack by sugar cane borers. They considered the increase due to lower parasitization by a *Trichogramma* sp., the activity of which was lowered by the 2,4-D. This type of altered interaction between parasite and prey is of concern to IPM programs. A similar potential disruption has been noted in orchards (Rock and Yeargan, 1973). Several herbicides were shown to be differentially toxic to mites. The herbicides were more toxic to the predatory mite than to the phytophagous mite. This is cause for concern if the predator/prey relationship is altered in favor of the problem mite; this type of interaction is only of concern, however, if the mites are actually sprayed. If they are on the trees this is unlikely; if they are on the orchard floor weeds, then they might be sprayed.

There can be positive interactions between herbicides and insect control. Direct insecticidal effects have been noted for some herbicides; EPTC has been shown to be a selective female chemosterilant for the Egyptian cotton leafworm (El-Ibrashy, 1971). Several sugarbeet herbicides, including cycloate, triallate and diallate caused increased mortality of the Collembollan *Onychiurus fimatus* Gisin and appeared to disturb the food searching mechanisms (Ulber, 1979). The attack against the sugarbeet seedlings was reduced. Rieckmann (1979) reported transitory reductions in the colonization of sugarbeets by alate aphids following treatment with phenmediapham; the insects settled less often during the 72 h following spraying, after which there was no effect. Although most herbicides have almost no insecticidal activity, it has been found that low concentrations of several herbicides acted synergistically to increase the activity of insecticides against *Drosophila melanogaster* (Lichtenstein et al., 1973). Activity increases ranged from 1.2-fold to nearly 9-fold. As these authors emphasized it is important that these types of interactions be known in integrated control programs. If the phenomenon is common it could be used as a control tactic for insects that might delay or reduce the build up of insecticide resistance (Fox, 1964). This has probably not been exploited, but might be of value in some situations.

Modification of the physiology of the host plant by herbicides has also been implicated in alteration of attack by insects. Increases in insects attacking, or thriving, following application of herbicides to crop plants include aphids on corn (Oka and Pimentel, 1976) by 2,4-D, increased wireworm damage in wheat by 2,4-D (Fox, 1948), and others. Plant growth regulators have also been shown to decrease aphid populations through altered host physiology (van Emden, 1964), and Williamson et al. (1979) have indicated that controlling the vigor of red raspberry canes by chemically pruning them with dinoseb can result in decreased disease and insect attack. It seems likely that contact herbicides would less often enter into these types of interactions as they kill plants or do not kill them; they rarely just alter the physiology of the plant like a growth regulator.

As a principle of pest management all the foregoing interactions between herbicides and insect control should be considered. It is apparent that our knowledge of the effects of herbicides on insects is limited; this is probably a reflection of the fact that direct herbicide effects on insect populations are not usually very drastic, otherwise the effects would have been much better investigated. As a general summary statement concerning the effects of herbicides of nontarget organisms it seems clear that the majority of the effects are not due to direct toxicity but reflect the habitat modification resulting from the weed control (Fox, 1964; Brooker and Edwards, 1974; Altieri, et al., 1977).

VI. INTERACTIONS BETWEEN NONWEED PEST CONTROL AND WEED MANAGEMENT

The control tactics employed to mitigate nonweed pests can have effects that interact with the management of weeds. Competitive ability of the crop can be altered by insect or pathogen attack which can then influence weed invasion. Rotations may be dictated for control of pests other than weeds; this can influence weed population dynamics. The use of pesticides can have direct effects through phytotoxicity to crops or weeds; this is especially true for mixtures of pesticides. Natural biological control of weeds can be perturbed by the measures used to mitigate a crop pest problem.

A. Interactions with Biological Weed Control

Any pest control practice that maintains the crop in as competitive condition as possible can be considered as a beneficial practice in terms of weed management. Conversely allowing a crop to be defoliated, even partially, by an insect or pathogen can substantially increase the severity of a weed problem. The author has recently demonstrated that uncontrolled attack of alfalfa by the Egyptian alfalfa weevil resulted in over a 50% increase in the quantity of yellow foxtail harvested later in the season (Norris and Schoner, 1975). It is apparent that if insect damage is allowed to occur because it is only causing "cosmetic damage" that this needs to be evaluted on the basis of the pest pentagon philosophy.

Biological control of weeds has traditionally focused on the introduction of organisms that utilize weeds as their food source, and that can cause dramatic reductions in the population of a particularly troublesome weed. Such managed biological control for weeds has been mainly successful against perennials; examples include control of prickly pear in Australia, Lanatana in Hawaii, and Klamath weed in California (Andres and Goeden, 1971; Frick, 1974). There are many projects currently under way for biological control of weeds (see Glass, 1975; Goeden et al., 1974; USDA, 1978). An approach that has seen increased emphasis in recent years has been the possibility of using plant pathogenic fungi for biological control of weeds. This topic was recently reviewed by Templeton et al. (1979), who coined the term "mycoherbicides" for sprays of pathogenic fungi that may be used to control weeds. There is reason to believe that this method will have utility in weed management in years to come. It seems unlikely that classical biological control of weeds (i.e., application or release of a specific control organism) will ever make major contributions to the control of annual weeds in row crops (Frick, 1974; Day, 1978). This is because a) weed problems are usually a complex of several species and thus removal of a single species, typical of a biological control program, will just allow the remaining weeds to grow better; b) there are selectivity problems as the insect or pathogen killing the weed might shift to the crop, as many common weeds are closely related to crops; c) annual weeds are ephemeral in nature and thus the food supply of the biological control agent is not always present, and d) results do not occur rapidly and thus serious competition could still occur in an annual crop.

The extent to which naturally occurring biological control plays a role in suppression of weeds has received little attention; it may, however, be a great deal more significant than currently recognized. Many weeds are attacked by endemic insects and mites, by pathogenic fungi, and by phytophagous nematodes. The level to which these organisms reduce the populations of weeds has not been investigated in most cases. The organisms involved may be the same pests that attack crops, examples include mites, omnivorous leaf roller and fleabeetles (authors personal observation), cutworms (Genung, 1959), and rice ring nematode (Hollis, 1972, 1977), or they may be fairly specific to the weed such as silver leaf nightshade nematode (Robinson et al. 1979), or they may be completely specific to a particular weed, such as the purslane sawfly to common purslane (Gorske et al., 1976). Any practice that is employed to mitigate pests attacking the crop may also disrupt this biological control of weeds. Kirkland and Goeden (1978) clearly demonstrated this when they utilized insecticides to determine the effectiveness of biological control of puncture vine. It is mainly through the use of pesticides that we are beginning to recognize the importance of natural biological control of weeds. The extent to which pesticides used for crop protection are also protecting weeds will be included under the section on pesticides in this chapter.

The effects of organisms that attack the aerial parts of plants are usually obvious. Phytophagous organisms also attack the plant parts in the soil. The significance of this predation/parasitism on the underground parts of plants cannot be overestimated; the regulation of weed seed populations is particularly critical. Weeds produce large numbers of seeds; many of these die in the soil without ever germinating. This is due to the activity of various soil-borne organisms, and is a form of natural biological control (see review on animal predation on seeds by Janzen, 1971). Any pest management practice that alters the activity of soil organisms could thus alter the size of the viable weed seed population. The significance of this is hard to judge due to lack of data. Carabid beetles have been reported to kill or damage seeds of many weed species in Indiana (Lund and Turpin, 1977); the beetles did not damage morningglory seeds but they killed almost 100% of the redroot pigweed, common chickweed, and crabgrass seeds present. Likewise Luff (1980) has reported that ground beetle larvae ate many grass and lambsquarters seeds in a strawberry field in Northern England. The latter relationship emphasized the complexity of the weed/insect interaction. Harpalus rufipes is a beneficial ground beetle in the adult stage yet Luff (1980) noted that its larvae probably would not survive without weed seeds as a food source, but at the same time the larval feeding helps to keep the weed population under control. Den Boer (1977) suggested that Carabid beetles need a disturbed habitat in which to survive, which adds another level of complexity to the interactions that could be occurring in response to pest management decisions. The evidence that natural predation on weed seed populations plays a significant role in the natural attrition of the population is sufficient to suggest that the effects of nonweed pest control decisions should be considered at this level.

B. Interactions Between Other Pesticides and Weeds

There are instances of insects and insecticides affecting weed management; these too must be known in a program approach to pest management. Many weeds may be under partial biological control by insects in the natural state (Gorske and Hopen, 1976; Gorske et al., 1977). It should be expected, as pointed out previously, that pesticides applied to protect crops will also protect weeds. Preplant insecticides increased the survival and early growth of weed seedlings, as noted by Finlayson et al. (1975). Maceljski (1968) had earlier reported that lindane and aldrin applications had resulted in increased growth of weeds. Recent work by Norris (unpublished) has shown that applications of phorate, disulfoton, or aldicarb resulted in increased growth and seed production by common purslane; competition by the weed against sugarbeets was increased when aldicarb was applied as a preplant treatment to the crop. The increased growth of the purslane was attributed to control of the purslane sawfly by the insecticides. Hollis (1972, 1977) reported a similar interaction between nematicide treatment and weeds. Applications of Nemacur to rice in the presence of weeds resulted in increased weed growth and reduced rice yield; control of the rice ring nematode, which had been providing partial control of the weeds was considered the reason for the response. The reason for these types of interactions is thus almost certainly due to organisms that normally provide partial biological control of the weeds being killed by the broad spectrum pesticide as suggested by Finlayson et al. (1975).

The converse can also be true! If an insect or disease attack defoliates a crop then much of the canopy is removed and the understory weeds then receive more light and grow better. Insecticidal control of the Egyptian alfalfa weevil in California has been shown to decrease subsequent growth of yellow foxtail (Norris and Schoner, 1975; Schoner and Norris, 1975). Wheat bulbfly attack has, likewise, been shown to allow both scentless mayweed and blackgrass to grow better in wheat due to lowered competition from the crop (Lawes Agricultural Trust, 1974). On a theoretical basis, it should be expected that any disease or insect attack that weakens the crop plant will make weeds grow better and make their control more difficult. An axiom of weed management is that a healthy, vigorous, competitive crop is the best method of weed control; therefore, control of insects and diseases that attack the crop will usually be beneficial in a weed management program.

A final interaction between weed control and insect control that should be noted is the fact that weed control may make chemical pest control easier. The weed canopy may intercept pesticide sprays, and stop the chemical from reaching the target pest(s). This has been noted for control of Egyptian alfalfa weevil in California alfalfa in relation to weediness of the crop. Controlling weeds may also mean that the insects are concentrated on the crop, and thus improve the efficiency of insecticide use; such an effect has been noted in establishing ryegrass using paraquat for weed control (Dixon and Davison, 1975). It would seem that this type of interaction would occur whatever type of weed control was practiced; as such "targeted" insecticide applications might mean less insecticide was needed. The control of weeds could then take on another dimension in pest management programs.

C. Interactions Between Pesticides

When a program calls for tank mixing of two pesticides, either the same or different types, it is possible that an interaction can occur that will affect the activity of one or both compounds. Herbicide antagonism and synergism is fairly well recognized (see review by Putnam and Penner, 1974), and is allowed for, or even used to advantage in weed management programs.

The reason for mixing two pesticides of different types include: may save time, may save energy due to less equipment used in the application, may cause less soil compaction, and may be required in order that two types of pests are controlled so that the anticipated yield gain can be achieved. The latter idea was recently pointed out by Abivardi and Altman (1978) who showed that nematodes needed to be controlled in sugarbeets if benefits from cycloate were to be derived fully. Likewise Merrill and Kistler (1978) showed that in order to obtain optimum growth of pines it was necessary to use fungicides for control of needlecast and to use herbicides for weed control. The data of Hollis (1972, 1977) for control of rice ring nematode also strongly suggest that control of weeds and nematodes in rice was necessary for optimum crop yield. In many cases, there is no undesired effect (Putnam and Penner, 1974; Thompson et al., 1977).

Probably the greatest problem with tank mixes, from the standpoint of weed management, has been decreased selectivity when herbicides are mixed with other pesticides. The herbicide becomes more toxic to the crop when applied in the mixture than when it is used alone. Propanil has been noted to be more toxic to rice when applied with carbofuran (Smith and Tugwell, 1975). Hammill and Penner (1973a, 1973b, 1973c) noted several interactions between insecticides and herbicides; carbofuran increased the activity of alachlor on barley, but did not alter the activity on corn (1973a); carbofuran increased the activity of chlorbromuron on both barley and corn (1973b); and the insecticide increased the activity of butylate on barley but not corn (1973c). These authors felt that the latter interaction might be useful for weed control in corn. Steyvoort (1975) reported on increased activity of propham when aldicarb was used in sugarbeets. Carbaryl has also been shown to enhance the activity of linuron (Del Rosario and Putnam, 1973). A recent report by Weierich et al. (1977) showed that the insecticide fonofos decreased the metabolism of terbacil in mint, with a concomitant increase in injury. Most of the above interactions can be traced to either altered uptake of the herbicide, or changes in the rate at which the tolerant species can metabolize it. These types of interactions must be known if the most effective use of pesticide combinations is to be made in pest management programs.

VII. INTERACTIONS AT THE LEVEL OF ECONOMIC THRESHOLDS

The concept of thresholds is central to all current IPM programs. According to Glass (1975) the economic threshold is defined as "that pest population density, or damage level, at which control measures should be taken to prevent an economic injury level from being attained." This implies that the control will cost less than the loss that would have occurred had nothing been done. This balance is so lopsided in favor of weed control that thresholds have not been worked out in many instances.

The concept of thresholds is basic to most insect management programs (Smith and Reynolds, 1966; Glass, 1975; Smith and Pimental, 1978; Thomason, 1978). It is now beginning to be used for certain disease management programs; for example, fireblight in pears (Thomson et al., 1977) and in several other crops (see Krause and Massie, 1975). The idea has not been employed to any degree at present in weed control except in a few cases. Some of the reasons for this include the following:

a) Low densities of weeds can cause yield loss. Barnyard grass at one per m^2 in rice can cause significant yield loss; one barnyard grass in a meter of sorghum row can lower yield by almost 10% and one grass plant per meter of sugarbeet row can cause a 50% yield loss. Most fields will have a population of this density and thus can economically justify being treated.

b) The dynamics of weed (plants in general) populations are long, but explosive. There is usually only one generation per year, but each generation is capable of almost explosive increase (e.g., one pigweed plant can produce in excess of 250,000 seeds) (Stevens, 1957; Harper, 1977). This contrasts with insect populations that can build with several generations in a year, and crash in the same year. Plasticity of plants also makes population prediction difficult (pigweed plants grown under adverse conditions may produce less than 100 seeds). All these factors have led to little attempt to model weed population dynamics, except in very approximate terms (see Harper, 1977; Sagar and Mortimer, 1977; Auld et al., 1978).

c) Weed problems are usually mixed populations of species. Predictions of one species will probably be of very little help in predicting what will happen to the whole complex. This is especially true when using selective herbicides.

d) The lack of effective postemergence treatments in many situations has meant that preplanting weed control must be practiced. If there is no economical way of controlling an emerged weed problem, but there is satisfactory preemergence control, then thresholds become even harder to establish. Many growers do attempt to decide whether a postemergence treatment is worthwhile on the basis of the yield of the current crop; this does not take into account, however, the long-term effect of allowing a few weeds to go to seed. In sorghum, Norris (unpublished) has estimated that roughly 400 seeds of barnyardgrass are returned to each square meter of soil from an infestation level of one grass plant per meter of row. In order to maintain the "status quo" it requires that about 99.8% of those seeds either die of natural causes or are killed by man. Most current weed control practices are not that effective.

e) Most thresholds refer to numbers of individuals; when a perennial weed like Johnsongrass, quackgrass, or nutsedge are considered there is a difficult question regarding what constitutes an individual as all the plants in a field could theoretically be a single clone. This dilemma was pointed out by Sagar and Mortimer (1977) with reference to modelling population dynamics of perennial weeds.

f) Another problem with weeds is the longevity of propagules, and their dormancy. Seeds of many weeds lie dormant in the soil for from one to many years (20 or more in some cases). This prolongs the time that they will germinate, and makes population predictions very difficult for any one season, unless the total seed population is known in conjunction with the dormancy status. To make matters more complicated, cultural and environmental factors can alter dormancy and germination (Mayer and Poljakoff-Mayber, 1975). These phenomena led to the saying "one year's seeding equals seven year's weeding"; this is a dilemma for the pest manager.

g) A serious problem with thresholds for weeds is the lack of any easy way of determining populations. This is a drawback to implementing any pest management program; there must be easy ways to determine the population level if thresholds are to be used in practice. Accurate determinations of weed seed populations are difficult, require considerable time and laboratory space, and require a high level of expertise (to recognize the seeds and seedlings). Even determining the current population for a postemergence treatment requires identifying and counting the weed seedlings in several locations in the field as weed populations are never uniform.

For these, and probably other reasons, it seems unlikely that "thresholds" will be much used for weed control in the immediate future. If valid computer models can be developed for long-term population dynamics it may be possible in the future to make predictions of population shifts in relation to particular weed control practices. Glass (1975) points out a further dilemma for a weed management program; is the threshold for a mixed population the same as that for a single species? As weeds usually occur in mixed stands, it again seems doubtful that thresholds will play much of a role in weed control in the immediate future.

The problem of mixed pest attack can be a further complicating factor in determining thresholds. If thresholds are viewed from the "pest pentagon" philosophy (Fig. 2) then it becomes apparent that thresholds may vary depending on how other pests interact with the one that is being controlled. As pointed out earlier, one of the anticipated developments for insect management will be increased use of computers to predict population increases, with the idea that the time at which a threshold will be attained can be better anticipated. The key to these computer predictions is accurate knowledge of the temperature environment of the insect. Weeds and their control alter that temperature (e.g., zero-till soil temperature) and thus can alter the accuracy of the computer predictions. There are currently a few examples of this type of interaction that have been documented. The author is working on relationships between weeds and Egyptian alfalfa weevil in California alfalfa (Norris and Cothran, 1974; Norris and Schoner, 1975; Norris and Burton, 1979). It is becoming clear that damage done by the weevil larval feeding is much worse if the field is weedy; conversely, if the field is weed free the alfalfa appears to be able to tolerate a higher weevil attack. This seems to suggest than in an IPM system that the threshold for the weevil should be varied depending on whether or not the field is weedy. Computer predictions of weevil population dynamics are proving to be inaccurate when weeds are present in the field (Summers, personal communication). Merrill and Kistler (1978) recently reported that needlecast disease of Scots pine was much more severe if weeds were present; this would alter the threshold for weed control depending on the presence or absence of the disease. Many of the examples of weed/insect interactions, and weed/disease interactions noted in the section on habitat diversification could be viewed in the light of the above discussion. When pests interact with each other, as this chapter has documented, then development of reliable, useful, economic thresholds are going to be extremely difficult. Attempts to develop simplistic thresholds may have the potential for creating even greater crop losses due to lack of recognition of such interactions.

VIII. SUMMARY

Weed management is an essential component of the crop production system. Weed control, regardless of method employed, involves removal or destruction of vegetation, which is a modification of the ecosystem. This agro-ecosystem modification causes changes in the microhabitat that can have affects on all other pest organisms. The alterations in the habitat by weeds can be through changes in the microclimate, through changes in shelter, and through changes in the food supply for pest or beneficial organisms. The damage done by other pest organisms can modify the crop growth which can then alter the severity of a weed problem; the control tactics applied against these other pests may also modify the growth of weeds. The changes in the microclimate should thus be judged on the basis of feedback reactions, as microenvironment alteration by one pest may result in changes in another pest, which then alter the situation for the original pest (Fig. 4). For these reasons it is essential that pest management develop an integrated approach that includes the "pest pentagon" philosophy (Fig. 2).

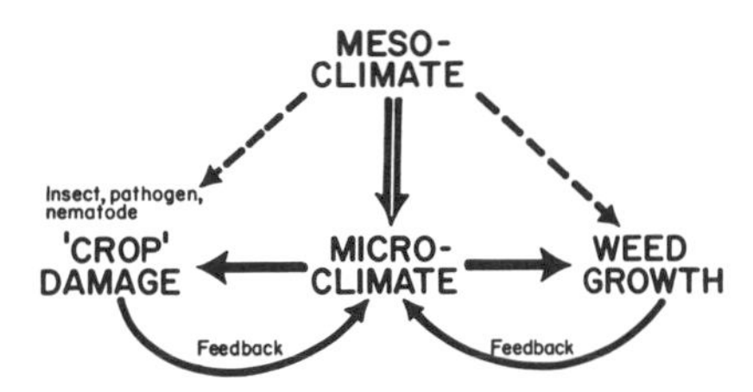

Fig. 4. Interactions and feedback between weeds and microclimate, and nonweed pests and microclimate.

The interactions between weeds and other pests can be beneficial, of no significance, or may be detrimental. It seems probable that no general statements can be made about such effects; each situation will have to be judged on its own merit. When herbicides are used to obtain the needed weed control, then there are additional interactions; these can be mediated through habitat modification, or they may be a direct consequence of the herbicide action. Herbicides can alter the physiology of the host plant, which then alters the way insects, pathogens and nematodes will react to the plant. Herbicides can also have direct effects on nontarget organisms. The above effects can again be useful, of no consequence, or may be detrimental in a unified pest management system. It is apparent that weed control is central to crop production systems, and that effects of weed management extend beyond the immediate response of removing undesired vegetation.

ACKNOWLEDGMENTS

I wish to thank Dr. M. A. Sall for suggestions during the preparation of the manuscript and for critical review of its contents. I also wish to thank Dr. R. M. Nowierski and Dr. D. E. Bayer for review of the manuscript.

REFERENCES

Aase, J. K., and Siddoway, F. H. (1980). Agric. Meteorol. 21:1-20.
Abivardi, C., and Altman, J. (1978). Weed Sci. 26:161-162.
Adams, J. B. (1960). Can J. Zoo. 38:285-288.
Adkisson, P. L. (1972). In "Implementing Practical Pest Management Strategies," Proc. Nat. Ext. Insect-Pest Management Workshop, Purdue University. pp. 37-50.
Alder, E. F., Klingman, G. C., and Wright, W. L. (1976). Weed Sci. 24:99-106.
All, J. N., and Gallaher, R. N. (1977). J. Econ. Entomol. 70:361-365.
All, J. N., Gallaher, R. N., and Jellum, M. D. (1979). J. Econ. Entomol. 72:265-268.
Altieri, M. A., van Schoonhover, A., and Doll, J. D. (1977). PANS 23:195-205.
Altieri, M. A., and Whitcomb, W. H. (1979a). Hort. Sci. 14:12-18.
Altieri, M. A., and Whitcomb, W. H. (1979b). Prot. Ecol. 1:185-202.
Altman, J., and Campbell, C. L. (1977a). Ann. Rev. Phytopath. 15:361-385.
Altman, J., and Campbell, C. L. (1977b). Phytopath. 67:1163-1165.
Altman, J., and Campbell, C. L. (1979). Z. Pflanzenkrank. Pflanzenschutz 86:290-302.
Anderson, J. R. (1978). In "Pesticide Microbiology," (I. R. Hill and S. J. L. Wright, eds.), pp. 313-534. Academic Press, New York, N.Y.
Anderson, J. R., and Drew, E. (1976). Zbl. Bakt. Abt. II 131: 247-258.
Andres, L. A. and Goeden, R. D. (1971). In "Biological Control," (C. B. Huffaker, ed.), pp. 143-164. Plenum Publ. Corp., New York.
Audus, L. J. (1970). Proc. 10th Brit. Weed Control Conf. pp. 1036-1051.
Auld, B. A., Menz, K. M., and Medd, R. W. (1978). Agro-ecosystems 5:69-84.
Baeumer, K., and Bakermans, W. A. P. (1973). Adv. Agron. 25: 77-123.
Bains, S. S., Jhooty, J. S., Sokhi, S. S., and Rewal, H. S. (1978). Plant Dis. Rep., 62:143.
Barnes, M. M. (1970). J. Econ. Entomol. 63:1462-1463.
Beck, S. D., and Reese, J. C. (1976). In "Biochemical Interactions Between Plants and Insects," (J. W. Wallace and R. L. Mansell, eds.), 10:41-92. Rec. Adv. Phytochem.
Bendixen, L. E., Reynolds, D. A., and Reidel, R. M. (1979). Res. Bull. 1109, Ohio Agric. Res. Dev. Center, Wooster, 64 p.
Benhamou, N., Giannotti, J., and Louis, C. (1978). Acta Phytopathol., 12:107-119.
Bennett, O. L. (1977). J. Soil Water Cons., 32:9-12.

Beyer, R., and Mey, S. (1976). Beitr. Trop. Landwirtsch. Vetrinarmed. 14:297-302.
Bird, G. W., and Hogger, C. (1973). Plant Dis. Rept. 57:402.
Boon, W. R. (1967). Endeavour 26:27-32.
Boursalis, M. G., and Doupnik, B., Jr. (1976). Bull. Entomol. Soc. Am. 22:300-302.
Brooker, M. P., and Edwards, R. W. (1974). Freshwater Biol., 4:311-335.
Brooks, D. H., and Dawson, M. G. (1968). Ann. Appl. Biol. 61: 57-64.
Brown, A. W. A. (1978). "Ecology of Pesticides." John Wiley and Sons, New York, 525 p.
Browne, L. T. (1975). Proc. 27th Annu. Calif. Weed Conf. 94 p.
Buczacki, S. T. (1973). Ann. Appl. Biol. 75:25-30.
Campbell, C. L., and Altman, J. (1977). Phytopath. 67:557-560.
CAST. (1977). Council Agric. Sci. Technol., Rept. #68, 28 p.
Chumakova, B. M. (1960). Trud. vsesoyii Inst. Zashch. Rast. 15:57-70.
Clay, D. V., and Davison, J. G. (1976). Proc. Brit. Crop. Prot. Conf. – Weeds. pp. 249-258.
Clayton, E. E., Shaw, K. J., Smith, T. E., Gaines, J. G., and Graham, T. W. (1944). Phytopath. 34:870-883.
Cooper, J. I., and Harrison, B. D. (1973). Ann. Appl. Biol. 73:53-66.
Crafts, A. S. (1975). "Modern Weed Control," University of California Press, Berkeley, pp. 1-20.
Croft, B. A., and McGroarty, D. L. (1977). Res. Rept. 33, Michigan State University, Agr. Expt. Sta., East Lansing, MI.
Cromartie, W. J., Jr. (1975). J. Appl. Ecol. 12:517-533.
Curry, J. P. (1970). Pedobiologia 10:329-361.
D'Arcy, C. J., and De Zoeten, G. A. (1979). Phytopath. 69:1194-1198.
Day, B. E. (1978). In "Pest Control Strategies," (E. H. Smith and D. Pimentel, eds.), pp. 203-213. Academic Press, New York.
Del Rosario, D. A., and Putnam, A. R. (1973). Weed Sci. 21: 465-468.
Dempster, J. P. (1969). J. Appl. Ecol. 6:339-345.
Dempster, J. P., and Coaker, T. H. (1974). In "Biology in Pest and Disease Control," (D. Price-Jones and M. E. Solomon, eds.), pp. 106-114. Blackwell, Oxford.
Den Boer, P. J. (1977). Misc. Paper, Landbouwhogeschool, Wageningen, 14:1-190.
Dhingra, O. D., and da Silva, J. F. (1978). Plant Dis. Rep. 62:513-516.
Dixon, G. M., and Davison, R. H. (1975). Proc. New Zealand Grassland Assoc. 37:143-151.
Doutt, R. L., and Nakata, J. (1973). Environ. Entomol. 2: 381-386.
Duffus, J. E. (1971). Ann. Rev. Phytopath. 9:319-340.
Edwards, C. A. (1970). Proc. 10th Brit. Weed Control Conf., pp. 1052-1062.

Edwards, C. A. (1975). Outl. Agric. 8:243-244.
Edwards, C. A., and Stafford, C. J. (1979). Ann. Appl. Biol. 91:132-137.
Ehlers, W. (1975). Soil Sci. 119:242-249.
Eijsackers, H., and van der Drift, J. (1976). In "Herbicides: physiology, biochemistry, ecology," (L. J. Audus, ed.), 2:149-174. Academic Press, New York.
El-Ibrashy, M. T. (1971). Experientia 27:808-809.
Feeny, P. (1976). In "Biochemical Interactions Between Plants and Insects," (J. Wallace and R. Mansell, eds.), 10:1-49. Rec. Adv. Phytochem.
Finlayson, D. G., Campbell, C. J., and Roberts, H. A. (1975). Ann. Appl. Biol., 79:95-108.
Fischl, G. (1974). Növényvédeleem, 10:242-245.
Flaherty, D. L. (1969). Ecol. 50:911-916.
Fowler, D. L., and Mahan, J. N. (1978). The pesticide review 1977. USDA, ASCS, Washington, D.C.
Fox, C. J. S. (1964). Can. J. Plant Sci. 44:405-409.
Fox, W. B. (1948). Sci. Agric. 28:423-424.
Franklin, M. T. (1970). Proc. 10th Brit. Weed Control Conf. pp. 927-933.
Frick, K. E. (1974). In Proc. Summer Inst. Biol. Control Plant Insects and Dis., (F. G. Maxwell and F. A. Harris, eds.), pp. 204-223. University Press of Mississippi, Jackson.
Fryer, J. D., Smith, P. D., and Hance, R. J. (1980a). Weed Res. 20:103-110.
Fryer, J. D., Smith, P. D., Ludwig, J. W., and Hance, R. J. (1980b). Weed Res. 20:111-116.
Genung, W. G. (1959). Fla. State Hort. Soc. Proc. 72:163-167.
Genung, W. G., and Orsenigo, J. R. (1970). Fla. State Hort. Soc. Proc. 83:161-165.
Gerawean, R. M. (1975). Proc. 27th Annu. Calif. Weed Conf., pp. 98-99.
Glass, E. H. (Coordinator). (1975). Entomol. Soc. Am., Spec. Publ. 75-2. pp. 141.
Goeden, R. D., Andres, L. A., Freeman, T. E., Harris, P., Pienkowski, R. L., and Walker, C. R. (1974). Weed Sci. 22:490-495.
Gorske, S. F., and Hopen, H. J. (1976). Proc. North Cent. Weed Contr. Conf. 30:30-31.
Gorske, S. F., Hopen, H. J., and Randell, R. (1976). Hort Sci. 11:580-582.
Gorske, S. F., Hopen, H. J., and Randell, R. (1977). Ann. Entomol. Soc. Sm. 70:104-106.
Goss, M. J., Howse, K. R., and Harris, W. (1978). J. Soil Sci. 29:475-488.
Grau, C. R. (1977). Phytopath. 67:551-556.
Greaves, M. P. (1979). Ann. Appl. Biol. 91:129-132.
Greene, M. G. (1976). Chem. Ind. 1976:641-646.
Gregory, W. W., and Musick, G. J. (1976). Bull. Entomol. Soc. Am. 22:302-304.

Griesbach, E., and Eisbein, K. (1975). Zentralbl. Bakteriol. Parasitenkd. Inkeftionskr. Hyg. II Naturwiss. Abst., 130: 745-760.
Griffith, D. R., Mannering, J. V., and Moldenhauer, W. C. (1977). J. Soil Water Cons. 32:20-28.
Grossbard, E. (1976). In "Herbicides: physiology, biochemistry, ecology," (L. J. Audus, ed.), 2:99-147. Academic Press, New York.
Grossbard, E., and Harris, D. (1976). 28th Int. Symp. Crop Prot., part 1., Mededel. Fakult. Landbouwwetenschappen, Ghent. 41:693-702.
Hagen, K. S. (1974). Proc. Tall Timb. Conf. Ecol. Anim. Contr. Habitat Manag. 6:221-261.
Hamill, A. S., and Penner, D. (1973a). Weed Sci. 21:330-335.
Hamill, A. S., and Penner, D. (1973b). Weed Sci. 21:335-338.
Hamill, A. S., and Penner, D. (1973c). Weed Sci. 21:339-342.
Harper, J. L. (1977). "Population Biology of Plants." Academic Press, Inc., London, 892 p.
Harvey, R. G., Hagedorn, D. J., and Deloughery, R. L. (1975). Crop Sci. 15:67-71.
Haynes, R. J. (1980). Agro-Ecosystems 6:3-32.
Heathcote, G. D. (1970). Proc. 10th Brit. Weed Control Conf. pp. 934-941.
Heitefuss, R. (1970). Z. Pflkrankh. Pflschutz. Sonderheft 5: 117-127.
Heitefuss, R. (1972). Z. Pflkrankh. Pflschutz. Sonderheft 6:79-87.
Hepperly, P. R., Kirkpatrick, B. L., and Sinclair, J. B. (1980). Phytopath. 70:307-310.
Hill, I. R., and Wright, S. J. L. (eds.). (1978). "Pesticide Microbiology." Academic Press, New York, 844 p.
Hollis, J. P. (1972). Plant Dis. Rep. 56:420-424.
Hollis, J. P. (1977). Nematologica 23:71-78.
Holm, L. (1969). Weed Sci. 17:113-118.
Holm, L. (1971). Weed Sci. 19:485-490.
Hopkins, A. R., Taft, A. M., and James, W. (1972). J. Econ. Entomol. 65:870-872.
Hsia, Y., and Christensen, J. J. (1951). Phytopath. 41:1011-1020.
Huffaker, C. B., and Croft, B. A. (1976). Env. Health Persp. 14:167-183.
Huffaker, C. B., and Croft, B. A. (1978). Calif. Agric. 32:6-7.
Hull, R. (1978). Pestic. Sci. 9:239-244.
Ingram, J. W., Byrum, E. K., and Charpentier, L. J. (1947). J. Econ. Entomol. 40:745-746.
Janzen, D. H. (1971). Annu. Rev. Ecol. Syst. 2:501-508.
Johnson, W. M. (1977). Diss. Abst. Int. B. 37:4250.
Jordan, L.S., McLaughlin, J., and Russell, R. C. (1973). Proc. 25th Annu. Calif. Weed Conf., pp. 109-113.
Kalmbacher, R. S., Minnick, D. R., and Martin, F. G. (1979). Agron. J. 71:365-368.

Katan, J., and Eshel, Y. (1973). Res. Rev. 45:145-177.
Katan, J., and Eshel, Y. (1974). Phytopath. 64:1186-1192.
Kavanagh, T. (1969). Sci. Proc. Royal Dublin Soc. Ser. B-2, 179-190.
Kavanagh, T. (1974). Sci. Proc. Royal Dublin Soc. Ser. B-3, 251-265.
Kelley, A. D., and Bruns, V. F. (1975). Weed Sci. 23:486-493.
Kelley, W. D., and South, D. B. (1980). Weed Sci. 28:599-602.
Kennedy, J. M., Talbert, R. E., and Morris, J. R. (1979). J. Am. Soc. Hort. Sci. 104:713-716.
Khattat, A. R., and Stewart, R. K. (1980). Ann. Entomol. Soc. Am. 73:282-287.
Kilgore, W. W. and Doutt, R. L. (1967). "Pest Control: Biological, Physical, and Selective Chemical Methods." Academic Press, New York, 477 p.
Kirkland, R. L., and Goeden, R. D. (1978). Env. Entomol. 7: 349-354.
Krause, R. A., and Massie, L. B. (1975). Annu. Rev. Phytopath. 13:31-47.
Larson, W. E., Triplett, G. B., Jr., Van Duren, D. M., Jr., and Musick, G. J. (1970). Crops Soils 23:14-20.
Lawes Agricultural Trust. (1974). Rothamsted Experimental Station Guide, 1974. Harpenden, Herts, U.K.
Lee, M., and Lockwood, J. L. (1977). Phytopath. 67:1360-1367.
Leeper, J. R. (1974). Proc. Hawaiian Entomol. Soc. 21:403-412.
Leigh, T. F. (1961). J. Econ. Entomol. 54:120-122.
Leigh, T. F., Grimes, D. W., Yamada, H., Stockton, J. R., and Bassett, D. (1969). Proc. Tall Timb. Conf. Ecol. Anim. Contr. 1:71-83.
Lichtenstein, E. P., Liang, T. T., and Anderegg, B. N. (1973). Science 181:847-849.
Liebee, G. L., and Horn, D. J. (1979). Env. Entomol. 70:485-486.
Lindeman, R. L. (1942). Ecol. 23:399-418.
Luff, M. L. (1980). Ann. Appl. Biol. 94:153-164.
Lund, R. D., and Turpin, F. T. (1977). Env. Entomol. 6:695-698.
Maceljski, M. (1968). Anz. Schädlingsk. 41:81-88.
MacKenzie, D. R., Cole, H., and Ercegovich, C. D. (1968). Phytopath. 58:1058.
Magyarosy, A. C., and Duffus, J. E. (1977). Plant Dis. Rep. 61:248-251.
Mantle, P. G., Shaw, S., and Doling, D. A. (1977). Ann. Appl. Biol. 86:339-351.
Mayer, A. M., and Poljakoff-Mayber, A. (1975). "The Germination of Seeds." Pergamon Press Inc., Elmsford, New York.
Mayse, M. A., and Price, P. W. (1978). Agro-Ecosystems 4:387-405.
Mellanby, K. (1980). "The Biology of Pollution," Edward Arnold Ltd., London, 68 p.
Merrill, W., and Kistler, B. R. (1978). Plant Dis. Rep. 62: 200-202.
Messenger, P. S. (1971). Proc. Tall Timb. Conf. Ecol. Anim. Contr. 3:97-114.

Miller, P. M., and Ahrens, J. F. (1969). Plant Dis. Rep. 53:642-646.
Miller, J. H., Carter, C. H., Garber, R. H., and DeVay, J. E. (1979). Weed Sci. 27:444-449.
Moffett, J. O., Morton, H. L., and MacDonald, R. H. (1972). J. Econ. Entomol. 65:32-36.
Morton, H. L., and Moffett, J. O. (1972). Env. Entomol. 1:611-614.
Morton, H. L., Moffett, J. O., and MacDonald, R. H. (1972). Env. Entomol. 1:102-104.
Myers, L. F., and Moore, R. M. (1952). J. Aust. Inst. Agric. Sci. 18:152-155.
Nalewaja, J. D. (1980). Proc. West Soc. Weed Sci. 33:5-15.
Nilsson, H. E. (1977). Phytopath. Z. 90:361-363.
Norris, R. F., and Burton, V. E. (1979). Weed Sci. Soc. Am., Abst. #79. p. 38.
Norris, R. F., and Cothran, W. R. (1974). Proc. West. Soc. Weed Sci. 27:74.
Norris, R. F., and Schoner, C. A., Jr. (1975). Proc. Calif. Alfalfa Symp. 5:104-112.
Oka, I. N., and Pimentel, D. (1976). Science 193:239-240.
Okafor, L. I., and deDatta, S. K. (1976). Weed Sci. 24:43-46.
Olkowski, H., and Olkowski, W. (1976). Bull. Entomol. Soc. Am. 22:313-317
Orum, T. V., Bartels, P. G., and McClure, M. A. (1979). J. Nematol. 11:78-83.
Periera, H. C. (1975). Outl. Agric. 8:211-212.
Perrin, R. M. (1975). Ann. Appl. Biol. 81:289-297.
Perrin, R. M. (1977). Agro-Ecosystems 3:93-118.
Phillips, R. E., Blevins, R. L., Thomas, G. W., Frye, W. W., and Phillips, S. H. (1980). Science 208:1108-1113.
Phillips, S. H., and Young, H. M., Jr. (1973). "No-tillage farming." Reimann Assoc., Milwaukee, WI, 224 p.
Pimentel, D. (1961). Ann. Entomol. Soc. Am. 54:76-86.
Pimentel, D., and Goodman, N. (1979). In "Pest Management in Transition," (P. de Jong, Coordinator), pp. 1-18. Westview Press, Boulder.
Pinckard, J. A., and Standifer, L. C. (1966). Plant Dis. Rep. 50:172-174.
Porter, D. M., and Rud, O. E. (1979). Phytopath. 69:1042.
Potts, G. R. (1970). Bird Study 17:145-166.
Powell, D. M., and Wallis, R. L. (1974). Proc. Tall Timb. Conf. Ecol. Anim. Contr. Habitat Manag. 6:87-97.
Putnam, A. R., and Penner, D. (1974). Res. Rev. 50:73-110.
Rieckmann, W. (1979). Z. Pflanzenkrank. Pflanzenschutz 86:755-762.
Robinson, A. F., Orr, C. C., and Abernathy, J. R. (1979). J. Nematol. 11:73-77.
Rock, G. C., and Yeargan, D. R. (1973). J. Econ. Entomol. 66:1342-1343.
Romney, R. K., Anderson, J. L., and Griffin, G. D. (1974). Weed Sci. 22:51-54.

Root, R. B. (1973). Ecol. Monogr. 43:95-124.
Sacher, R. F., Hopen, H. J., and Jacobsen, B. J. (1977). Down to Earth 32:16-21.
Sagar, G. R., and Mortimer, A. M. (1977). Appl. Biol. 1:1-47.
Sarai, D. S. (1969). J. Econ. Entomol. 62:1507-1508.
Schoner, C. A., Jr., and Norris, R. F. (1975). Proc. Calif. Alfalfa Symp. 5:41-48.
Scott Russell, R. (1977). "Plant Root Systems: Their Function and Interaction With the Soil." McGraw-Hill Book Co., (UK) Ltd., London, 298 p.
Sewell, G. W. F., and Wilson, J. F. (1974). Ann. Appl. Biol. 76:37-47.
Sherrod, D. W., Shaw, J. T., and Luckman, W. H. (1979). Env. Entomol. 8:191-195.
Smith, E. H., and Pimentel, D. (eds.). (1978). "Pest Control Strategies." Academic Press, New York, 334 p.
Smith, J. G. (1969). Ann. Appl. Biol. 63:326-330.
Smith, J. G. (1976a). Ann. Appl. Biol. 83:1-13.
Smith, J. G. (1976b). Ann. Appl. Biol. 83:15-29.
Smith, J. R., and Ferry, B. W. (1979). Ann. Bot. 43:93-99.
Smith, R. F. (1954). J. Econ. Entomol. 47:205-210.
Smith, R. F., and Reynolds, H. T. (1966). Proc. FAO Symp. Integ. Pest Control 1:11-17.
Smith, R. J., Jr., and Tugwell, N. P. (1975). Weed Sci. 23: 176-178.
Soil Conservation Society of America. (1973). "Conservation tillage." Soil Cons. Soc. Am., Ankeny, Iowa, 241 p.
Soil Conversation Society of America. (1977). "Conservation tillage: problems and potentials." Spec. Publ. 20. Soil Cons. Soc. Am., Ankeny, Iowa. 75 pp.
Stam, P. A., Clower, D. F., Graves, J. B., and Schilling, P. E. (1978). J. Econ. Entomol. 71:477-480.
Stedman, O. J. (1976). Rept. 1975, #258, Rothamsted Expr. Sta., 1976, part 1. pp. 371.
Steiner, H. (1966). Proc. FAO Symp. Integr. Pest Control. 3:13-20.
Stern, V. M., Smith, R. F., van den Bosch, R., and Hagen, K. S. (1959). Hilgardia 29:81-101.
Stevens, J. B. T. (ed.). (1975). "Energy in Agriculture." Span 18:1-44.
Stevens, O. A. (1957). Weeds 5:46-55.
Steyvoort, L. van. (1975). 3rd Int. Meeting Sel. Weed Contr. Beet Crops. pp. 197-200.
Syme, P. D. (1975). Env. Entomol. 4:337-346.
Tamaki, G. (1975). Env. Entomol. 4:958-960.
Tamaki, G., Fox, L., and Chauvin, R. L. (1980). Env. Entomol. 9:62-66.
Tamaki, G., Moffitt, H. R., and Turner, J. E. (1975). Env. Entomol. 4:274-276.
Tamaki, G., and Olsen, D. (1979). Env. Entomol. 8:314-317.

Teasdale, J. R., Harvey, R. G., and Hagedorn, D. J. (1978). Weed Sci. 26:609-613.
Teasdale, J. R., Harvey, R. G., and Hagedorn, D. J. (1979). Weed Sci. 27:467-472.
Templeton, G. E., TeBeest, D. O., and Smith, R. J., Jr. (1979). Annu. Rev. Phytopath. 17:301-310.
Theunissen, J., and Den Ouden, H. (1980). Entomol. Exp. & Appl. 27:260-268.
Thomason, I. J. (Tech. Coord.). (1978). "Integrated Pest Management." Calif. Agr. 23:1-36.
Thompson, A. R., Price, J. L., and Roberts, H. A. (1977). Ann. Appl. Biol. 86:415-422.
Thomson, S. V., Schroth, M. N., Moller, W. J., Reil, W. O., Beutel, J. A., and Davis, C. S. (1977). Calif. Agric. 31:12-14.
Tomlinson, J. A., and Carter, A. L. (1970). Ann. Appl. Biol. 66:381-386.
Tomlinson, J. A., Carter, A. L., Dale, W. T., and Simpson, C. J. (1970). Ann. Appl. Biol. 66:11-16.
Triplett, G. B., Jr. (1976a). Entomol. Soc. Am. Symp.: "Crop Production with Reduced Tillage Systems." Bull. Entomol. Soc. Am. 22:289-291.
Triplett, G. B., Jr. (1976b). Entomol. Soc. Am. Symp.: "Crop Production with Reduced Tillage Systems." Bull. Entomol. Soc. Am. 22:298-299.
Triplett, G. B., Jr. (1978). Agron. J. 70:577-581.
Triplett, G. B., Jr., and Lytle, G. D. (1972). Weed Sci. 20:453-457.
Tucker, W. (1978). Harper's 257:43-58.
Ulber, B. (1979). Z. Ang. Entomol. 87:143-153.
Unger, P. W., and Phillips, R. E. (1973). In "Conservation Tillage," Soil Conserv. Soc. Am. pp. 42-54.
U.S.D.A. (1978). "Biological agents for pest control. Status and prospects." U.S. Govt. Printing Office, Washington, D.C.
van Emden, H. F. (1964). Nature 201:946-948.
van Emden, H. F. (1965a). Sci. Hort. 17:121-136.
van Emden, H. F. (1965b). J. Appl. Ecol. 2:171-196.
Van Emden, H. F. (1970). Proc. 10th Brit. Weed Control Conf. III:917-919.
van Emden, H. F., and Williams, G. F. (1974). Annu. Rev. Entomol. 19:455-475.
Vengris, J. W., Colby, C., and Drake, M. (1955). Agron. J. 47:213-216.
Vickerman, G. P. (1974). Proc. 12th Brit. Weed Control Conf. III:929-939.
Volz, M. G. (1977). Agro-Ecosystems 3:313-323.
Way, M. J. (1968). Proc. 9th Brit. Weed Control Conf. pp. 989-994.
Way, J. M., and Chancellor, R. J. (1976). In "Herbicides: physiology, biochemistry, ecology," (L. J. Audus, ed.), 2:345-372. Academic Press, New York.

Weinbaum, Z., and Milbrath, G. M. (1976). Plant Dis. Rep. 60:469-471.
Weierich, A. J., Nelson, Z. A., and Appleby, A. P. (1977). Weed Sci. 25:27-29.
Wicks, G. A., and Smika, D. E. (1974). Abst. #107, Weed Sci. Soc. Am.
Wilkinson, V. (1969). New Phytol. 68:701-708.
Wilkinson, V., and Lucas, R. L. (1969). New Phytol. 68:709-719.
Williamson, B., Lawson, H. M., Woodford, J. A. T., Hargreaves, A. J., Wiseman, J. S., and Gordon, S. C. (1979). Ann. Appl. Biol. 92:359-368.
Wright, S. J. L. (1978). In "Pesticide Microbiology," (I. R. Hill and S. J. L. Wright, eds.), pp. 535-602. Academic Press, New York.
WSSA. (1979). "Herbicide Handbook." Weed Sci. Soc. Am. Champaign, Illinois, 479 p.
Wyse, D. L., Meggitt, W. F., and Penner, D. (1976). Weed Sci. 24:16-21.
Yarham, D. J. (1975). Outl. Agric. 8:245-247.
Yarris, L., and Wharton, W. J. (1978). News, SEA, USDA, WR-66-78. 2 p.
Yarwood, C. E. (1968). BioScience 18:27-30.
Yarwood, C. E. (1970). Science 168:218-220.
Yeates, G. W., Stout, J. D., Ross, D. J., Datch, M. E., and Thomas, R. F. (1976). New Zealand J. Agric. Res. 19:51-61.
Zadoks, J. C., and Schein, R. D. (1979). "Epidemiology and Plant Disease Management," Oxford University Press, 427 p.
Zandstra, B. H., and Motooka, P. S. (1978). PANS 24:333-338.
Zimdahl, R. L. (1979). Proc. West. Soc. Weed Sci. 32:13-30.

EFFECTS OF LIGHT AND TEMPERATURE ON WEED/CROP GROWTH AND COMPETITION

David T. Patterson
U.S. Department of Agriculture
Southern Weed Science Laboratory
Stoneville, Mississippi

I. INTRODUCTION

Weeds are among the most important of all crop pests. In the United States, annual losses in crop yield and quality due to weeds, combined with the costs of weed control, exceed the costs of insects, plant diseases, and nematodes (Ennis, 1976; Shaw, 1978). Unlike other pests, weeds compete directly with crops for the environmental resources upon which plant growth depends.

Although many definitions of "competition" have been proposed (Zimdahl, 1980), for our purposes it can be considered the simultaneous demand by more than one plant for the same resources, when the immediate supply of the resources is below the combined demand of the plants (Donald, 1963).

The primary resources for which weeds and crops compete are water, light, and mineral nutrients (Harper, 1977; Zimdahl, 1980). Competition also may occur for oxygen in the soil and carbon dioxide in the air. Competition for physical space also may occur, as for germination microsites and room for tuber and root development. Competition for agents of pollination and seed dispersal may occur in some agroecosystems (Risser, 1969).

The overall detrimental effects of weeds on crops ["interference" as defined by Harper (1961)] may include allelopathy in

ISBN 0-12-332850-0

addition to competition. Allelopathy, the detrimental effect of one plant on another caused by the synthesis and release of toxic or inhibitory substances (Rice, 1974) is clearly related to competition, because competition-induced stress may increase the production of allelopathic substances, and growth inhibition caused by allelopathy may reduce the competitive ability of the affected plant.

Zimdahl (1980), in a comprehensive review of weed-crop competition, summarized potential weed-related yield losses in various crops. Of course, these yield losses vary greatly with the species of weed and crop, the densities of the weed and crop plants, the duration of weed competition, soil type, fertility level, management practices, and general climatic conditions. Maximum yield losses can exceed 90% of the potential yield in many crops, but actual losses in farmers' fields are generally much lower.

Various workers have attempted to delineate the characteristics that endow a plant with competitive ability or competitiveness. Clements et al. (1929), in their classic treatment of plant competition, stated:

> "It is evident that practically all the advantages or weapons of competing species are epitomized in two words - amount and rate. Greater storage in seed or rootstock, more rapid and complete germination, earlier start, more rapid growth of roots and shoots, taller and more branching stems, deeper and more spreading roots, more tillers, larger leaves, and more numerous flowers are all the essence of success."

The ability to compete effectively with associated crop species is only one of many factors contributing to the success and impact of a particular weed in an agroecosystem. In fact, many of the most devastating agronomic weeds might be considered poor competitors in the classical sense (Baker, 1974; Grime, 1979; Harper, 1977), because they thrive only in intensively disturbed or managed habitats – often supplied with supplemental water and nutrients – and have little ability to successfully invade and establish themselves in natural plant communities. For example, Sharp (1976), in a phytosociological survey of weed communities in North Carolina, reported that agronomically important weeds, many of which are exotic and/or annuals, rarely persist after the abandonment of cultivated fields, but are displaced by presumably more competitive perennial plants characteristic of the "old-field" ecosystem.

As Harper (1965) pointed out, generalizations about the competitive ability of a particular species are risky. Competitiveness is not an intrinsic property of an individual species or plant. It is a relative term, depending on the association of plants with which an individual competes. Nevertheless, according to Harper (1965), success in mixed stands of competing species generally is associated with early and rapid establishment from large seeds, rapid canopy development, and rapid root growth.

Factors contributing directly or indirectly to the success of agronomic weeds include the following:

1. Physiological and/or morphological similarity to the crop.
2. Vegetative reproduction.
3. Efficient and rapid uptake of water and nutrients through rapid development of exploitive root systems.
4. Rapid partitioning of plant biomass into leaf area (Potter and Jones, 1977; Patterson et al. 1978b) and consequent rapid development of canopy shade over crop.
5. Efficient photosynthetic uptake of CO_2 [C_4 versus C_3 photosynthetic pathway (Black et al., 1969)].
6. Low transpiration/photosynthesis ratio resulting in more efficient use of water.
7. Rapid rate of growth and high initial relative growth rate (Grime and Hunt, 1975).

8. Freedom from environmental constraints, conferred by broad tolerance to environmental variables and/or capacity for acclimation and adaptability to changing environmental factors.

Several of the attributes of Baker's (1974) "ideal weed" also may contribute to success in agronomic weeds. These include internally-controlled dormancy mechanisms, nonspecific germination requirements, copious seed production, phenotypic plasticity, ability to produce at least some seed in a wide range of environmental conditions, and special abilities for interspecific interference (allelopathy, viny habit, smothering growth). Bazzaz (1979) provided further examples of characteristics of successful weeds.

A survey of various lists of important agronomic weeds reveals many common characteristics (McWhorter and Patterson, 1980). For example, of the 37 worst soybean weeds in the United States 38% are monocots, 32% are perennials, 35% have some form of vegetative reproduction, 19% produce rhizomes, 38% have the C_4 photosynthetic pathway, 55% are exotic to the United States, and 55% have been reported to have allelopathic properties (Wax, 1976). Of the world's worst 18 weeds (Holm et al., 1977), 72% are monocots, 44% are perennials, 61% reproduce vegetatively, 33% produce rhizomes, and 78% have the C_4 pathway (McWhorter and Patterson, 1980). Among these characteristics perhaps the most outstanding are the tremendous overrepresentations of C_4 plants and monocots as important agronomic weeds in proportion to their occurrence in the world's flora.

II. ENVIRONMENTAL FACTORS INFLUENCING GROWTH AND COMPETITIVENESS

Temperature and light, together with water and nutrient availability, are the primary factors which govern plant growth and hence, the potential competitiveness of crop plants and their associated weeds. In this review, we will consider the effects of light and temperature only.

The general effects of light and temperature on plant growth have been reviewed extensively (Blackman, 1960; Daubenmire, 1974; Evans, 1963; Went, 1957). The environmental responses of nine major crops are summarized in Evans (1975).

Various workers have described the light and/or temperature responses of a number of important weeds: johnsongrass [*Sorghum halepense* (L.) Pers.] (McWhorter and Jordan, 1976); purple nutsedge (*Cyperus rotundus* L.) (Wills, 1975); ragweed parthenium (*Parthenium hysterophorus* L.) (Williams and Groves, 1980); quackgrass [*Agropyron repens* (L.) Beauv.] and *Agrostis gigantea* Roth (Williams, 1971); emex (*Emex australis* Steinh) and spiny emex (*Emex spinosa* Campd.) (Weiss and Simmons, 1977); yellow nutsedge (*Cyperus esculentus* L.) (Keeley and Thullen, 1978); velvetleaf (*Abutilon theophrasti* Medic.), spurred anoda [*Anoda cristata* (L.) Schlecht], prickly sida (*Sida spinosa* L.), common cocklebur (*Xanthium pensylvanicum* Wallr.), johnsongrass and redroot pigweed (*Amaranthus retroflexus* L.) (Potter and Jones, 1977); hemp sesbania [*Sesbania exaltata* (Raf.) Cory], velvetleaf, and redroot pigweed (Patterson et al., 1978b); itchgrass (*Rottboellia exaltata* L.f.) (Patterson, 1979 and Patterson et al., 1979); and cogongrass [*Imperata cylindrica* (L.) Beauv.] (Patterson, 1980b and Patterson et al., 1980).

Comparative studies of the light and temperature responses of weeds and crop plants are few. Potter and Jones (1977) reported that relative growth rates (R_w) and relative leaf area expansion rates (R_a) of corn (*Zea mays* L.), cotton (*Gossypium hirsutum* L.), soybean [*Glycine max* (L.) Merr.], and six associated weed species were greater at day/night temperatures of 32/21C than at 21/10C or 38/27C. At all three temperatures, the weeds had higher R_w and R_a than the crops. Noguchi and Nakayama (1978a) reported that, in Japan, cool temperatures favored the growth of corn, upland rice (*Oryza sativa* L.), soybean, and peanut (*Arachis hypogaea* L.) over that of several weeds, early in the growing season. However, weed growth was accelerated by warmer temperatures 40 to 50 days after planting, and weed competition then became more severe. In another study of the effects of temperature on weed-crop competition in Queensland, Australia, Hawton (1979) found that air temperatures of 23 C or greater favored the germination and early growth of goosegrass [*Eleusine indica* (L.) Gaertn.] over that of millet (*Setaria anceps* Stapf. cv. Nandi). He suggested that the competitive balance could be shifted in the direction of the crop by seeding it earlier in the season.

Patterson et al. (1978b) found that shading greatly reduced R_w and R_a of cotton and three associated weeds. However, the R_w and R_a of the weeds were higher than those of cotton, regardless of the light level during growth. From studies of the effects of artificial shade on the growth of yellow nutsedge, Keeley and Thullen (1978) concluded that corn, alfalfa (*Medicago sativa* L.), and other rapidly developing crops would effectively suppress the weed through competition for light. In similar studies, Noguchi and Nakayama (1978b) concluded that, of several crop plants, soybean and peanut were most effective in suppressing Henry crabgrass [*Digitaria adscendens* (H.B.K.) Henr.], chufa (*Cyperus microiria* Steud.), and common purslane (*Portulaca oleracea* L.).

Other studies summarized below from the author's work, serve as further examples. Such studies of the comparative light and temperature responses of weeds and crop plants contribute to our understanding of the weed-crop interactions that occur in agroecosystems. They may also aid in the prediction of future weed problems and in the development of integrated weed management systems to meet future needs.

Investigations of the light and temperature responses of weeds and crop plants require the use of controlled environmental facilities in which environmental variables can be studied singly or in combination. Most of the work summarized here has been conducted in the Duke University Phytotron Unit of the Southeastern Plant Environment Laboratory (Kramer et al., 1970).

The technique of mathematical growth analysis (Kvet et al., 1971; Hunt, 1978) provides a convenient means of examining the processes of total dry matter production and leaf-area expansion that are important in determining a plant's vegetative growth and potential competitiveness under a variety of environmental conditions. The basic formulas used in mathematical growth analysis are summarized in Table I.

Table I. Growth analysis definitions and formulas.

$T_2 - T_1$ = length of harvest interval (days) = ΔT

W_1, W_2 = dry wt at beginning and end of harvest interval

A_1, A_2 = leaf area at beginning and end of harvest interval

$\Delta W = W_2 - W_1$; $\Delta A = A_2 - A_1$

R_w = Relative Growth Rate (g/g/day) = $(\ln W_2 - \ln W_1)/\Delta T$

R_a = Relative Leaf Area Growth Rate (dm^2/dm^2/day) = $(\ln A_2 - \ln A_1)/\Delta T$

NAR = Net Assimilation Rate (g/dm^2/day) = $(\Delta W/\Delta A) \times (\ln A_2 - \ln A_1)/\Delta T$

LAI = Leaf Area Index = dimensionless ratio of leaf area to land area

CGR = Crop Growth Rate (g/dm^2 land surface/day) = NAR x LAI

LWR = Leaf Weight Ratio (g/leaf wt/g tot wt)

LAD = Leaf Area Duration (dm^2 days) = $[\Delta A/(\ln A_2 - \ln A_1)] \times (\Delta T)$

BMD = Biomass Duration (g days) = $[\Delta W/\ln W_2 - \ln W_1)] \times (\Delta T)$

SLW = Specific Leaf Weight (g/dm^2)

SLA = Specific Leaf Area (dm^2/g), SLA = 1/SLW

LAR = Leaf Area Ratio (dm^2/g tot wt), LAR = LWR x SLA

This technique requires destructive harvests of plant material at appropriate intervals during an experiment. The basic data on leaf area, total plant dry weight, and dry weights of plant parts can then be used to calculate rates of dry matter production per unit leaf area (net assimilation rate or NAR), relative growth rates, relative leaf-area expansion rates, and partitioning coefficients for plant biomass and leaf area allocation.

Several growth analysis parameters are particularly relevant to studies of potential competitiveness. For example, the leaf area duration (LAD) or total amount of leaf area present during a particular interval serves as one indicator of the potential impact of an individual plant in a competitive situation, since leaf area is important in competition for light. The amount of dry matter produced by an individual plant is an indicator of its overall utilization of the resources available for plant growth. Dry matter production is the product of average NAR and LAD. The maximum relative growth rate (R_w) of individual plants was used by Grime and Hunt (1975) as an overall indicator of potential competitive ability. The R_w is the product of the NAR and the leaf area ratio (LAR). The LAR or amount of leaf area per unit of total plant weight is a measure of the relative leafiness of a plant. The LAR in turn is the product of the leaf-weight ratio (LWR = leaf weight/total plant weight) and the specific leaf area (SLA = leaf area/leaf weight). The LWR is a measure of the partitioning of plant biomass into the leaf component, and the SLA is a measure of the allocation of leaf biomass as leaf area.

These growth analysis relationships are particularly useful in evaluating the response of plants to irradiance during growth. For example, in a study of the effects of shading on the growth of the exotic noxious weed itchgrass, Patterson (1979) found that shading to 60, 25, or 2% of full sunlight markedly reduced dry-matter production. Leaf-area production was less severely reduced, even though the leaves produced in the shade were much thinner (indicated by their greater SLA values). The partitioning of plant biomass into the leaf component (indicated by changes in LWR) increased significantly with each increase in the intensity of shading. The combined increases in SLA and LWR, in response to increasing shade, resulted in great increases in LAR also. These increases in LAR represent an adaptation to shading at the whole plant level since the investment of plant biomass in photosynthetic light-harvesting structure is increased.

Adaptation to growth irradiance also occurs at the level of individual leaves and chloroplasts. Thus, Patterson et al. (1978a) found that growth irradiance influenced the photosynthetic capacities of cotton and the associated malvaceous weed velvetleaf primarily through effects on leaf mesophyll volume, chlorophyll content, and photosynthetic unit (PSU) density per unit leaf area. Velvetleaf had a maximum photosynthetic rate per unit mesophyll volume about twice as great as that of cotton, indicating an advantage for the weed in terms of potential productivity per unit of leaf material.

In studies of the effects of shading on the aquatic weed water hyacinth [*Eichornia crassipes* (Mart.) Solms], maximum photosynthetic capacity was closely correlated with leaf mesophyll conductance and protein content per unit area (Patterson and Duke, 1979). Photosynthetic light adaptation characteristics observed in water hyacinth may be of adaptive significance in a floating aquatic plant that may be exposed to a range of irradiances during growth.

Adaptation to shading may also occur at the population level through the development of genetically distinct ecotypes (Bjorkman, 1973; Gauhl, 1976; Patterson, 1980a). However, in a study of local populations of the exotic noxious weed cogongrass from shaded and exposed habitats, Patterson (1980b) found little evidence of ecotypic differentiation. Plants from all three populations grew much more vigorously in full sunlight than in 56 or 11% of full sunlight. Adaptation to shading was evident in increased LWR, SLA, and LAR in all populations. Based on its observed responses to artificial shading, cogongrass should be capable of surviving and producing rhizomes under the canopy shade of common row crops.

Temperature as an environmental variable affects both the amount and duration or periodicity of weed and crop growth. Temperatures early in the growing season are particularly important because of their effects on seed germination and on the development rates of weed and crop seedlings. Growth advantages established by either the crop seedling or associated weeds may be critical during early season competition.

For example, velvetleaf and spurred anoda, malvaceous weeds of increasing importance in cotton production in the South, have been reported to be more competitive with cotton following abnormally cool periods early in the growing season (Chandler, 1977; Oliver and Lambert, 1974). In controlled-environment experiments, we compared the responses of cotton, velvetleaf, and spurred anoda to simulated naturally occurring chilling events typical of those occurring during the first month of the growing season for cotton in the Yazoo-Mississippi Delta (Patterson and Flint, 1979a). We found that a three-day period of chilling at 17/13C day/night temperatures reduced dry matter and leaf-area production in all three species, in comparison to control plants maintained at a simulated average May temperature of 26/21C. However, the chilling treatment reduced the growth of cotton more than that of the two weeds. The growth-inhibiting effects of chilling persisted even after a 15-day recovery period (seven days at 26/21C followed by eight days at 29/23C simulating early June temperatures).

At the end of the second recovery period, 42 days after planting, weed/crop ratios for both leaf area and total dry weight were significantly greater when both the weed and the crop had been subjected to the chilling treatment. The persistent inhibitory effects of chilling on cotton were primarily associated with reduced leaf area expansion and LAD. Evaluations of the effects of chilling on photosynthesis and stomatal conductance in both cotton and velvetleaf showed that both of these parameters were significantly reduced by chilling, but complete recovery to the level of the unchilled controls occurred within 94 hours after the termination of the chilling treatment (Patterson and Flint, 1979a).

Studies of the temperature responses of exotic weeds can aid in predicting their potential geographic range in new environments. In studies of itchgrass, an exotic weed of tropical origins currently present in the United States only in southern Florida and southern Louisiana, we found by growing the plant in 36 combinations of day and night temperature, it could grow vigorously and set seed in temperature regimes simulating summer conditions as far north as Wisconsin (Patterson et al., 1979).

In subsequent experiments, we compared the chilling responses of itchgrass with those of the adapted varieties of corn and soybean with which it would be expected to compete, in temperature regimes simulating the first five weeks of the growing season for corn and soybean at Madison, Wisconsin; for soybean at Carbondale, Illinois; for corn at Waycross, Georgia; and for soybean at Baton Rouge, Louisiana (Patterson and Flint, 1979b). We found that the cool early growing season temperatures and superimposed simulated naturally occurring chilling events typical of the first three locations severely reduced the growth of itchgrass relative to that of the adapted crop varieties. Thus, we concluded that itchgrass is unlikely to be a serious early season competitor with corn or with soybeans outside the South.

Similar studies with another exotic weed, cogongrass, have shown that its sensitivity to cool temperatures reduces its chances of becoming a serious pest outside the Gulf Coast states (Patterson et al., 1980).

III. CONCLUSIONS

Studies of the comparative environmental responses of weeds and crops are essential to the development of effective integrated pest management systems. Knowledge of weed responses to environmental factors can help predict when and whether a particular weed will constitute a problem. For example, the relative responses of a weed and a crop to temperature extremes may indicate whether the weed will be more important in early or late season. The relative responses to irradiance may indicate whether canopy shading by the developing crop will effectively suppress the weed or vice versa.

Differential responses to water stress and limited nutrients, although not considered in this review, may likewise be extremely important in weed-crop interactions.

Controlled environment studies can aid in the prediction of the potential range and importance of recently introduced exotic noxious weeds. Such information is essential for the development of effective quarantine and eradication programs for exotic weeds and for the setting of priorities for such programs.

Weed phenology models developed from controlled-environment experiments may aid in determining critical stages in weed life cycles, in scheduling timely application of control practices, and in synchronizing biological control methods. Likewise, weed productivity models, which may ultimately interface with crop productivity models, should aid in the prediction of the competitiveness of specific weeds with specific crops and in the establishment of economic threshold levels for weed infestations.

Finally, increased basic knowledge of the competitiveness and environmental responses of successful weeds may facilitate the development of more competitive and productive crop varieties, or of altogether new crops. These improvements could contribute significantly to freeing agronomic crops from the environmental constraints that so severely limit crop production.

REFERENCES

Baker, H. G. (1974). Ann. Rev. Ecol. Syst. 5:1-24.
Bazzaz, F. A. (1979). Ann. Rev. Ecol. Syst. 10:351-371.
Bjorkman, O. (1973). Photophysiol. 8:1-63.
Black, C. C., Chen, T. M., and Brown, R. H. (1969). Weed Sci. 17:338-344.
Blackman, G. E. (1960). In "Growth in Living Systems" (M. X. Zarrow, ed.), pp. 525-556. Basic Books, Inc., New York.
Chandler, J. M. (1977). Weed Sci. 25:151-158.
Clements, F. E., Weaver, J. E., and Hanson, H. C. (1929). "Plant Competition - An Analysis of Community Function." Carnegie Inst., Washington, D.C.
Daubenmire, R. F. (1974). "Plants and Environment." John Wiley & Sons, New York.
Donald, C. M. (1963). Advan. Agron. 15:1-118.

Ennis, W. B., Jr. (1976). In "World Soybean Research" (L. D. Hill, ed.), pp. 375-386. Interstate Printers and Publishers, Inc., Danville, Illinois.
Evans, L. T., ed. (1963). "Environmental Control of Plant Growth." Academic Press, New York.
Evans, L. T., ed. (1975). "Crop Physiology - Some Case Histories." Cambridge University Press, London.
Gauhl, E. (1976). Oecologia 22:275-286.
Grime, J. P. (1979). "Plant Strategies and Vegetation Processes." John Wiley & Sons, Chichester.
Grime, J. P. and Hunt, R. (1975). J. Ecol. 63:393-422.
Harper, J. L. (1961). In "Mechanisms in Biological Competition" (F. L. Milthorpe, ed.), pp. 1-39. Oxford University Press, Cambridge.
Harper, J. L. (1965). In "Genetics Today" (S. J. Geerts, ed.), 2:465-482. Pergamon Press. Oxford.
Harper, J. L. (1977). "Population Biology of Plants." Academic Press, London.
Hawton, D. (1979). Weed Res. 19:279-284.
Holm, L. G., Plucknett, D. L., Pancho, J. W., and Herberger, J. P. (1977). "The World's Worst Weeds. Distribution and Biology." University Press of Hawaii, Honolulu.
Hunt, R. (1978). "Plant Growth Analysis." Edward Arnold, London.
Keeley, P. E., and Thullen, R. J. (1978). Weed Sci. 26:10-16.
Kramer, P. J., Hellmers, H., and Downs, R. J. (1970). BioSci. 20:1201-1208.
Kvet, J., Ondok, J. P., Necas, J., and Jarvis, P. G. (1971). In "Plant Photosynthetic Production. Manual of Methods." (Z. Sestak, J. Catsky, and P. G. Jarvis, eds.), pp. 343-391. W. Junk, The Hague.
McWhorter, C. G., and Jordan, T. N. (1976). Weed Sci. 24:88-91.
McWhorter, C. G. and Patterson, D. T. (1980). In "Proc. World Soybean Res. Conf. II" (F. T. Corbin, ed.), pp. 371-392. Westview Press, Boulder, Colorado.
Noguchi, K. and Nakayama, K. (1978a). Jap. J. Crop Sci. 47:48-55.
Noguchi, K. and Nakayama, K. (1978b). Jap. J. Crop Sci. 47:381-387.
Oliver, D. and Lambert, B. (1974). Weeds Today 5:22.
Patterson, D. T. (1979). Weed Sci. 27:549-553.
Patterson, D. T. (1980a). In "Predicting Photosynthesis for Ecosystem Models" (J. D. Hesketh and J. W. Jones, eds.), I:205-235. CRC Press, Inc., Boca Raton, Florida.
Patterson, D. T. (1980b). Weed Sci. 28:735-740.
Patterson, D. T., and Duke, S. O. (1979). Plant & Cell Physiol. 20:177-184.
Patterson, D. T., and Flint, E. P. (1979a). Weed Sci. 27:473-479.
Patterson, D. T., and Flint, E. P. (1979b). Weed Sci. 27:645-650.

Patterson, D. T., Duke, S. O., and Hoagland, R. E. (1978a). Plant Physiol. 61:402-405.
Patterson, D. T., Meyer, C. R., and Quimby, P. C., Jr. (1978b). Plant Physiol. 62:14-17.
Patterson, D. T., Meyer, C. R., Flint, E. P., and Quimby, P. C., Jr. (1979). Weed Sci. 27:77-82.
Patterson, D. T., Flint, E. P., and Dickens, R. (1980). Weed Sci. 28:505-509.
Potter, J. R., and Jones, J. W. (1977). Plant Physiol. 59:10-14.
Rice, E. L. (1974). "Allelopathy." Academic Press, New York.
Risser, P. G. (1969). Bot. Rev. 35:251-284.
Sharp, D. (1976). Vegetatio 31:103-136.
Shaw, W. C. (1978). Proc. Southern Weed Sci. Soc., New Orleans. 31:28-47.
Wax, L. M. (1976). In "World Soybean Research" (L. D. Hill, ed.), pp. 420-425. Interstate Printers and Publishers, Inc., Danville, Ilinois.
Weiss, P. W. and Simmons, D. M. (1977). Weed Res. 17:393-397.
Went, F. W. (1957). Chron. Bot. 17:1-343.
Williams, E. D. (1971). Weed Res. 11:159-170.
Williams, J. D., and Groves, R. H. (1980). Weed Res. 20:47-52.
Wills, G. D. (1975). Weed Sci. 23:93-96.
Zimdahl, R. L. (1980). "Weed-Crop Competition - A Review." International Plant Protection Center, Oregon State Univ., Corvallis, Oregon.

MICROHABITAT VARIATION IN RELATION TO WEED SEED GERMINATION AND SEEDLING EMERGENCE

Raymond A. Evans and James A. Young
USDA/SEA-AR
Reno, Nevada

I. INTRODUCTION

Weedy species have invaded and dominated large areas of degraded rangelands of the western United States. Of almost 40 million hectares of sagebrush (*Artemisia*) rangelands in the West, less than 5% are in good condition (U.S. Forest Service, 1972). In the Humboldt River basin, which is representative of Northern Nevada, only 1% of the rangelands are in high condition (Anon., 1966). The remainder of these rangelands is characterized by low densities of perennial grasses and other forage plants, high cover of shrubs of low palatability, and infestations of alien annual weeds. Plant communities within the big sagebrush (*Artemisia tridentata*)/bunchgrass vegetation type are extraordinarily subject to invasion by alien annual species (Jardine and Anderson, 1919). Apparently, no highly competitive native annual species were evolved to dominate low seral plant communities created by past grazing abuses.

The invasion by alien annual species into sagebrush rangelands is fairly recent. Downy brome (*Bromus tectorum*) was not known to be present in Western Nevada at the turn of the century (Kennedy and Doten, 1901). Today, downy brome and associated annual species characterize the landscape on millions of hectares in the Great Basin (Klemmendson and Smith, 1964).

ISBN 0-12-332850-0

A seral continuum of alien annual species exists on degraded rangelands and dominates these rangelands where sagebrush has been removed by fire or other means (Piemeisel, 1951). The seral continuum is composed of Russian thistle (Salsola iberica), tumble mustard (Sisymbrium altissimum), tansey mustard (Descurainia pinnata), and downy brome. In specific habitats, other annual species, such as medusahead (Taeniatherum asperum) and scotch thistle (Onopordium acanthium), extend this continuum (Fig. 1).

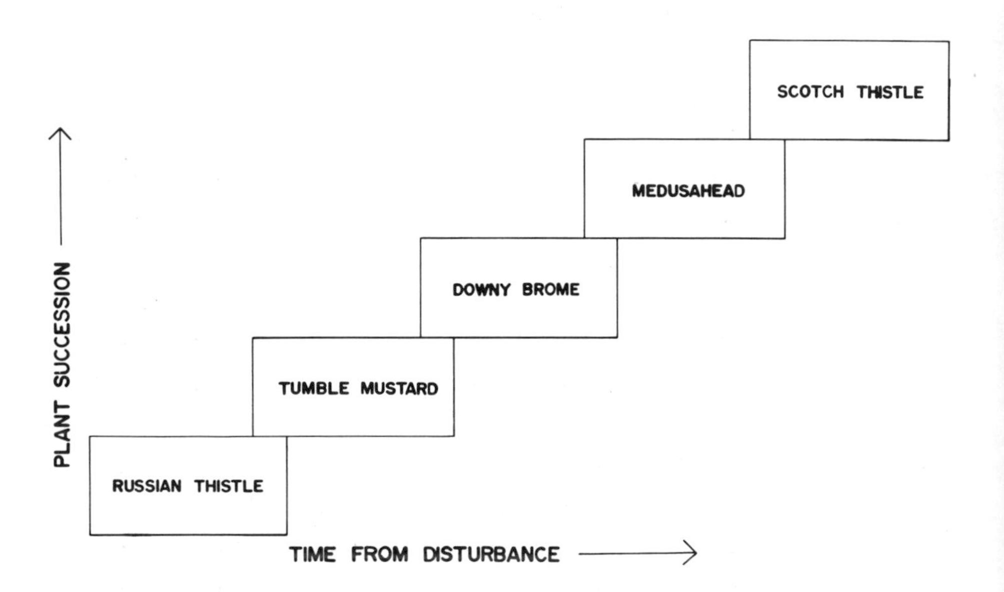

Fig. 1. Seral continuum of annual species on degraded big sagebrush rangelands.

Botanical composition and dominance by specific plant species are rapidly changing phenomena after disturbance in this annual community. The order of dominance of species is predictable and is based on reproductive and growth characteristics. These characteristics are: seed production, seed dispersal, seed germination, earliness of maximum growth, earliness of maturity, and competitive ability (adapted from Piemeisel, 1951). The first annual to invade and dominate disturbed lands is Russian thistle; next are the mustards, and downy brome is the last. The succession of these annuals is relatively rapid because every year they involve the processes of seed dispersal, seed germination, seedling growth, and plant maturation.

Recent studies of these communities indicate that relationships of inherent characteristics of seed germination and plant growth of these species interact with soil surface characteristics to determine dominance of one species over another. The effects of plant litter cover and microtopography on germination and establishment of annual species were studied by Evans and Young (1970, 1972). The objective of these studies was to develop an understanding of community dynamics among annual species to form a basis for range improvement technology.

Soil surface characteristics of the seedbed, which alter the environmental potential to allow germination, emergence and establishment of the alien annual species, are plant litter cover and microtopography (Evans and Young, 1970, 1972). These seedbed characteristics tend to create safe microsites for establishment of specific plant species (Harper et al., 1965).

II. PLANT LITTER

Plant litter is defined as the procumbent stems, leaves, and inflorescences of plants, mainly herbaceous from this community, from previous years' growth. Standing, dried, annual plants, especially grasses, are usually pressed down by winter snow following a particular growing season. Plants of some species remain standing for longer periods; others, such as Russian thistle, retain their original shape for sometime. However, they all contribute to the litter of the plant community.

In a study by Evans and Young (1970), the role of plant litter in modifying microsites in the seedbed environment and its effect on establishment of annual alien plants was investigated.

In trials of this study, litter of different annual species was placed on the soil surface with varying depths of coverage (0, 5, 7.5, and 10 cm). Before placement of litter, caryopses of medusahead and downy brome were broadcasted and raked into the soil surface. Frequency, density, and yield of these two species and those of tumble mustard and other broadleaf weeds were determined in relation to litter coverage. In addition, air temperature (3 cm above soil surface), soil surface temperature, soil temperature 1.5 cm below the surface, relative humidity near the soil surface (0 to 3 cm), insolation at soil surface level, and soil moisture at depths of 1, 3, and 9 cm were monitored in relation to litter cover treatments during selected times of germination.

Establishment and growth of downy brome and medusahead were increased and those of tumble mustard and other broadleaf weeds were decreased by litter cover. There was an interacting effect of litter coverage and favorable microrelief of the soil surface (Table 1). Both tended to moderate the harsh conditions on a bare soil surface and allow germination and establishment of plants.

Average diurnal temperatures during the spring growing period at 3 cm above the soil surface and above the litter indicated higher temperatures at noon and somewhat lower temperatures at night with litter than over bare soil (Fig. 2). Trends of daily maximum and minimum temperatures during the germination period also indicated increased maximum air temperature with litter cover at 3 cm. Minimum air temperatures were not significantly affected by cover of litter at this height above the soil surface.

Soil surface temperatures were greatly modified by litter cover in the spring germination period (March 20 to April 17). Average day temperatures were 9°C lower and night temperatures were 5°C higher under litter cover than on bare soil (Fig. 3). By litter coverage, maximum soil surface temperatures were lowered as much as 15°C and minimum temperatures raised 10°C during the germination period (Fig. 4). Soil temperatures at 1.5 cm below

Table 1. Frequency percentages of downy brome and tumble mustard relation to litter-covered soil surface, one with vàriations in microrelief, and to smooth soil surface (Evans and Young, 1970).

Species	Characteristic of soil surface	Date of sampling		
		April 7	April 16	April 28
Downy brome	Litter covered	50	34	48
	With microrelief	43	60	46
	Smooth	7	6	6
Tumble mustard	Litter covered	13	0	0
	With microrelief	20	0	10
	Smooth	67	100	90

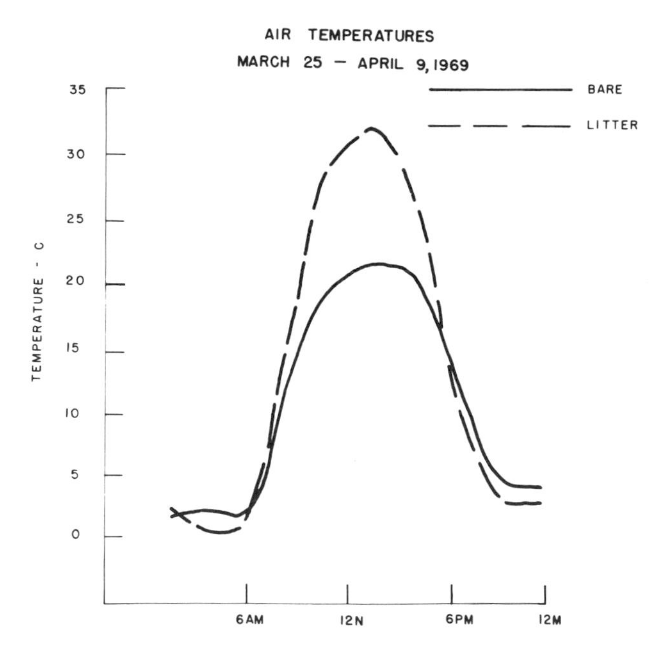

Fig. 2. Diurnal pattern of air temperatures 3 cm above the bare soil surface and that covered with plant litter averaged from values taken by continuous monitoring in the spring growing season of 1969 (Evans and Young, 1970).

the surface followed similar trends as did soil surface temperatures but with less magnitude.

Litter covering the ground acted as an insulating layer absorbing and trapping insolation during the daylight hours. At night, the litter held the warm air on the soil surface. The processes together moderated temperatures on or just below the soil surface, where most of the seeds of the annual species are located (Young et al., 1969). Probably most important in terms of germination and seedling growth is that litter caused the minimum temperatures to remain above the critical level for germination (Table 2). Also important is that litter cover decreased extreme fluctuation between minimum and maximum daily temperatures. Most seeds germinate better under moderate than widely fluctuating temperature regimes (Evans et al., 1981).

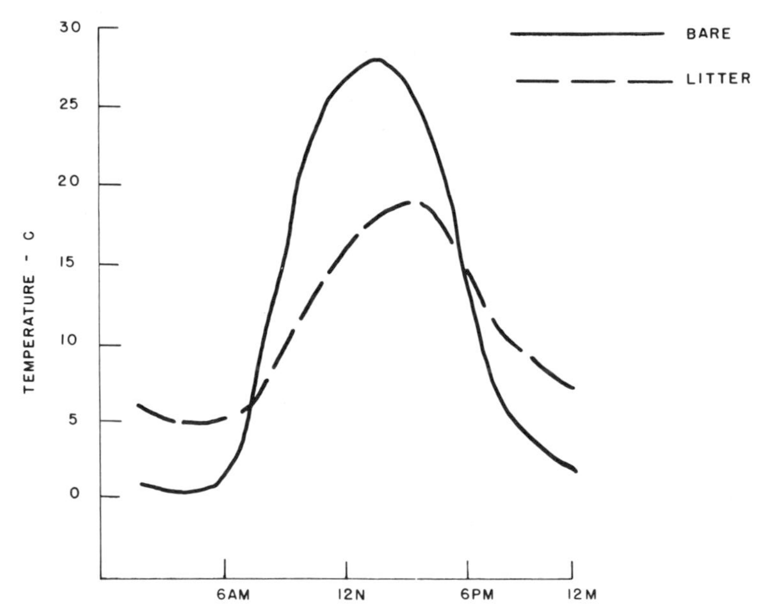

Fig. 3. Diurnal pattern of air temperatures on the surface of bare soil and that covered with plant litter from values taken by continuous monitoring in the spring growing season of 1969 (Evans and Young, 1970).

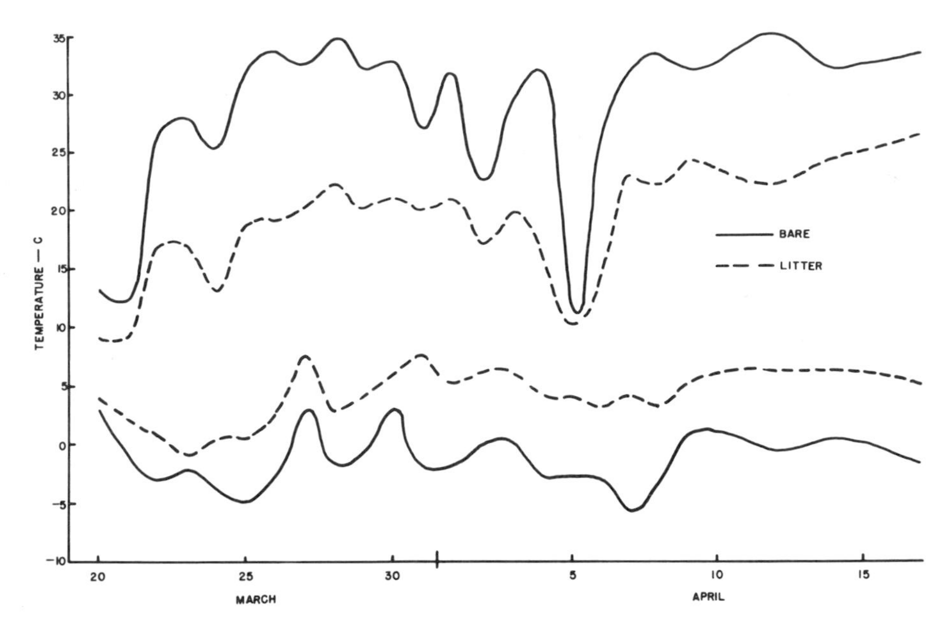

Fig. 4. Maximum and minimum air temperatures on the surface of bare soil and soil covered with plant litter during the spring growing season of 1969 (Evans and Young, 1970).

Moisture relationships on the soil surface and immediately beneath the surface are probably more important than the relationship of temperature to germinating seeds. Diurnal pattern of relative humidity indicated that seeds on the surface of bare soil were subjected to severe drying conditions during the day (Fig. 5). This moisture imbalance occurred when the most favorable temperature conditions prevailed for germination. Atmospheric moisture conditions under litter were much more stable and favorable than those on bare soil. Under litter, relative humidity varied only between 60 and 95% in its diurnal range.

In combining moisture and temperature modification due to litter cover, we see that not only are the minimum temperatures more conducive for germination, but, as the temperature rises during the day, moisture conditions also remain more favorable. With temperatures being restrictive during the night and moisture during the day, temperature-moisture conditions for seed germination on bare soil are beyond the range conducive to germination most of the time. These conditions would be modified during the periods of cloudy, wet weather. At such times, relative humidity and temperature of exposed microsites (those on bare, smooth soil) tend to remain more constant with less diurnal fluctuation.

Table 2. Germination percentages as indicated by extension of radicles and coleoptiles or cotyledons of downy brome, medusahead, tumble mustard, and Russian thistle after two weeks of incubation at ten temperatures (Evans and Young, 1970).

	Downy brome		Medusahead			Russian[b]
Temperature	Radicle	Coleoptile	Radicle	Coleoptile	Mustard[a]	thistle
3	0	0	0	0	0	0
0	2	0	0	0	0	26
3	51	0	8	0	10	43
5	99	24	80	5	45	56
7	100	93	79	70	57	88
9	100	99	81	79	79	78
10	100	100	83	83	85	88
15	100	100	82	82	100	78
20	100	100	90	90	100	66
25	100	100	61	61	100	29

[a] If the radicle and cotyledons of the tumble mustard seedlings were emerged from the seed coat, the seeds were considered germinated.

[b] The Russian thistle seeds were considered germinated if the embryo was completely uncoiled.

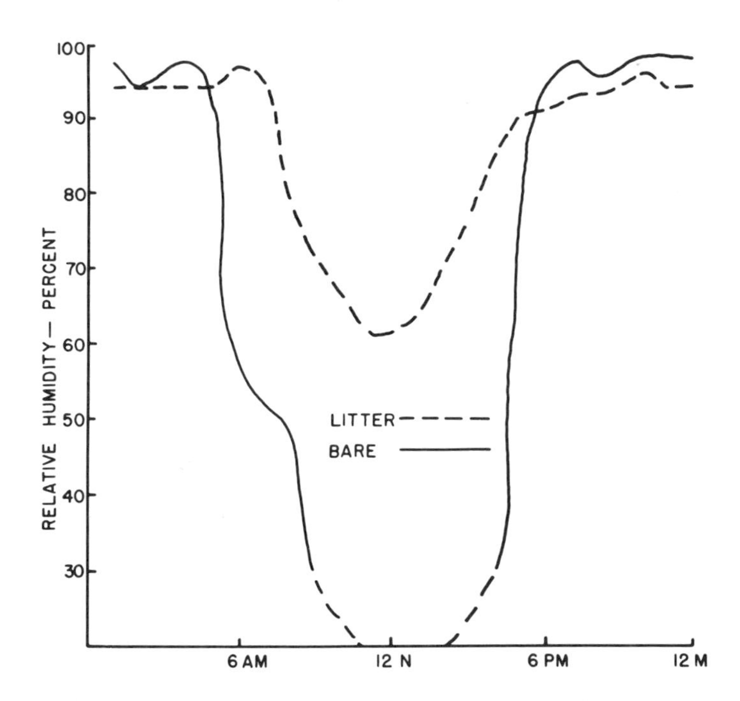

Fig. 5. Diurnal pattern of relative humidity at soil level (0 to 3 cm) for bare soil and for soil covered with plant litter as shown by average values taken in March and April, 1969. Relative humidity values below 30% are extrapolated (Evans and Young, 1970).

During the monitoring period, very little precipitation occurred, with virtually no water recharge of the soil profile. Under these conditions, soil moisture depletion from the surface to 1 cm was less rapid under litter than on the bare soil, but by the end of the spring germination period, both were near -15 bars of tension (Fig. 6). Litter cover would tend to maintain favorable moisture conditions and create germination "safe sites" for longer periods between rains during the germination period.

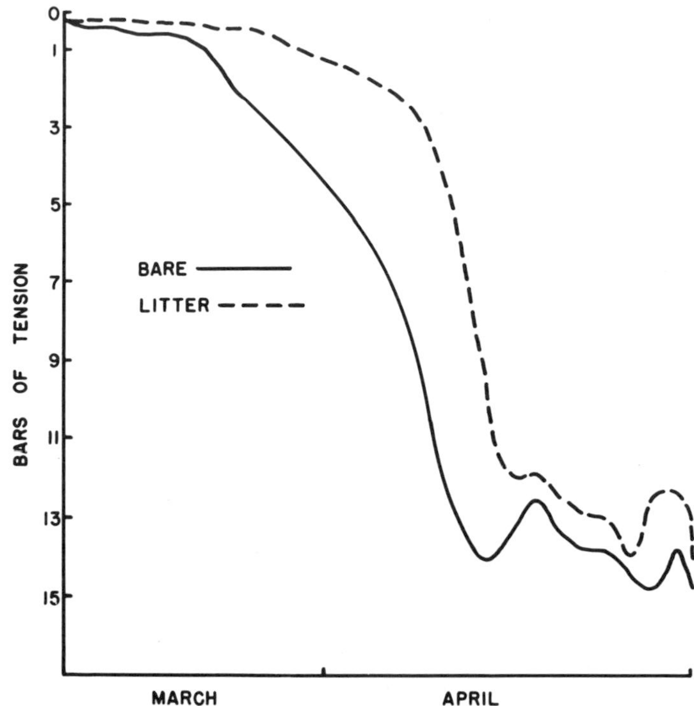

Fig. 6. Soil moisture depletion from the surface to 1 cm below the surface of bare soil and that covered with plant litter during the spring growing period in 1969 (Evans and Young, 1970).

Litter cover and the establishment of downy brome and other alien annual grasses form a ubiquitous and characteristic relationship on millions of acres of rangelands in the Western United States. Further, this relationship is an important factor controlling plant succession in these communities. The succession from broadleaf plants to downy brome is accompanied by an accumulation of litter. In areas where downy brome dominates, a litter residue exists from year-to-year.

III. SOIL MICROTOPOGRAPHY

Depressions on the soil surface (rough versus smooth microtopography) increase germination of seeds and caryopses and the establishment of seedlings of annual plants by altering their physical environment in two basic ways. The first is that the depressed sites retain soil moisture at the surface longer and have more favorable atmospheric moisture and temperature regimes than the smooth soil surface. The second is that conditions are created for more adequate soil coverage of the seeds and caryopses which, in turn, further modify their microenvironment.

Trials were established comparing emergence and growth of several annual species in relation to various degrees of microtopography of the soil surface (Evans and Young, 1972). These trials were conducted during spring and fall growing seasons. On the plots, pits were dug 3 by 6 cm and 3, 6, or 9 cm deep. Seeds of the annuals were distributed on the soil surface or uniformly covered with 1 cm of soil both on the smooth soil surface and in the pits. Plots were established on clay-textured and sandy-loam soils.

The microenvironmental monitoring conducted in this study was similar to that conducted in the plant litter study (Evans and Young, 1970). Air temperatures on the soil surface and 3 cm above and soil temperatures at 1 and 3 cm below the surface were continuously measured with thermistors. Relative humidity near the soil surface (0.3 cm height) was measured with electric hydrometric circuit elements. Soil moisture from the surface to 1 cm deep and at 3 cm below the soil surface was measured with gypsum blocks. The data recording system used was essentially the same as the one used in the litter study.

In the spring trial on clay soil, more than 100 times as many downy brome seedlings emerged from pits (9 cm deep) as from smooth soil (Fig. 7). Emergence increased more than 30 times when caryopses were buried 1 cm deep, in comparison to broadcast on the surface of smooth soil. Similar results were found with tumble mustard except initial emergence was more rapid and maximum emergence was considerably less (Fig. 8). Effects of seedbed differences persisted throughout the growth period and affected herbage production of the two species. On sandy loam soil,

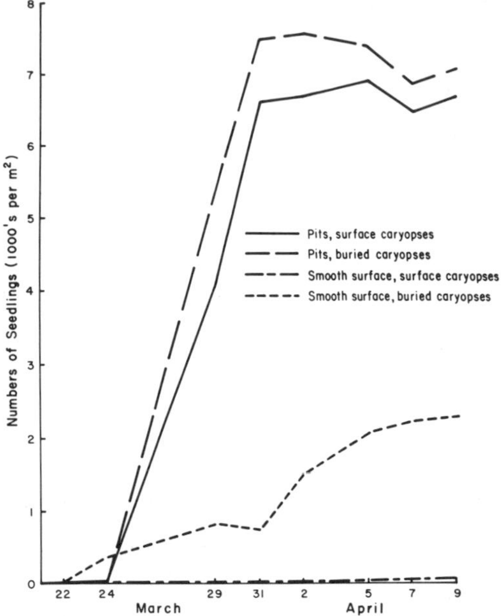

Fig. 7. Seedling emergence of downy brome during early spring on a clay soil seedbed in relation to microtopography and seed burial (Evans and Young, 1972).

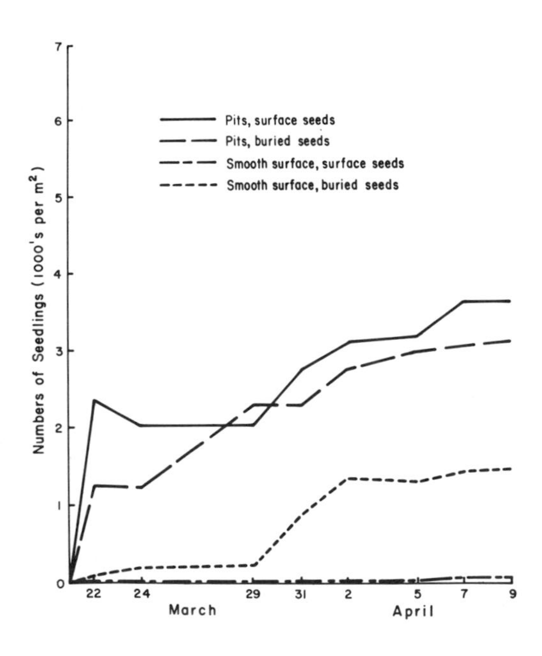

Fig. 8. Seedling emergence of tumble mustard during early spring on a clay soil seedbed in relation to microtopography and seed burial (Evans and Young, 1972).

sloughing-in of the pits limited emergence and growth by burying seeds and emerging seedlings as deep as 7 cm.

In the fall study, emergence of downy brome and medusahead was again dramatically affected by microtopography of the soil surface. Three weeks after planting, 30 seedlings of downy brome per square meter emerged on the soil surface where the caryopses had been unburied, compared with 10,300 seedlings per square meter in 9-cm pits (Evans and Young, 1972). In the same comparison, there were 40 medusahead seedlings per square meter on the surface compared with 6,010 in the 9-cm pits. When the caryopses of both species were covered with soil, differences of seedling emergence between smooth surface and pits were either negated or drastically reduced.

Soil movement and its subsequent filling in of holes, pits, and crevices materially aids in seed burial creating a more favorable environment for germination and seedling growth. However, in some cases, burial may be too deep as in deep pits with loose soil, adversely affecting emergence. Instability and movement of surface soil are directly related to textural differences of soil and to type and amount of precipitation during the period of seed germination, seedling emergence, and growth.

About 1% of the caryopses and seeds of the annual species seeded on the smooth soil surface resulted in seedling establishment. In contrast, in the 9-cm pits, the number of emerging seedlings of downy brome and medusahead was over 100% of the number of seeded caryopses.

Emergence rate of tumble mustard varied from 20 to 72% in pits. The foregoing indicated that the functions of microtopography, in the case of these two annual grasses, involve both dispersal and establishment. Because of the small size and high density of tumble mustard seeds, modification of the seedbed of this rather gross scale affected only its establishment. Perhaps topography on a microscale, as found in nature on soil surfaces, would also affect disperal of tumble mustard seeds.

Maximum and minimum air temperatures on the soil surface were much modified by microtopography during the spring growing period (Fig. 9). Temperatures on the smooth soil surface varied from -9 to 40 C in the 9-cm pits, maximum daily temperatures were as much as 17 C lower during the germination period. Minimum soil surface temperatures in the pits were at or above 0 C during the growing season except for four days when they dipped to a minimum of -3 C.

Diurnal patterns of soil surface temperatures further emphasize differences due to microtopography and demonstrate that an eight hour day temperature alternating with a 16 hour night

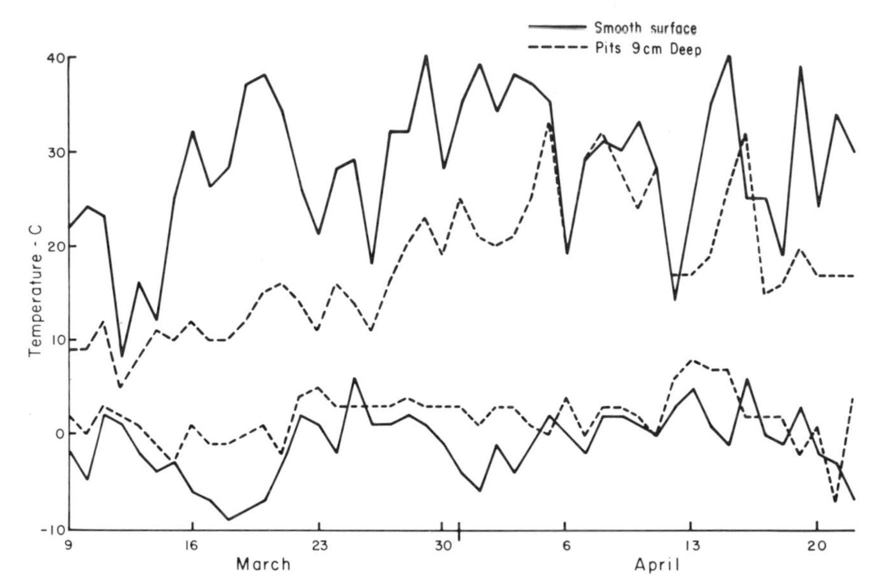

Fig. 9. Maximum and minimum air temperatures on the surface of smooth soil in the clay seedbed and of soil at the bottom of 9-cm deep pits during early spring 1971 (Evans and Young, 1972).

temperature generally portrays the temperature regime in the field (Fig. 10).

Temperature regimes were relatively constant over the fall germination period in this study, with diurnal fluctuation of soil surface temperatures ranging from 0 to 20 C. Modification of temperature attributable to microtopographic differences was

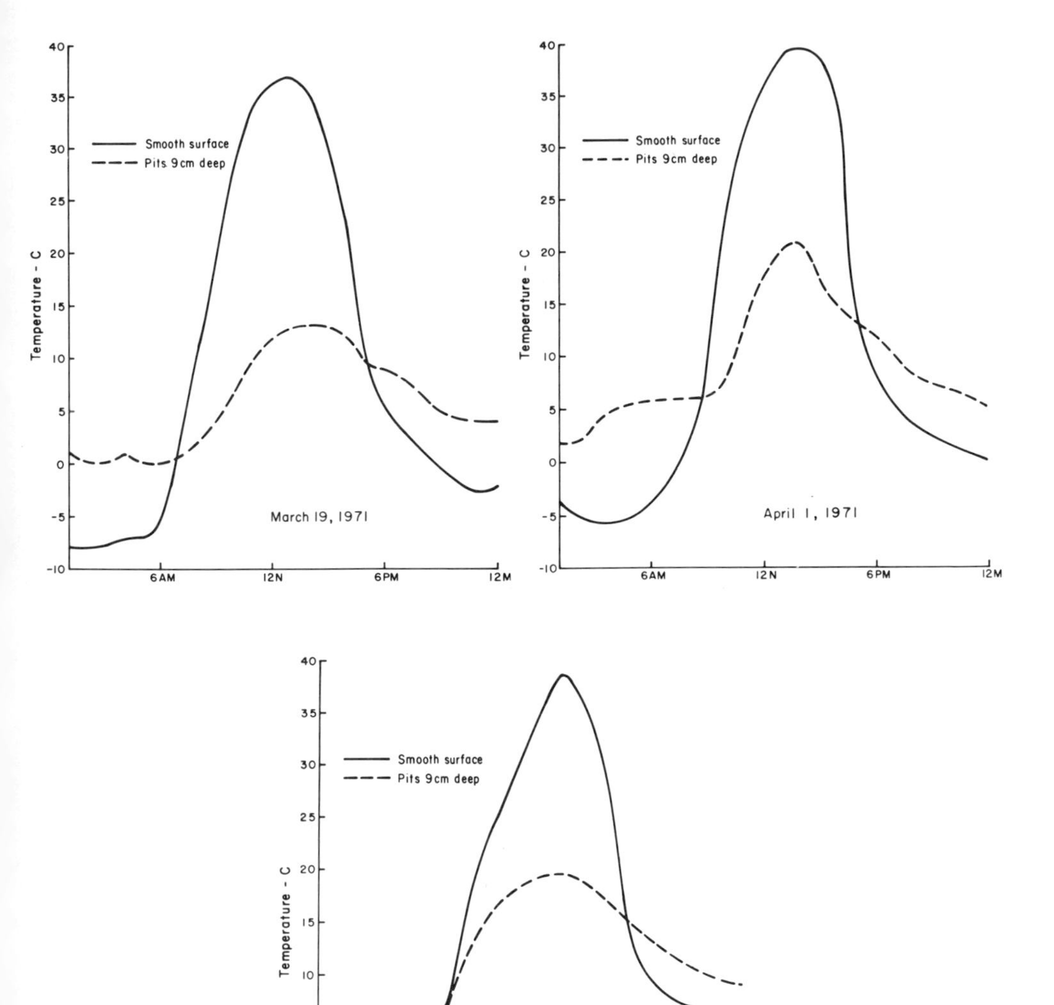

Fig. 10. Diurnal patterns of air temperatures at the surface of smooth soil in the clay seedbed compared with those at the bottom of 9-cm pits at three dates in the spring trial (March 19, April 1, and April 19) (Evans and Young, 1972).

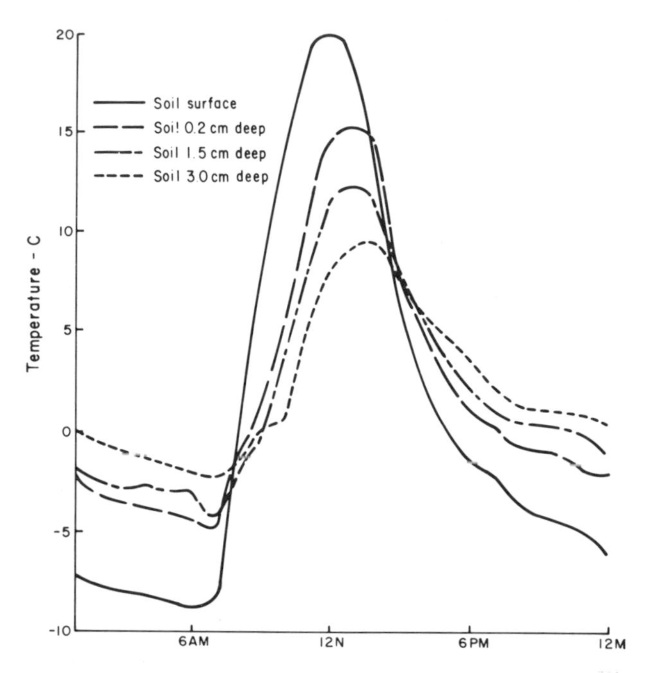

Fig. 11. Mean diurnal ranges of temperatures in relation to soil depth in a sandy loam soil comparing those from the soil surface to 3 cm deep for one week in the fall of 1969 (Evans and Young, 1972).

slight. The effect of soil coverage on temperature during this period was depth-related but was nonlinear (Fig. 11).

Depletion of available water in the surface layer of the soil was drastically altered by favorable microtopography. In both clay and sandy loam soils, water was unavailable midway through the spring growing period on the smooth surface; in the pits, water remained available during all this period (Fig. 12). Also, in the sandy loam on the smooth surface, a short period of rapid water depletion occurred at the time of initial seedling emergence.

In laboratory studies of germination and seedling growth of these annual species, threshold values were established in relation to temperature and water stress for downy brome, medusahead and tumble mustard (Evans and Young, 1932). These values for

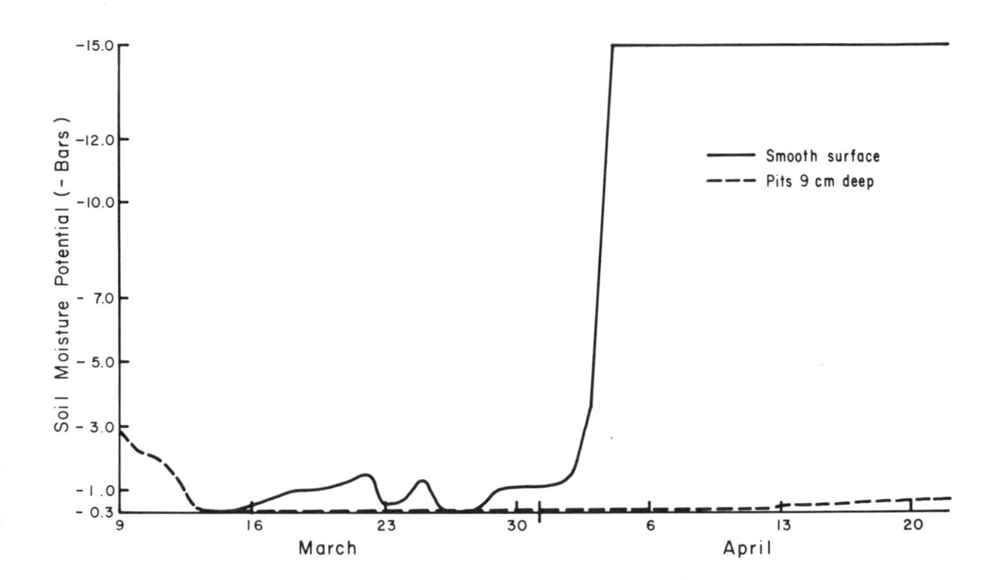

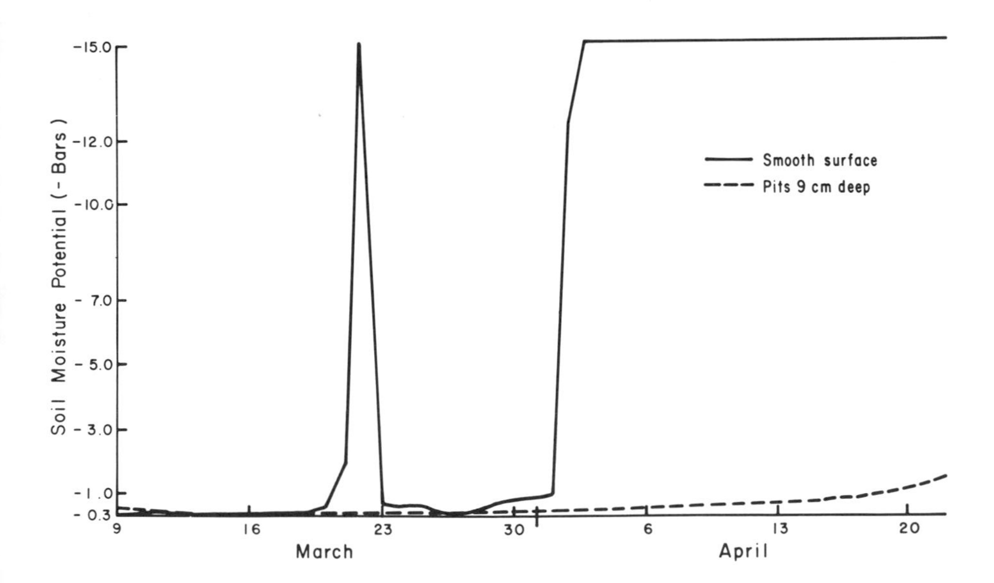

Fig. 12. Soil moisture potential in relation to microtopography of clay-soil (top) and sandy loam (bottom) seedbeds in spring 1971 (Evans and Young, 1972).

temperature and available water, when compared to field values, show the importance of microtopography for establishment of these species.

IV. SEED CHARACTERISTICS

A. Seed Size and Dispersal

The classic work of E. J. Salisbury (1942) established and illustrated the general tendency for plants in successional habitats to have small seeds, in intermediate habitats to have larger seeds, and in woodlands and forests to have large seeds. Small seed size often interacts with dispersal mechanism and number of seeds produced. Plants of successional habitats that produce an abundance of small seeds often have a highly developed seed dispersal system.

These relationships can be illustrated in the degraded sagebrush grasslands by the continuum of weedy species, Russian thistle, tumble mustard, and downy brome. Russian thistle is the first species to invade disturbed places in sagebrush grasslands. This species produces an abundance of relatively small and simple seeds. These seeds are not dispersed from the plant until the parent plant breaks off at the soil surface and tumbles across the landscape, beating the seeds free and ensuring their wide dispersal (Young and Evans, 1972).

Tumble mustard is a winter-early-spring annual in contrast to the spring-summer life cycle of Russian thistle which it easily out competes. Tumble mustard also tumbles, widely dispersing a multitude of small dense seeds. The two species appear to have evolved similar seed production and dispersal strategies with the winter-early-spring annual preempting the environmental potential before the spring-summer annual gets a chance to establish.

The highest species in this successional continuum is downy brome. This annual grass is a highly competitive winter-early-spring annual that out competes both Russian thistle and tumble mustard. Downy brome usually produces far fewer seeds per plant than either of the lower successional stage dominants. The seeds of downy brome are much larger and are complex caryopses.

The same trend is apparent among the dominant native species in this environment. Most subspecies of big sagebrush, the landscape dominant, produce an abundance of wind-dispersed achenes (McDonough and Harness, 1974), ensuring rapid repopulation of disturbed areas. Most of the native herbaceous perennials in this ecosystem are relatively long lived and replace themselves only at infrequent intervals (West et al., 1978). Many of the dominant perennial grasses and valuable browse species have complex germination strategies which reflect this infrequent need to recruit seedlings to the population.

B. Interaction of Dormancy Systems

Inherent dormancy systems interact with seed numbers, size and dispersal systems to modify germination-establishment strategies. A good example of this interaction is Russian thistle. Seedlings of this species are not frost tolerant and are usually killed if they germinate in the fall. This would appear to be a severe handicap for a species that broadcasts its small seeds to become established only on disturbed areas. Russian thistle seeds have an inherent temperature after-ripening requirement that limits germination soon after maturity to a restricted number of relatively warm temperatures (Young and Evans, 1972). Gradually over winter, the after-ripening requirements breakdown and by late spring Russian thistle will germinate at virtually any seedbed temperature from 0 to 50 C, including warm temperatures alternating with 0 C.

Seeds of both tumble mustard and downy brome are usually completely free of after-ripening requirements when mature. This freedom from inherent dormancy appears to be a nearly universal characteristic of weedy annuals that colonize numerous habitats.

C. Seeds and the Physical Environment of Seedbeds

Germination on a specific seedbed is controlled by inherent characteristics of seeds, or in some cases, through modification of the physical environment by the seed itself. If we again go to the seral continuum of annuals on degraded sagebrush rangelands, a diverse array of characteristics and mechanisms allowing germination can be demonstrated.

Russian thistle, with its multitude of seeds and efficient dispersal system, does not overtly modify potential seedbeds through the presence of its seeds. The simple seeds of this species take advantage of brief transitory periods, such as rainstorms, that permit germination (Wallace et al., 1968). To accomplish this type of germination, the physiological system of the seed must be totally oriented toward rapid germination. In the case of Russian thistle, rapid germination is enhanced by foregoing a well-developed seedcoat and allowing the nearly naked embryo to uncoil as soon as it is moistened. This is a high-risk type of germination which leads to mortality if the radicle does not penetrate the seedbed and immediately start supplying the seedling with moisture.

Tumble mustard modifies the seedbed environment through the production of mucilage on the seedcoat when moistened. In botanical terminology this trait is known as "myxospermy." A large number of species in numerous families have various forms of seed mucilage (Young and Evans, 1973). Among the Angiospermae, myxospermatic species belong to 49 families (Grubert, 1974). Undoubtedly, mucilage serves different functions among this large number of species. In some species it may serve as a dispersal agent (Salisbury, 1942), in others, to provide seeds with a moisture-supplying substrate or to cement seeds to the moisture supplying substrate (Grubert, 1974).

In the case of tumble mustard the existence of mucilage produced by epidermal cells on the seedcoat appears to impart the clear-cut ecologic advantage of being able to germinate on the soil surface (Young et al., 1970). Seeds on the soil surface must be able to take up moisture from the seedbed faster than they lose it to the atmosphere. Several factors enter into how steep the loss gradient is from the seed to the atmosphere and the rate of moisture uptake. If the relative humidity near the soil surface is at or very near saturation, obviously the loss gradient is not very steep. The moisture content of the seedbed and the degree of contact between the seedbed and the seed influence hydraulic conductivity and the rate of moisture flow from the soil to the seed. A fine-textured seedbed will provide more points of contact than a coarse-textured soil. Harper and Benton (1966) use an analogy to heat flow physics to explain the advantage seed mucilage imparts to allow germination on the soil surface. This analogy implies that the drying mucilage on the parts of the seed exposed to the atmosphere retards moisture loss from the seed while the moist mucilage under the seed provides continuous points of contact between the seed and the soil particles in the seedbed. This analogy is logical but remains to be proven.

The phenomenon is very real and can be easily demonstrated with seeds of tumble mustard and downy brome. If both are planted on the soil surface or buried with as little soil coverage as possible, the mustard seeds will germinate on the soil surface and occupy the pots to the exclusion of the downy brome. In the slightly buried treatment, the downy brome seedlings will establish first and occupy the pots to the virtual exclusion of the tumble mustard seedlings (Young et al., 1970).

Downy brome caryopses lack mucilage, but their presence in sufficient numbers will influence the physical seedbed as a form of litter to allow caryopses to germinate if they are protected by a mulch of other caryopses (Evans and Young, 1970).

D. Environmentally Induced Dormancy

The inherent physiological system of seeds interacts with the physical and biological environment of the seedbed not only to produce germination, but also to induce dormancy. Caryopses of downy brome are highly germinable when mature or soon after maturity. Some of the caryopses germinate, some are picked up by predators, and a large proportion of each season's production acquires an environmentally induced dormancy (Young et al., 1969). The seeds that find an acceptable environment for germination germinate and the population reaps the rewards of simultaneous germination and the preemption of environmental potential from other species. Those that acquire dormancy provide a reserve of viable, but nongerminable seeds. This dormancy breaks down slowly over time and germination is enhanced by such stimuli as nitrate ions or gibberellin (Evans and Young, 1975). If a prolonged period of available moisture extends well into spring, nitrification produces available nitrate, and a flush of germination from previously dormant downy brome caryopses can be expected. Similarly, the accumulation of nitrate nitrogen in a fallowed seedbed will enhance the germination of dormant downy brome caryopses which escaped the fallowing operation by being dormant. The acquisition of dormancy allows downy brome populations to enjoy the fruits of continuous germination from a renewable reserve.

E. Temperature and Germination

Working in environments where temperatures permitting germination and suitable moisture conditions are largely out of phase, our attention was soon focused on the critical importance of seedbed temperatures in controlling population dynamics. In order to comparatively evaluate the germination response of various weed and crop species, we have developed a 55 constant and alternating temperature regime test. Constant temperatures consist of 0, 2, 5, and 5 degree increments through 40 C. Alternating temperatures consist of 16 hours at each constant temperature and eight hours at each possible higher temperature. For example, 0 C alternates with 2, 5, 10, 15, 20, 25, 30, 35, and 40 C while 35 C alternates with 40 C only. This temperature has to be extended for some species to 50 and 60 C or to -2, -4, and -6 C for others. The 8/16 hour alternating regimes are roughly based on the diurnal regime in field seedbeds in the early spring and late fall when germination occurs on sagebrush rangelands.

The effects of constant and alternating temperatures on an individual selection being tested are analyzed using a quadratic response surface. This quadratic response surface is composed of a series of regression equations, one for each cold-period (16 hour) temperature through the series of warm-period temperatures (8 hour) with calculated values and their confidence limits (Evans et al., 1980).

A single profile can be very informative but the temperature profiles are most valuable on a comparative basis such as when a multitude of species growing in the same community are compared (Young et al., 1973). However, when one compares a number of profiles, with a few exceptions, the feeling is that when you have seen one, you have seen them all. Differences between profiles can be quite subtle and the mass of data makes it difficult to make visual comparisons.

We have developed a series of parameters derived from the quadratic response surfaces to make comparisons more meaningful. These parameters include mean of the entire profile, mean of those regimes that have some germination, percentage of regimes with some germination, mean of regimes with optimum germination, and percentage of regimes with optimum germination. Our definition of "optimum germination" includes those means that are not lower than the maximum and its confidence interval. In addition, it is often useful to breakdown the germination response into percentiles of <10, 10-25, 26-50, 51-75, 76-90, and >90%.

Germination data for these profiles are collected at the end of one, two, and four weeks of incubation. By comparing germination rates at the end of one week, from one to two weeks, and two to four weeks from various temperatures, it is possible to evaluate the degree of temperature stress on the germination process. Through this process we have determined, for example, that for cool season grass species, colder, warmer, and widely fluctuating temperatures compared with optimum temperatures can delay the rate of germination as well as reduce total germination (Evans and Young, 1981).

F. Seedbed Temperatures

The germination profiles for temperature are of their greatest value when compared to seedbed temperatures. Based on the microenvironmental monitoring of field seedbeds (Evans et al., 1970), we have developed a discriminate breakdown of seedbed temperatures for comparison to germination temperature profiles (Fig. 13).

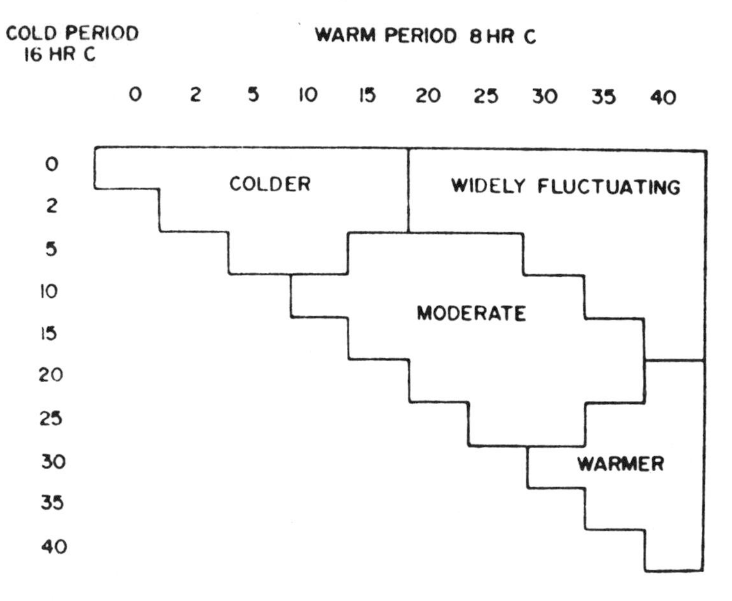

Fig. 13. Discriminate breakdown of seedbed temperatures into germination categories of moderate, colder and warmer than moderate, and widely fluctuating.

Using this type of analysis, it is possible to make comparisons between the inherent response of species and seedbed temperatures. For small-seeded species of fescue and bluegrass that have been notoriously difficult to establish on rangelands, we found that the seeds had very poor germination at widely fluctuating temperatures (Young et al., 1971). Because of their small size, these species have to be seeded at shallow depths where they are exposed to widely fluctuating temperatures.

G. Moisture and Seed Germination

Germination rate and total germination of seeds are influenced by soil water matric potential, osmotic potential, and hydraulic conductivity (Collis-George and Hector, 1966).

Water uptake by a seed is an essential prerequisite of germination. Three stages in moisture relationships can be distinguished during seed germination: a) imbibition, during which water absorption into the seed takes place; b) a pause during which enzymatic transformation and initiation of meristematic activities takes place; and c) the start of growth with radicle elongation and emergence through the seedcoat (Mayer and Poljackoff-Mayber, 1963). This sequence of stages is governed by water uptake from the external substrate, namely the soil or solution (Hadas, 1976).

Considerable information concerning available moisture in field seedbeds has been gathered by microenvironmental monitoring. Most of the available literature on laboratory studies deals with varying osmotic potentials with little information on the influence of matric potential or hydraulic conductivity. The reason for this preponderance is the relative ease of varying osmotic potentials with aqueous solutions of different solutes. The soil is intentionally bypassed in order to eliminate various complications inherent with soil-water media.

A capsule summary of water relations and seed germination is rather obvious in that germination rate decreases with decreasing external water potential and for each species, there is a critical value of water potential below which germination will not occur.

H. Interaction of Moisture Stress and Temperature

The most comprehensive studies of the interaction of moisture stress and temperature on seed germination have been done by Sharma (1976). He concluded that for most species the effects are additive. Seeds will be in more stress at a given moisture potential at extremely warm or cold incubation temperatures. This effect is exemplified by the germination of spiny hopsage (*Grayia spinosa*) (Wood et al., 1976).

I. Germination in Simulated Litter

Litter is such an important factor in seedbed quality in terms of modifying environmental parameters that it is easy to overlook the fact that litter also serves as a site of germination. Caryopses of most annual grasses will germinate in simulated litter without the tip of the caryopses in contact with moisture-supplying substrate (Young et al., 1971). Often the germination of these species in simulated litter is considerably different from their germination in petri dishes. Caryopses that germinate in litter have to have roots that can withstand drying or can reinitiate growth from lateral buds after the primary root has dried and died.

V. INTEGRATED WEED CONTROL-SEEDING TECHNOLOGY

A. Atrazine Fallow

The atrazine-fallow technique (Eckert and Evans, 1967) for herbaceous weed control preparatory to seeding on downy brome rangelands is successful because of the principles brought out in this chapter. The effects of litter cover and bare soil and various microtopographies of the soil surface on seed germination and seedling growth are keys to success of weed control and establishment of grasses and other forage species.

In the atrazine-fallow technique, atrazine [2-chloro-4-(ethylamino)-6-(isopropylamino)-*s*-triazine] is applied in the fall and a fallow is created the next growing season. In the year after application of atrazine, no annual plants can grow because of herbicidal activity in the soil and the litter accumulation from past years is broken down at an accelerated rate. The area is seeded to wheatgrass (Agropyron sp.) or other forage species one year after herbicide application. In seedling year of the forage species, downy brome is controlled by a litter-free seedbed which preempts germination because of unfavorable moisture and temperature regimes. Usually at that time the atrazine is not active in the surface layers of the soil, so weed control is dependent upon microenvironmental parameters of the seedbed. Seedling success of forage species is enhanced by seeding in the bottom of furrows made with a modified rangeland drill (Asher and Eckert, 1973). The furrow provides a favorable microenvironment for seedling establishment (Evans et al., 1970) and removes any herbicide residues on the soil surface to the area between the furrows.

B. Weed Control Systems

For degraded big sagebrush communities in which perennial grasses are gone and downy brome has invaded the shrub understory, it is possible to vertically integrate, combine in sequence, 2,4-D [(2,4-dichlorophenoxy) acetic acid] and atrazine-fallow treatments into a weed-control system (Evans and Young, 1977) for brush and herbaceous weed control in preparation for seeding forage species.

REFERENCES

Anonymous. (1966). Water and related land resources. Humboldt River Basin, Nevada–Basinwide Report, No. 12. Econ. Res. Serv., Forest Service. Soil Cons. Serv. and College of Agric., University of Nevada, Reno. 120 p.

Asher, J. E., and Eckert, Jr., R. E. (1973). J. Range Manage. 26:377-379.

Collis-George, N., and Hector, J. B. (1966). Austr. J. Soil Res. 4:145-164.

Eckert, Jr., R. E., and Evans, R. A. (1967). J. Range Manage. 20:35-41.

Evans, R. A., and Young, J. A. (1970). Weed Sci. 18:697-703.

Evans, R. A., and Young, J. A. (1972). Weed Sci. 20:350-356.
Evans, R. A., and Young, J. A. (1975). Weed Sci. 23:354-357.
Evans, R. A., and Young, J. A. (1977). J. Range Manage. 30:331-336.
Evans, R. A., and Young, J. A. (1981). J. Range Manage. (in press).
Evans, R. A., Holbo, H. R., Eckert, Jr., R. E., and Young, J. A. (1970). Weed Sci. 18:159-162.
Evans, R. A., Young, S. A., Easi, D. A., and Book, D. N. (1980). 61st Annu. Meeting, Western Soc. Crop Sci., Davis, CA. pp. 37.
Grubert, M. (1974). Acta. Biol. Venez. 8(3-4):315-551.
Hadas, A. (1976). J. Exp. Bot. 77:480-489.
Harper, J. L., and Benton, R. A. (1966). J. Ecol. 54:151-166.
Harper, J. L., Williams, L. T., and Sagar, G. R. (1965). J. Ecol. 539:273-286.
Jardine, J. T., and Anderson, M. (1919). U.S. Dept. Agric. Bull. 790. 98 p.
Kennedy, P. B., and Doten, S. B. (1901). Nevada State University Agric. Exp. Sta. Bull. 51. 54 p.
Klemmendson, J. O., and Smith, J. G. (1964). The Bot. Review 30:226-262.
Mayer, A. M., and Poljackoff-Mayber A. (1963). Pergamon Press, Oxford. 236 p.
McDonough, W. T., and Harness, R. O. (1974). Northwest Sci. 48:17-20.
Piemeisel, R. L. (1951). Ecol. 32:53-72.
Salisbury, E. J. (1942). In "The Reproductive Capacity of Plants." Bell, London. 330 p.
Sharma, M. L. (1976). Agron. J. 68:390-394.
U.S. Forest Service. (1972). In "The nation's range resources—a forest-range environment study (FRES)." USDA Forest Res. Rept. 19. 147 p.
Wallace, A., Rhods, W. A., and Frolich, E. F. (1968). Agron. J. 60:76-78.
West, N. E., Tausch, R. J., Rea, K. H., and Tueller, P. T. (1978). J. Range Manage. 31:87-92.
Wood, M. K., Knight, R. W., and Young, J. A. (1976). J. Range Manage. 29:53-56.
Young, J. A., and Evans, R. A. (1972). Agron. J. 64:214-218.
Young, J. A., and Evans, R. A. (1973). Weed Sci. 21:52-54.
Young, J. A., Evans, R. A., and Eckert, Jr., R. E. (1969). Weed Sci. 17:20-26.
Young, J. A., Evans, R. A., Gifford, R. O., and Eckert, Jr., R. E. (1970). Weed Sci. 18:41-47.
Young, J. A., Evans, R. A., and Kay, B. L. (1971). Agron. J. 63:551-555.
Young, J. A., Evans, R. A., and Kay, B. L. (1973). Agron. J. 65:656-659.

UTILIZING METEOROLOGICAL DATA FOR MODELING CROP AND WEED GROWTH

J. L. Anderson and E. A. Richardson
Departments of Plant Science and Soil Science
and Biometeorology
Utah State University
Logan, Utah

I. INTRODUCTION

A model (according to Webster) is a small copy or imitation of an existing object, or a preliminary representation that serves as a plan for construction of a usually larger object. In reality these are definitions of models of physical objects in general. The term, as commonly used in many present day sciences, has different connotations.

Mathematicians have been using a form of modeling for hundreds of years. Each mathematical representation of a curve or of a scientific phenomena constitutes an example of today's use of the term, model. The equations developed by astronomers to define Kepler's laws of motion of the planets about the sun, the law of gravitation and many other physical principles that can be expressed in mathematical terms fit within the generalized term of models.

Economists, using statistical regression equations, have developed predictive models to analyze and forecast the behavior of the stock market and the economy in general. More recently, agricultural economists have utilized similar linear and multiple regression techniques to predict agricultural crop production. Models also allow them to identify the implications of such crop

ISBN 0-12-332850-0

stress factors as soil moisture deficits, the influence of cold spring temperatures, and hot summer temperatures on the economic structure of the agricultural community.

Extensions of these statistical models can be found in the literature of many disciplines. Biologists and plant scientists, for example, have attempted, with varying degrees of success to relate selected environmental factors to such physiological processes as photosynthesis, respiration, and growth. Agricultural engineers have tried to relate the water use of plants to irrigation scheduling by means of multiple regression equations and models.

Another form of modeling, which was developed about the same time as the thermometer, relates various environmental factors to physical processes and has been given the name of climatic or phenoclimatography modeling.

II. MODELS OF WEEDS GROWN UNDER CONTROLLED CONDITIONS

Weeds are often grown in controlled environment greenhouses or growth chambers and observed for their morphological or physiological responses to a specific set of environmental conditions. When the resultant growth data are analyzed, models can be constructed that will simulate growth of that particular weed population under the field conditions approximated in the model. Wherever possible, models developed from observations under controlled environmental conditions should be field tested and refined before being applied to predict population performance.

Patterson et al. (1979) analyzed the growth of itchgrass (*Rottboellia exaltata* L. f.) under varying ranges of day and night temperatures and predicted that the weed could reach 75 to 100% of its maximum potential growth in the Gulf Coast states, the lower Midwest, the Southern Atlantic states, and the Southwest. Since itchgrass is not native to the United States and occurs here only in limited areas, it was not practical to test the model by introducing this noxious weed to these potential problem regions.

Further studies which imposed the three days of chilling temperatures that commonly occur during the early growth season of corn (*Zea mays* L.) and soybean (*Glycine max* L.) showed that itchgrass was more sensitive to growth reduction from chilling than was either crop (Patterson and Flint, 1979b). It was predicted, therefore, that itchgrass was unlikely to become an important early-season competitor with corn or soybean in regions outside of the South because of chilling probability. This second study demonstrated the limitations of extrapolating results from controlled environmental studies into predictions of population responses under natural conditions without verification from field studies. In a companion study (Patterson and Flint, 1979a), the weeds velvetleaf (*Aubtilon theophrastic* Medic.) and spurred anoda (*Anoda cristatu* (L.) Schlect.) were shown to recover faster and more completely from a three-day chilling period than did cotton (*Gossypium hirsutum* L.). The results helped explain the field observation that these two weeds are especially competitive with cotton following abnormally cool periods early in the growing season.

Patterson (1979) further determined the effect of shading on itchgrass growth and photosynthetic activities. Mathematical growth analysis indicated that shading, 60%, 25% or 2% sunlight, markedly reduced itchgrass biomass production. But even though dry-weight production was greatly reduced by shading, itchgrass maintained its capacity for high photosynthetic activity and high growth rates when subsequently exposed to high irradiance. These characteristics were interpreted as explaining its competitiveness with crop species and perhaps why it is ranked by Holm et al., (1977) as the eighteenth worst weed in the world.

Orwick et al. (1978) at Purdue University developed a simulation model for robust white foxtail (*Setaria viridus* var. *robusta-alba* Schreiber) or robust purple foxtail (*S. viridus* var. *robusta-purpurea* Schreiber). This model, SETSIM, is patterned after the basic plant growth simulation of an earlier alfalfa carbohydrate balance model, SIMED (Holt et al., 1975). SETSIM uses a compartmental approach to represent carbon flow within a foxtail population. These models require 30 input factors to regulate nine physiological rates in their projections. SETSIM can predict the most active period of robust foxtail growth and hence, the time of its greatest susceptibility to a systemic postemergence herbicide. Further studies of the competitive effects of foxtail and other weeds on soybean growth and development (Hagood et al., 1980; Orwick and Schreiber, 1979) should lead to the development of a simulation model that will predict soybean growth under weed-free conditions and in competition with weeds.

Simulation models based on photosynthesis and translocation rates often provide a broader scientific understanding of population growth and development. These models, however, tend to be site specific. Furthermore, because their input factors are not commonly available, except to scientists, the models often have limited applications.

A. Physical Models

Richardson, at Utah State University, has pioneered the development of what he calls physical models in contrast to the statistical or physiological models referred to previously. In developing these more generalized phenoclimatography models, the objective has been to utilize only readily available weather information as input factors, yet have predictive models with sufficient accuracy for use in agricultural decisionmaking.

The first fruit tree model developed under this concept was the chill unit model in which environmental temperatures were related to the time of rest completion of peach trees (Richardson et al., 1974). This weighted-temperature model is based on a "chill unit" concept wherein one chill unit equals one hour exposure at 6 C. As environmental temperatures depart from this optimum value, their contributions toward meeting rest requirements of the tree decrease until (at -1.1 and 15.6 C) the chill unit accumulation is 0. As temperatures climb above 15.6 C, the contributions toward meeting the chilling requirements of a tree become negative (Erez and Lavee, 1971).

Bud temperatures in an orchard often vary significantly from ambient air temperature as measured in a standard instrument shelter. With calm winds during daylight hours, the bud temperature may exceed the ambient air temperature by 10 to 15°C or more. Similarly, at night under clear, calm conditions, radiative losses from a bud may reduce its temperature 3 to 5°C below that of the ambient air. Winds of 7 m/s or higher will maintain bud temperatures quite near that of the ambient air. As bud temperature data are not readily available, curves were developed relating shelter temperature to effective bud temperature (Fig. 1).

The chill unit concept was developed to explain the manner in which fruit buds respond to temperatures during the winter. Curve A in Fig. 1 represents the relation between actual bud temperature and the rate of meeting the rest or dormancy requirements of the buds. Since most of the temperatures available to users are measured in a standard instrument shelter (Stevenson Screen), it was necessary to develop a method for relating these shelter temperatures, as represented by Curve B, to the actual bud temperatures. Analysis of the influence of wind and radiation on the actual temperature of the buds enabled the establishment of Curve C which represents the effective bud temperature.

The effective bud temperature curve was used in developing the chill requirements of various fruit cultivars. Curve A should be used to compare the temperatures of buds being grown in a growth chamber or greenhouse since wind and radiative effects are not present. Chill unit requirements have been calculated for other deciduous fruit by a statistical method (Ashcroft et al., 1977). The values have been tabulated in Table 1.

Environmental temperatures influence both the breaking of rest in deciduous plants and their subsequent phenological development. Early concepts regarding the relationship between

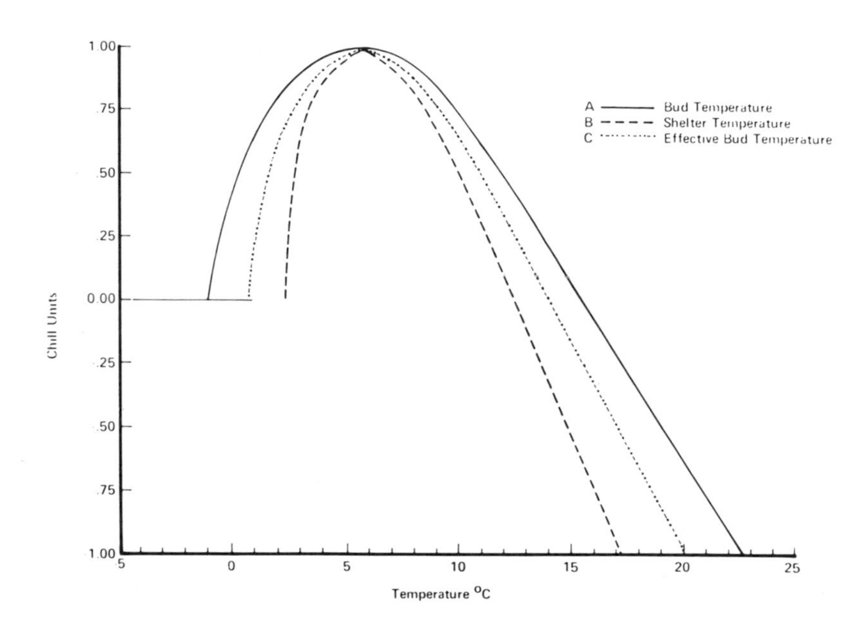

Fig. 1. Chill unit development curves.

environmental conditions and plant growth and development, especially fruit tree blossoming, have been reviewed elsewhere (Lombard and Richardson, 1979; Wang, 1963). A growing degree hour (GDH) fruit tree model was developed wherein the date of fruit blossoming could be predicted (Richardson et al., 1975). In the model, one GDH is defined as 1 h at 1°C above 4.5 C with the maximum accumulation for 1 h occurring at 25 C. The equation for calculating the daily accumulation of GDH is:

$$\text{GDH (C)} = \sum_{i=1}^{24} (Ti - 4.5)$$

Table 1. Chill unit requirements to the end of rest for various fruit trees.

Variety-Cultivar	Chill Units (CU in °C)
Apple "Red Delicious"	1234
Apricot "Tilton"	720
Cherry "Bing"	880
"Montmorency"	954
Grape "Concord"	860
Peach "Elberta"	800
"Redhaven"	870
Pear "Bartlett"	1210
Prunes "Italian"	788

where Ti = 4.5 C for $T_h < 4.5$ C; Ti = 25 C for $T_h > 25$ C; and Ti = T_h for 4.5 C $\leqq T_h \leqq$ 25 C, in which T_h is the temperature in C at hour h. Table 2 lists the GDH accumulation to certain phenological stages after chilling has resulted in rest completion. A computer program was developed to predict bud phenology based on date of rest completion, GDH accumulation to date, and long-term averages of temperatures (normal temperature) at a given site (Anderson et al., 1975). Hourly temperatures from sites not recording hourly temperatures were synthesized from daily maximum and minimum temperatures recorded in instrument shelters at a test orchard.

The predictive capabilities of the chill unit and GDH models were field tested for two years. Cooperators at 30 orchard sites in 11 states (which had long-term temperature histories) recorded daily maximum and minimum temperatures and mailed them to the

Table 2. Growing degrees hour accumulated (GDH) after chilling completion through various stages of fruit bud development in tree fruits using linear model (23).

Variety Cultivar	GDH at various season stages (GDH °C)			
	Bud swell	First pink	First bloom	Full bloom
Apple "Red Delicious"	2,061	4,856	6,172	6,933
Apricot "Tilton"	1,444	3,039	3,533	4,111
Cherry "Bing"	2,722	4,661	5,328	6,072
"Montmorency"	1,556	4,780	6,196	6,920
Grape "Concord"	4,139	–	16,056	17,778
Peach "Elberta"	2,167	3,717	4,239	5,117
"Redhaven"	1,411	3,710	4,174	4,926
Pear "Bartlett"	1,794	4,244	5,044	5,644
Prunes "Italian"	2,533	5,417	5,683	6,411

laboratory each week. The temperature inputs were used to refine computer predictions of date of rest completion.

Following accumulation of sufficient chill units to satisfy rest requirements, bloom dates were predicted based on long-term temperature averages. These predictions of bloom date were refined weekly as field data arrived and updated predictions were made on site temperature data to date and long-term averages for the remainder of the bud development season. The standard deviation of predicted dates using actual temperatures for all experimental sites was ± 3 days. The method of estimating hourly temperatures produced a GDH accumulation within two percent of actual for continental areas where no routine changes of air mass occur. At sites with maritime climates along the coast or sites in canyons where routine changes of wind direction result in regular changes of the air mass GDH estimation error was much greater. The fruit tree models have been refined recently to improve their accuracy.

Since spring bud development could be estimated so accurately by GDH accumulation, it seemed logical that bud development could be delayed by lowering bud temperature after rest was completed. Therefore, a program of overhead sprinkling was initiated to reduce bud temperature by evaporative cooling. The chill unit model was used to estimate the date of rest completion and thereafter the trees were sprinkled to cool the buds whenever the instrument shelter temperature exceeded 7 C (Anderson et al., 1976). By estimating the stage of sprinkled bud development at any desired time and assuming normal temperatures for the remainder of the spring season, dates of phenology development of the sprinkled bud were made. Using the models to control sprinkling, a delay in full bloom of nearly three weeks was achieved. Such bloom delay has protected treated orchards from late spring freeze damage.

B. Tomato and Weed Models

With the ultimate purpose of developing a model to estimate annual weed competition with tomatoes, Anderson and coworkers have developed a series of phenological models for tomato and four annual weeds common in tomatoes. These models were patterned after the Utah fruit tree models in that the input parameters are commonly available meteorological observations.

Richardson (unpublished data) tested four different methods of calculating GDH on tomato phenology data collected from seven sequential plantings in California. The average difference between the predicted and observed dates of first bloom among the plantings ranged from 4.3 days, using a linear response model with no upper temperature limits, to 1.8 days using an asymmetric curvalinear set of equations. The latter became the basis of a tomato model (Anderson et al., 1981). This model, based on standard instrument shelter data, was used to predict the growth and development of sequential plantings of tomatoes in Utah and California during 1978 and 1979. When tomato plants were grown weed free, one per square meter plot, under fertilized and irrigated conditions where plant competition, nutrients and moisture were not limiting, temperature appeared to be the growth controlling variable. The model predicted the dates of various phenological stages about as accurately as variability among plants in the test population would allow. Average differences between predicted and observed dates of yellow bud stage and first red fruit among all plantings were less than two and five days, respectively. The correlation between energy accumulation as defined by the model and biomass development was 94.7%.

Results with preliminary models developed for black nightshade (*Solanum nigrum* L.) and hairy nightshade (*Solanum sarrachoides* Sendt.) were similar to those of the tomato model except for differences in the cardinal temperatures (Hinkley, unpublished data). The optimum temperature for tomato was 26 C, for black nightshade 24 C and for hairy nightshade, 28 C. The critical temperatures, temperatures beyond which little or no growth will occur but no irreversible damage is done to the plant, for the three species were assumed to be 36 C, 40 C, and 38 C, respectively (Table 3).

A phenoclimatography model was developed for redroot pigweed (*Amaranthus retroflexus* L.) relating phenology, plant height, volume and biomass to GDH accumulation (Slack, 1981). The cardinal temperatures for the pigweed model were 10, 32, and 40 C. GDH accumulation from seedling emergence to anthesis was dramatically affected by photoperiod (Anderson and Salisbury, 1977). Plants emerging during photoperiods of 14 h or longer required twice the GDH to reach anthesis as plants emerging during photoperiods of less than 14 h. Daylength at emergence appeared to be the critical trigger for the photoperiod response. Seedling populations

Table 3. Cardinal temperatures for curvilinear asymmetric models of tomato and associated weed species.

Plant	Cardinal temperatures (°C)		
	Base	Optimum	Critical
Tomato (VF-145)	10	26	36
Barnyardgrass	10	32	40
Redroot pigweed	10	32	40
Black nightshade	10	24	40
Hairy nightshade	10	28	38

emerging 10 days apart bracketing the natural 14 h daylength threshold had a two fold difference in energy requirements to reach anthesis. Under greenhouse conditions, plants emerging during photoperiods of less than 14 h developed only four to five leaves prior to anthesis whereas plants emerging during photoperiods greater than 14 h had 14-16 leaves by anthesis (Koller, 1974).

In a companion study, Palmer amaranth (*Amaranthus palmeri* S. Wats.) also showed a photoperiod response (Slack and Anderson, 1979). The response, however, appeared to be cumulative, gradually intensifying as the season progresses. Redroot pigweed in contrast had a narrow window following seedling emergence through which to sense daylength. The differences in photoperiod response and GDH accumulation to seed maturity may partially explain the adaptability and distribution of the two related amaranths.

A preliminary barnyardgrass [*Echinochloa crusgali* (L) Beauv.] model has the cardinal temperatures of 10, 32, and 40 C (Anderson and Richardson, 1979) Barnyardgrass The world's third worst weed (Holm et al., 1977) also has a definite photoresponse: the longer the daylength, the greater the GDH accumulation prior to anthesis. Not only does barnyardgrass have a wider photoperiod window than redroot pigweed, but the window is operative later in its growth cycle. Evidently, barnyardgrass does not sense daylength until about 2500 GDH have accumulated, or when the plant is in the 3-4 leaf stage. A similar response is present in some of the perennial range grasses (Richardson, 1981; Richardson and Leonard, 1981).

III. SUMMARY

Few attempts have been made to model weed species. Weeds are generally more difficult to model than crop species because of the wide genetic and phenotypic variability that occurs in native populations. Modeling of individual weedy plants grown under controlled conditions has provided some basic biological information on the respective species but will be of limited usefulness to agriculture unless it can predict population distribution [such as Patterson (1979) has been able to do] or competition with crops. Models that predict the phenological development of weeds that serve as alternate hosts of insects and other plant pests would be of value in integrated pest management strategies. Upon completion of individual phenological models for weeds that commonly compete with tomatoes, we intend to develop a model that will predict the competitive effects of various populations of barnyardgrass, redroot pigweed, and nightshades on tomato growth and production.

REFERENCES

Anderson, J. L., Ashcroft, G. L., Richardson, E. A., Alfara, J. F., Griffin, R. E., Hanson, G. R., and Keller, J. (1975). J. Am. Soc. Hort Sci. 100:229-231.

Anderson, J. L., Ashcroft, G. L., Griffin, R. E., Richardson, E. A., Seeley, S. D., and Walker, D. R. (1976). Final Report. Utah State Agric. Exp. Sta. Proj. 562-366-084.

Anderson, J. L., and Richardson, E. A. (1979). Abstr. Weed Sci. Soc. Am. p 91.

Anderson, J. L., Richardson, E. A., and Ashton, F. M. (1981). Hort Sci. (In press.)

Anderson, J. L., and Salisbury, F. B. (1977). Proc. West. Soc. Weed Sci. 30:15-17.

Ashcroft, G. L., Richardson, E. A., and Seeley, S. D. (1977). HortSci. 12:347-348.
Erez, A., and Lavee, S. (1971). J. Am. Soc. Hort. Sci. 96: 711-714.
Hagood Jr., E. S., Bauman, T. T., Williams Jr., J. L., and Schreiber, M. M. (1980). Weed Sci. 28:729-734.
Holm, L. G., Plucknett, D. L., Panchio, J. V., and Herberger, J. P. (1977). The world's worst weeds. Distribution and biology. University Press of Hawaii, Honolulu. 609 p.
Holt, D. A., Bula, R. J., Miles, G. E., Schreiber, M. M., and Peart, R. M. (1975). Purdue University Agric. Exp. Sta. Res. Bull. 907. 26 p.
Koller, D. (1974). Final Report U.S.D.A. Project A10-CR-80. Hebrew University of Jersusalem, Israel. 208 p.
Lombard, P. B., and Richardson, E. A. (1979). In "Modification of the Aerial Environment of Plants," B. J. Barfield and J. F. Gerber, (eds.). ASAE Monograph No. 2, Am. Soc. Agric. Engrs. St. Joseph, Missouri. pp. 429-440.
Orwick, P. L., and Schreiber, M. M. (1979). Weed Sci. 26:655-674.
Orwick, P. L., Schreiber, M. M., and Holt, D. A. (1978). Weed Sci. 26:691-699.
Patterson, D. T. (1979). Weed Sci. 27:549-553.
Patterson, D. T., and Flint, E. P. (1979a). Weed Sci. 27:473-479.
Patterson, D. T., and Flint, E. P. (1979b). Weed Sci. 27:645-650.
Patterson, D. T., Meyer, C. E., Flint, E. P., and Quimby Jr., P. C. (1979). Weed Sci. 27:77-82.
Richardson, E. A. (1981). Contract UT-910-CTO-003, Bur. Land Mgmt., U.S. Dept. of Interior, Salt Lake City, Utah. 64 p.
Richardson, E. A., and Leonard S. G. (1981). Abstr. 15th Conf. Agric. and Forest Meteor. pp. 182-185. Am. Meteor. Soc.
Richardson, E. A., Seeley, S. D., and Walker, D. R. (1974). HortSci. 9:331-332.
Richardson, E. A., Seeley, S. D., Walker, D. R., Anderson, J. L., Ashcroft, G. L., (1975). Hort. Sci. 10:236-237.
Slack, E. M. (1981). The phenological development of redroot pigweed (*Amaranthus retroflexus* L.)., M.S. Thesis, Utah State University. Logan, Utah.
Slack, E. M., and Anderson, J. L. (1979). Proc. West Soc. Weed Sci. 32:92-96.
Wang, J. Y. (1963). Agricultural Meteorology. Pacemaker Press. 693 p.

INTERACTIONS AMONG WEEDS, OTHER PESTS AND CONIFERS IN FOREST REGENERATION

Steven R. Radosevich
University of California
Davis, California

and

Susan G. Conard
Department of Forest Science
Oregon State University
Corvallis, Oregon

I. INTRODUCTION

Forests occupy vast areas of North America. They range from the Boreal zone to the tropics and exist from semiarid to rain forest conditions. Forest vegetation also occupies a wide range of soil and topographic conditions at elevations from sea level to over 3500 meters. All of these factors make the forest a highly diverse and complex vegetation type. Forest vegetation is also exposed to an array of natural and man-caused disturbances which are often as diverse as the vegetation that results from them. Following such catastrophic events, some semblence of the original forest will return, in time. The regeneration goes through several seral stages in which species dominant in one stage are largely replaced by plant species common to the next. In this sequence it makes little difference if the initial disturbance occurred from a "natural" event or was man-caused. It appears

ISBN 0-12-332850-0

that reoccurring disturbance is common in most forests and "virgin stands" have simply had a long time to recover from the most recent catastrophe. This had lead Spurr (1964) to state "disturbance to tree development and growth are normal, instability of the forest is inevitable, and the changeless virgin forest is a myth."

A hypothetical example of early succession following disturbance in the coniferous zones of Northern and Western America is presented in Fig. 1. In this diagram, initial establishment of a pioneer community of herbaceous vegetation is replaced by a shrub dominated (brush) community. The intermediate shrub community will eventually be replaced by conifers in the absence of additional disturbance. However, the natural reestablishment of conifers usually proceeds slowly since they are readily suppressed by brush growth. The success of these early phases of succession in conjunction with site characteristics and regional climate, determine the time spans required for conifer dominance to reestablish. These early (herb and brush dominated) phases of succession have been the object of most forest regeneration efforts.

II. DOMINANCE POTENTIAL

The same environmental resources (light energy, CO_2, nutrients, and water) are fundamental to every forest system. Therefore, the relative success of any species in a seral stage will depend on its abilities to survive environmental site conditions and to preempt limited resources. Success of a species will depend on the resource requirements of that species, the availability of resources during periods of demand, and adaptations to partitioning of resources in time and space relative to other species in the community.

Newton (1973) has proposed the term "dominance potential" to relate the ability of an individual species to preempt resources. In defining this process, Newton (1973) states:

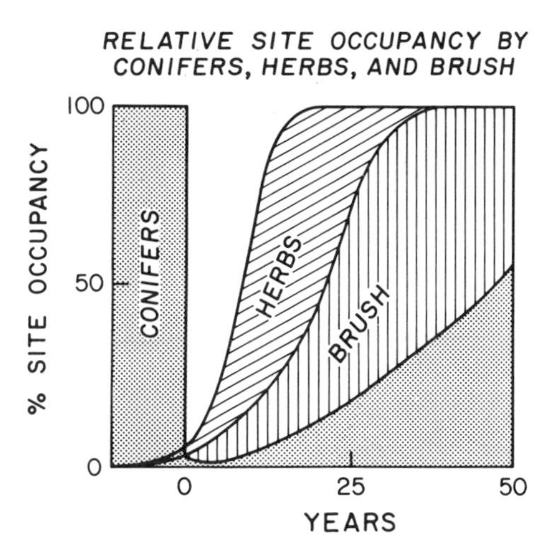

Fig. 1. A hypothetical example of early forest succession following disturbance.

"dominance potential (DP) is the summation of all features of a species relating to its ability to assume dominance (over a specific length of time) in a limited system. For a given site and given an equal start, a species with high DP will tend to replace or suppress a species with lower DP. A species with lower DP will prevail if it establishes first, but a species with high long-term DP will eventually displace a species with high short-term DP."

Thus, it appears that a series of communities will develop after disturbance, as in Fig. 1, with species of high short-term DP initially dominating. As succession proceeds, species of each community apparently maintain their dominance for longer periods of time. If disturbance occurs on a frequent basis, it is likely that succession will not proceed beyond the early stages (Fig. 2).

Fig. 2. Forest succession in the Sierra Nevada reverted to an earlier successional (shrub) stage as a result of frequent fire.

III. CONIFER REGENERATION IN BRUSHFIELDS

Over 22 million hectares of potentially productive forestlands in the Pacific Coast and Rocky Mountain regions are dominated by brush and weed-tree species (Gratkowski, et al., 1973). These brushfields often have resulted from disturbance to the natural forest, primarily by fire or logging activity. Many woody shrub species produce large quantities of seed which retain their viability for years and germinate when exposed to favorable conditions typical of over-story timber removal, fire, or soil scarification. Quick (1944) observed nearly 7 million viable brush seeds per hectare in the soil and litter of an over-mature stand of timber. Most of the common brush species also sprout readily from roots or stumps after removal of their above-ground parts. Roberts (1980), while evaluating the effectiveness of manual brush cutting for competition release of young Douglas fir (*Pseudotsuga menziesii*) in the Pacific Northwest, observed that all brush species sprouted vigorously after they were cut. After about 18 months of regrowth, brush had returned to near pretreatment levels (1.5 m) and no increase in Douglas fir height growth had occurred. Because of the rapid occupation of sites by shrub or other brush species some form of brush suppression has often been desirable for adequate conifer regeneration (Fig. 3). However, once conifers assume dominance, they should remain there due to their high long-term dominance potential, especially in the absence of other constraining factors like animals, insects, or disease. The need for brush suppression is often dependent on the species of conifer and shrub species involved and effectiveness and timeliness of site preparation in relation to planting (Schubert and Adams, 1971).

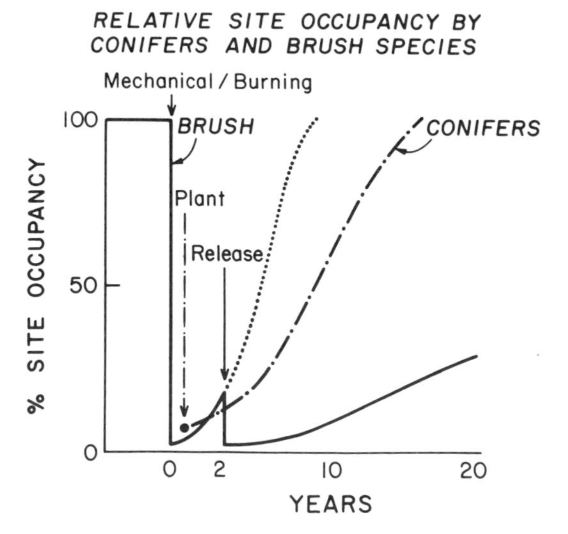

Fig. 3. A hypothetical example of succession following brush removal by site preparation. If shrub sprouting and rapid growth occurs (heavy dashed line) some form of brush suppression (release) is often necessary to allow conifer dominance.

When sites are totally occupied by shrub or undesirable tree species, some form of initial site preparation is required to establish a plant community in which conifers are dominant (Fig. 3). Procedures include mechanical site scarification, the use of mechanical shrub shredders, controlled fire, herbicides, or combinations of these tools. The result is usually a site in which bare mineral soil is exposed and shrub top growth is initially suppressed. These are conditions which enhance dominance of planted conifers since available niches for resource utilization are not fully occupied at that time (Fig. 3). The conifer growth will be best when the trees become established before the competitors, are well separated from the competitors, and the competitors are weak or have been weakened.

Considerable effort has been made to identify the effects of brush on conifer establishment and growth. Stewart (1980) has summarized over two hundred papers on this subject. Although the conditions and species varied in Stewart's summarizations, it was generally agreed that brush vegetation inhibited conifer growth. Small trees, those below or within the brush canopy, appeared to respond better than taller trees to brush control (Table 1, Fig. 4). Although most studies (Stewart, 1980) attributed increased coniferous growth to increased availability of light, soil moisture, or nutrients following brush control, the actual degree of resource limitation during those studies was unknown.

IV. ENVIRONMENTAL FACTORS

Species of early communities may initially mediate their microenvironment to enhance their own competitive ability and successional development. For example, crown fire or clearcut logging in the white fir (*Abies concolor*) zone of the Sierra Nevada (1200 to 2100 m elevation) initiates a seral sequence in

Table 1. Height growth of young Douglas-first for a five-year period after release from varnishleaf ceanothus (Gratkowski and Lauterbach, 1974).[1]

Release Treatment[2]	Height of trees when released (feet)					
	1	2	3	4	5	6
	Percent[3]					
None	100	100	100	100	100	100
Cut; stump sprayed	185	185	183	184	184	185
Basal sprayed	130	136	140	144	147	149
Aerial spray	255	217	200	188	181	171

[1] Height growth on the aerial spray area was for a six-year period after spraying.

[2] Treatment applied to varnishleaf ceanothus.

[3] Height growth as a percentage of growth of similar trees under live ceanothus.

which previously forested sites become fully occupied by sprouting (fire adapted) shrub species within ten years after disturbance (Skau, et al., 1970; Fowells and Schubert, 1951). White fir saplings in these brushfields may be quite high (1,000 to 10,000 stems/ha) within 20 or 30 years after a disturbance, but the trees

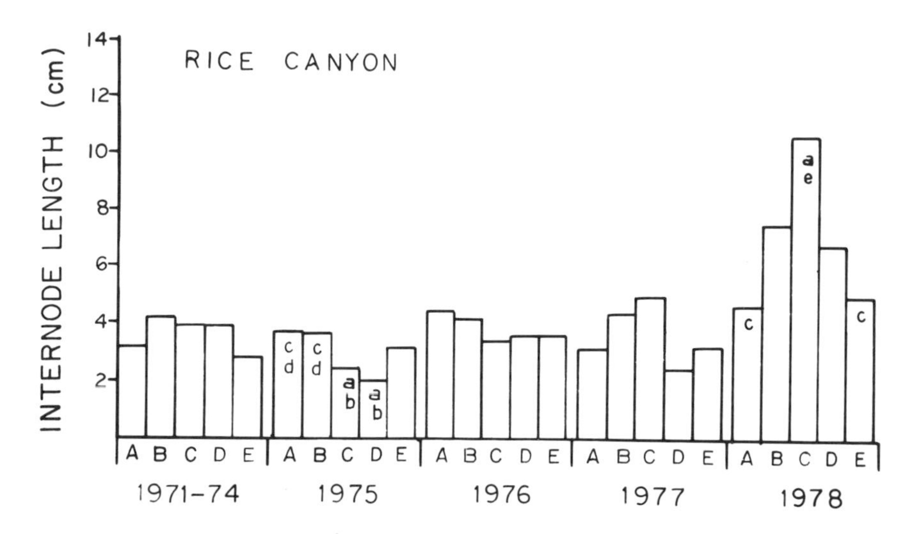

Fig. 4. Median internode growth of white fir saplings during the 1971 through 1978 growing seasons. Letters on each column indicate those treatments from which median growth is significantly different at an overall probability level of 0.05. Comparisons made were among treatments within years. A through E refer to treatments: (A) not disturbed; (B) shrub stems cut, removed, allowed to regrow; (C) shrub stems cut, herbicide treated, and shade provided with cut branches; (D) identical to C but shrub canopy removed from plot; (E) herbicide applied to undisturbed shrubs and trees. (Conrad and Radosevich, 1981b.)

often grow only 3 to 4 cm per year under the shrub canopy (Conrad, 1980). Conard and Radosevich (1981a) measured maximum and minimum air temperatures above and within the shrub canopy (Fig. 5). Air temperatures were 2 to 3°C higher within the shrub canopy, where most of the trees occurred, than above the canopy. Minimum air temperatures were also consistently lower (2 to 3°C) within the shrub canopy than above it. Figure 6 shows the response of net photosynthesis to temperature for white fir and the major shrub constituent in that study based on laboratory data. The narrow

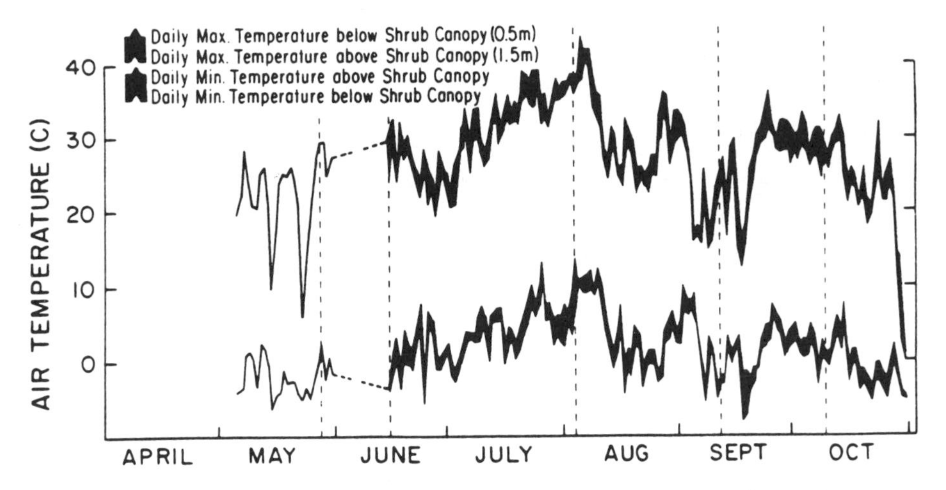

Fig. 5. Maximum and minimum air temperatures above (1.5 m) and in the shrub canopy (0.5 m) from May through October 1978 (Conrad and Radosevich, 1981a).

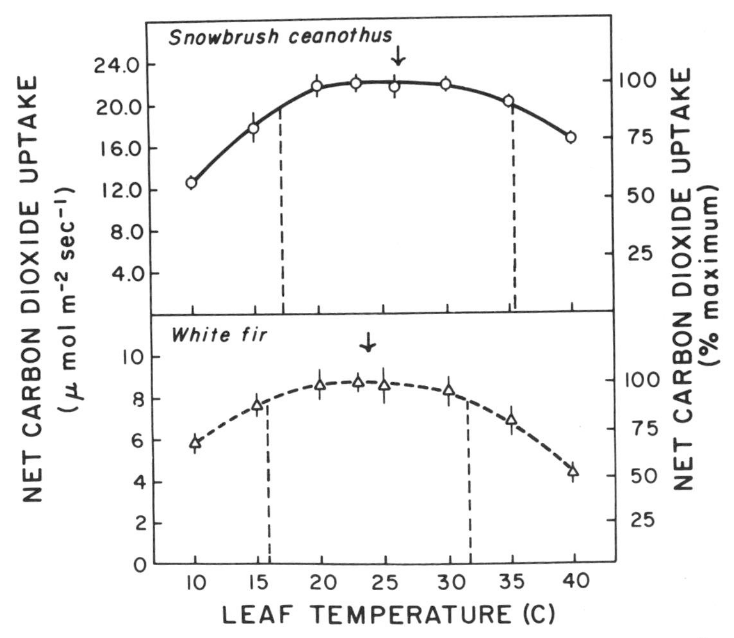

Fig. 6. Response of net photosynthesis to temperature for white fir and snowbush ceanothus (*Ceanothus velutinus*). Each point is the mean value of three plants. Vertical bars represent one standard deviation. Dashed vertical lines indicate the temperature range at which net photosynthesis is at least 90% of maximum. Arrows identify the midpoint of that range (Conrad and Radosevich, 1981a).

range of optimum temperature for white fir photosynthesis suggests that the wide temperature fluctuations within the shrub canopy may disfavor growth of white fir relative to snowbush ceanothus. In another study at the same site, Conard and Radosevich (1981b) observed that white fir growth was doubled by the fourth growing season when artificial shade was provided in conjunction with the removal of competing shrubs (Fig. 4). A smaller growth increase was observed when no shade was provided if shrub control was greater than 80 percent. They concluded that poor soil moisture availability (due to the presence of a rapidly transpiring shrub cover) was the most important factor for the suppression of white fir growth. The benefit of added shade was an effect of improvement of the water balance in shaded trees. These data suggest that on drought-prone sites dominance potential is enhanced by rapid early growth after disturbance and the ability to capture available soil moisture. Only those trees which can tolerate shade and economically use residual water supplies will survive to eventually attain long-term dominance. In contrast, the ability to outgrow competitors or survive their shade may be the overriding characteristic on moist sites (Newton, 1973).

The amount of conifer growth achieved will depend upon the interaction of numerous environmental factors, all of which may be influenced by competing vegetation. The effect of environment (as well as interactions between species competing for site resources) will also be dependent on physiological responses of individual species. It is important, therefore, to know the level of environmental resources or factors available and to understand the species responses to them. Several examples will illustrate the importance of environment in regeneration.

Maguire (1955) observed that soil surface temperatures can reach 72C on bare soil at low elevation in the Sierra Nevada. He reported significant increases in ponderosa pine (*Pinus ponderosa*) seedling survival from artificial shade and attributed this response to milder soil surface temperature rather than decreased transpiration. It seems likely, however, that both factors were occurring. Sunlight, although necessary for photosynthesis, may also be harmful to certain coniferous species if irradiance is too high. Severe solarization of Engelmann spruce (*Picea engelmannii*) was noted on high elevation exposed sites and has been attributed to plantation failure in the central Rocky Mountains (Ronco, 1975). Shade during the growing season markedly increased survival. Tappeiner and Helms (1972) observed higher first year survival of Douglas fir and white fir in shaded microsites rather than in full sunlight. They attributed seedling mortality to a number of causes, notably drought. These studies demonstrate that under conditions of high moisture stress or irradiance the presence of shade may be crucial for conifer survival. In these situations, the benefits of shading may outweigh any potential decrease in photosynthesis as a result of lowered light intensity or temperature. Both temperature and light are necessary for plant growth but both can be detrimental, if excessive.

Because environmental factors interact with physiological processes some workers have attempted to express microenvironmental requirements of early conifer growth in physiological terms (Cleary, 1970; Lindquist, 1977; Tappeiner and Helms, 1972). A study conducted by Radosevich and Conard concerning the comparative physiology of several coniferous species planted on an exposed site in the Sierra Nevada should demonstrate the importance of interspecific differences in response to environmental conditions. Five species of conifers were planted in 1975 on a previously brushraked and disked site so that the influence of competitors was removed. Seasonal and diurnal measurements of plant moisture status, leaf conductance to water vapor, and gross photosynthesis were determined during 1977 and 1978 with a Scholander pressure chamber (Scholander et al., 1965), diffusive resistance porometer, and a $^{14}CO_2$ exposure apparatus (Shimshi, 1969; Tieszon et al., 1974), respectively. Seasonal patterns of moisture stress (predawn xylem sap tension) for five coniferous species are presented in Table 2. As the normal summer drought

Table 2. Predawn xylem sap tension of five coniferous species planted on an exposed site in the Sierra Nevada. $LSD_{.05}$ for species, date, and species x date = 1.0, 1.2, and 2.0 bars, respectively.

Species	Predawn xylem sap tension (bars)								
	4/13/77	7/6/77	8/29/77	9/23/77	10/21/77	12/13/77	5/26/78	7/6/78	8/29/78
Pondoerosa pine	6.6	3.2	4.8	13.8	3.6	–	3.3	3.9	3.4
Jeffrey pine	8.3	2.4	3.4	7.9	3.6	–	3.0	3.7	4.0
Sugar pine	4.4	0.9	5.4	7.4	5.8	–	2.6	4.1	4.2
Douglas fir	7.2	4.4	6.6	9.1	6.6	–	4.1	4.9	7.8
White fir	5.3	4.5	5.8	12.1	8.4	–	2.6	3.9	8.5

that is common in the Sierra Nevada progressed through spring and summer in 1977 and 1978, predawn xylem sap tension (moisture stress) did not increase significantly for ponderosa pine and Jeffrey pine (*Pinus jeffreyii*). A similar response was noted for sugar pine (*Pinus lambertiana*) in 1978. Predawn xylem sap tension rose sharply for sugar pine, Douglas fir, and white fir in August 1977 and for Douglas fir and white fir in 1978. Highest levels of moisture stress for every species were observed in September 1977 (Table 2). While ponderosa pine and white fir had significantly higher levels of predawn moisture stress in September 1977 than the other three species, the recovery of ponderosa pine to low levels of xylem sap tension was complete by October 21, 1977. This is in contrast to the continued high stress of white fir. Pharis (1966) has compared seedling drought resistance of several conifer species and concluded, based on soil moisture content, that ponderosa pine, Douglas fir, and sugar pine were most, intermediate, and least tolerant to drought conditions, respectively. Tappeiner and Helms (1972) indicate that effects of drought are more important for white fir than Douglas fir for regeneration on exposed sites. These data suggest that Douglas fir and white fir are not as well adapted to efficient water utilization as ponderosa pine and Jeffrey pine (Table 2). By the second year of our study, sugar pine responses appeared similar to those of other pines. The 1977 patterns for sugar pine may reflect a lag in establishment for younger stock rather than lower drought tolerance.

Seasonal patterns of $^{14}CO_2$ fixation (gross photosynthesis) for the five coniferous species are presented in Table 3. In 1977 the $^{14}CO_2$ incorporation of all species, except white fir, increased from April to July. Thereafter, each species maintained relatively constant gross photosynthesis levels until October 1977. Gross photosynthesis of Douglas fir and white fir was lower than the other species through these summer months. $^{14}CO_2$ incorporation of all species except ponderosa pine was significantly

Table 3. Gross photosynthesis of five coniferous species planted on an exposed site in the Sierra Nevada. $LSD_{.05}$ for species, date, and species x date = 1.0, 1.3, and 3.2 mg $^{14}CO_2$ gdw^{-1} hr^{-1}, respectively.

Species	$^{14}CO_2$ Fixation (mg $^{14}CO_2$ gdw^{-1} hr^{-1})								
	4/13/77	7/6/77	8/29/77	9/23/77	10/21/77	12/13/77	5/26/78	7/6/78	8/29/78
Ponderosa pine	6.3	10.9	6.9	10.5	11.8	2.7	7.7	13.8	7.9
Jeffrey pine	4.5	8.2	8.7	9.2	12.0	3.1	4.6	9.5	9.2
Sugar pine	3.2	6.6	6.5	5.2	8.8	2.7	7.0	9.8	10.2
Douglas fir	3.8	4.5	4.2	3.6	12.0	2.3	3.9	6.3	4.3
White fir	5.7	4.3	3.4	3.9	8.0	2.5	4.2	4.8	4.5

higher in October 1977 than August. This increase in gross photosynthesis was accompanied by a decrease in predawn xylem sap tension (Table 2). The October 1977 sample date was preceded by several cm of rainfall and also cooler temperatures than earlier sample dates. In contrast to the four species, $^{14}CO_2$ incorporation (Table 3) of ponderosa pine remained high throughout the growing season and did not significantly respond to seasonal variation in plant water stress (Table 2). Gross photosynthesis of every species decreased in the winter. Helms (1972) and Cleary (1970) indicate that photosynthesis of ponderosa pine and Douglas fir is under environmental control. With both species, low environmental stress (saturation vapor pressure deficit or plant water stress) resulted in high levels of photosynthesis. Plant moisture stress exceeding 18 bars decreased Douglas fir photosynthesis to about 50% of its maximum rate in those studies. Since xylem sap tension will affect stomatal closure and photosynthesis (Waring and Cleary, 1967; Slatyer, 1967) high seasonal plant moisture stress may ultimately inhibit conifer growth. There is considerable evidence that height growth of most trees occurs at the expense of stored carbohydrates (Kozlowski, 1962; Brown, 1974). Such studies indicate that a current season's stem growth is dependent on carbohydrate reserves produced and stored the previous year(s). Although no effort was made to assess carbohydrate storage and translocation patterns, it is significant that ponderosa pine, which maintained high season-long levels of gross photosynthesis (Table 3), was also most productive in stem elongation (Table 4).

Diurnal patterns (May 26, July 6, August 29, 1978) in $^{14}CO_2$ fixation and leaf conductance to water vapor for ponderosa pine, Douglas fir, and white fir are presented in Fig. 7. On May 26, 1978, stomatal conductance of the three species remained constant and high throughout the day. In contrast, on July 6, 1978, stomatal conductance of Douglas fir and white fir was significantly less than on the earlier date. The mean

Table 4. Survival and stem growth of five coniferous species planted on an exposed site in the Sierra Nevada. $LSD_{.05}$ for tree survival and total stem growth is 19 percent and 3.8 cm, respectively.

Species	Survival %	Stem growth (cm)
Ponderosa pine	69	39.5
Jeffrey pine	66	25.4
Sugar pine	56	14.8
Douglas fir	52	22.1
White fir	59	16.8

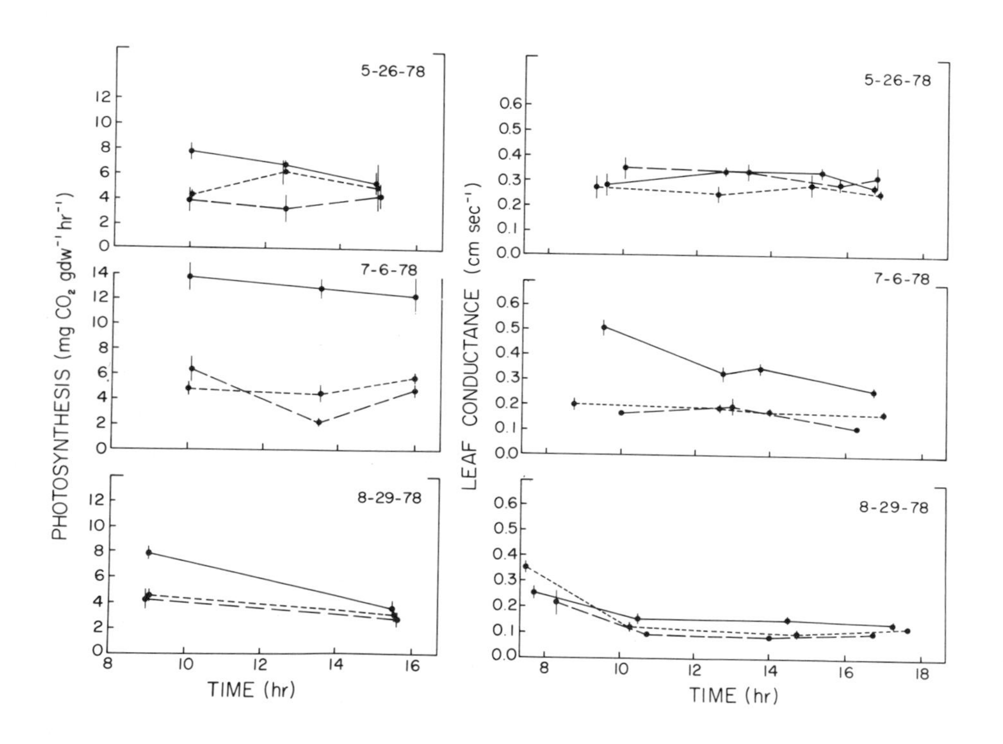

Fig. 7. Diurnal gross photosynthesis and leaf conductance of ponderosa pine, Douglas fir, and white fir. Vertical lines represent one standard deviation.

stomatal conductance of ponderosa pine on July 6 was similar to that on May 26, although there is evidence of a gradual stomatal closure throughout the day. Early morning conductance on August 28, 1978, for each species was higher than later times of the day. Since this date represents the most xeric condition of the three dates, it appears that considerable stomatal closure can occur by mid-morning to control transpiration loss. Photosynthesis values varied among species at each date. Ponderosa pine always fixed more $^{14}CO_2$ than the other two species. A trend of decreasing $^{14}CO_2$ incorporation was observed for each species as diurnal stomatal closure occurred. Ponderosa pine, which apparently is better adapted to xeric conditions (Tables 2 and 3) than Douglas fir and white fir (Fowells, 1965) maintained high stomatal conductance later into the season than did the other species. The diurnal variation in stomatal conductance and $^{14}CO_2$ incorporation is primarily a function of stomatal closure as temperatures warm and moisture becomes limiting. Similar diurnal responses of ponderosa pine, Douglas fir, and other tree species to diurnal environmental conditions have been reported (Helms, 1970, 1972; Cleary, 1970; Kramer, 1962).

These studies indicate the importance of the relationship between juvenile conifer growth and physiological responses to environmental conditions on exposed sites. Despite the value of site preparation in reduction of competing vegetation and increasing soil moisture availability, the exposed conditions created may prove favorable only to species adapted to manage high evaporative demands by balancing moisture uptake with transpiration loss. Ponderosa pine maintained consistently better relief of water stress, higher levels of photosynthesis, and faster growth rates than the other coniferous species. Although the other species varied in these responses, Jeffrey pine and Douglas fir seemed intermediate while white fir grew less and seemed least adapted to exposed sites. Sugar pine also grew poorly in this study, but physiological responses were often similar to other pines, especially during the last year of study. Sugar pine may be better physiologically adapted to the conditions of exposed sites than white fir or Douglas fir, but planting very young nursery stock may have decreased the ability of this species to initially establish and grow. These data agree with Pharis (1966) and Tappeiner and Helms (1972) and indicate that drought tolerant and rapid growing species should be planted soon after site preparation to obtain optimum conifer site capture on cut-over and manipulated areas.

Lanini and Radosevich (unpublished data) have compared the influence of shrub regrowth on soil moisture availability under three different site preparation regimes. Within the first growing season after site preparation, shrub regrowth had significantly reduced available soil moisture and by the second season xylem water potential of conifers was also reduced regardless of site preparation method. It is apparent that physiological status, environmental requirements, and resource needs of both coniferous and shrub species must be identified so that appropriate microsites can be created for maximum conifer survival and growth.

V. INFLUENCE OF VEGETATION MANAGEMENT ON OTHER PEST POPULATIONS

Harper (1977) believes that seedling establishment is probably the most critical time in the life cycle of a plant. He defines a "safe site" as a zone where conditions are favorable for germination, resources are available for consumption during growth, and hazards from predators, pathogens and competitors are absent. This definition can be extended for our purposes to include safe sites for planting as well as for germination. Conifer seedling survival often seems to be directly related to the availability of safe sites or favorable planting spots. Hermann and Chilcote (1965) examined the factors (soil condition and shade) which affected seedling establishment of Douglas fir after clearcut logging (Fig. 8). Major causes of seedling mortality the first year were heat, disruption by animals (mice, deer, and rabbits) and damping-off. Despite the wide range in causes of mortality, the overall variation among treatments was small. Up to a certain limit, if a seedling was not killed by one agent, it was killed by another. While this study does not identify the characteristics of a safe site, it does demonstrate their occurrence and the subtle interactions which may occur between various site factors.

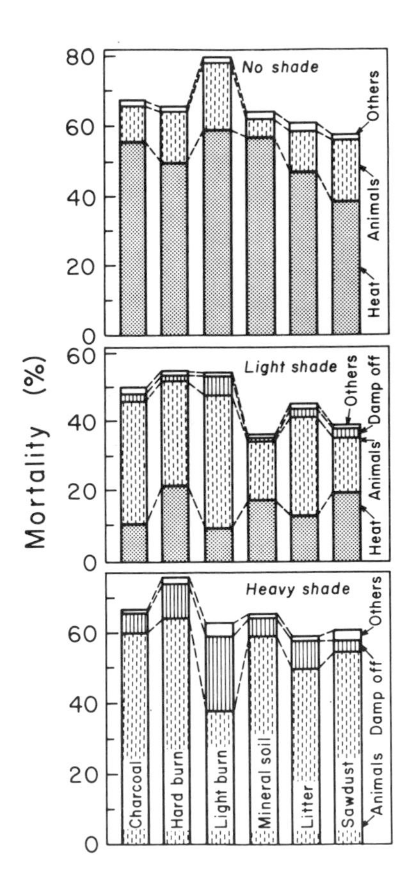

Fig. 8. The causes of mortality among seedlings of Pseudatsuga douglasii on six different substrates and under three shading regimes in a clear-cut forest in Oregon, USA (Herman and Chilcote, 1965).

VI. SMALL MAMMALS

Halves of three clearcuts in Western Oregon were treated by Borrecco, et al. (1979) in the springs of 1970 and 1971 with combinations of herbicides to control herbaceous vegetation. The vegetation and small mammal populations were sampled before and after herbicide application on both treated and untreated areas. Herbicide treatment significantly reduced the amount of grass and forb ground cover and increased growth and survival of Douglas fir. Growth of most shrubs was also better on treated than untreated areas. Reduction of herbaceous vegetation altered the species composition of small mammal communities as they responded according to their habitat preferences (Table 5). Species preferring grassy habitats were less abundant on treated than untreated areas.

Pocket gophers (Thomomys sp.) feed on roots, stems, and sometimes foliage of seedlings of most coniferous species. Their feeding activity usually results in tree mortality (Crouch, 1971). These animals can destroy new plantations within a year of planting, or annually kill a portion of the trees for ten years or more (Crouch and Hafenstein, 1977). For many years, gophers have been controlled by poisoning to reduce their numbers (Barnes, et al., 1970). However, there is apparently little documentation that poisoning with toxic baits by itself has resulted in successful protection of conifer seedlings (Crouch and Hafenstein, 1977), even though gopher poisoning has resulted in a dramatic reduction of damage to many agricultural crops.

Selective competition reduction between conifers and other plant species can be accomplished by herbicide application, tillage, or manual removal of undesirable vegetation in conifer plantations. Recent studies indicate substantial benefits to conifer seedlings from reduced competition of grass, forb, and brush species (Black and Hooven, 1977; Crouch and Hafenstein, 1977; Gratkowski and Lauterbach, 1974; Stewart and Beebe, 1974). There are also reports of declining gopher populations after infested range or forest land was treated with herbicides (Black and Hooven, 1977; Keith, et al. 1959; Howard and Childs, 1959; Hull, 1971). The population reduction was in response to the loss of grass and forbs which apparently constitute a primary source of the gopher food supply.

Table 5. Five principal species of small mammals removed from the three study areas in both years and their percent of composition of the small mammal communities (Borreco et al., 1979).

Area &	1970				1971			
	Untreated		Treated		Untreated		Treated	
Species	Number	%	Number	%	Number	%	Number	%
Carlson Creek								
Sorex vagrans	30	39.5	11	17.5	5	11.1	1	2.2
Sorex trowbridgii	4	5.3	15	23.8	3	6.7	7	15.2
Percomyseus maniculatus	18	23.7	33	52.4	20	44.4	35	76.1
Microtus oregoni	19	25.0	–	–	9	20.0	1	2.2
Zapus trinotatus	2	2.6	–	–	4	8.9	1	2.2
TOTAL	73	96.1	59	93.6	41	91.1	45	97.8
Williams Creek								
Sorex vagrans	39	21.1	38	33.0	5	16.1	3	8.1
Sorex trowbridgii	7	3.8	11	9.6	3	9.7	2	5.4
Peromyseus maniculatus	30	16.2	26	22.6	12	38.7	27	73.0
Microtus oregoni	94	50.8	35	30.4	5	16.1	–	–
Zapus trinotatus	4	2.2	–	–	2	6.5	–	–
TOTAL	174	94.2	110	95.6	27	87.1	32	86.5
Maxfield creek								
Sorex vagrans	40	14.3	14	12.4	28	23.7	7	11.5
Sorex trowbridgii	2	0.7	18	15.9	–	–	2	3.3
Peromyseus manicultus	25	8.9	33	29.2	2	1.7	13	21.3
Microtus oregoni	174	62.1	37	32.7	52	44.1	22	36.1
Zapus trinotatus	34	12.1	3	2.7	34	28.8	14	23.0
TOTAL	275	98.2	105	92.9	116	98.3	58	95.1

VII. INSECTS

Insect control is seldom considered in reference to vegetation management. However, this subject may deserve closer attention. The various methods of vegetation manipulation usually result in predictable shifts in plant species population which can affect both microenvironmental and habitat needs of insects. Recently, Orcutt and Radosevich (unpublished data) observed that about 90 percent reduction in herbaceous vegetation from herbicide application also reduced the grasshopper density in the treated plot by over 50 percent. Conifer mortality from grasshopper feeding in untreated areas was 98 percent while no grasshopper-caused mortality was evident where the herbicide was applied. A disked site preparation treatment was also included in that study. Disking also eliminated grasshoppers as a cause of conifer mortality even though grasshopper densities remained high. It is believed that disking caused the herbaceous vegetation which reoccupied that plot to be in a more palatable state by mid-season when most grasshopper damage occurred. Since grasshoppers apparently do not prefer to feed on conifers; less conifer mortality also resulted on that plot.

These data (Orcutt and Radosevich) and those of Black and Hooven (1977), Borrecco (1979), and Crouch and Hafenstein (1977) illustrate that the population dynamics of various organisms (pest and nonpest) can be manipulated through habitat management. The environmental requirements of these organisms and economic threshold levels of pests must be determined to effectively utilize this concept for conifer regeneration.

VIII. SUMMARY

The primary goal of vegetation management in forestry is to alter the competitive balance of early successional vegetation to favor conifer establishment and growth. Such alteration can result in conifers exercising dominance early and with long-term consequences. Since the same environmental resources are common to all forests, it is the ability of individual species to out-compete their neighbors in a limited system which determines succession. Coniferous, brush, or herbaceous species must, therefore, respond in a predictable manner to changes in environmental conditions. It is important for researchers to identify limiting resources, microenvironments (safe sites), and the physiological controls for both coniferous and brush growth. In this way predictive and realistic management schemes (models) can be developed for forest vegetation management and conifer regeneration. In addition to the predictable shifts of plant species which result from vegetation management, the dynamics of populations of other organisms associated with that vegetation can be affected. This affect can be positive (if additional pest damage is reduced), neutral, or negative in terms of conifer regeneration. It is important that the possibility for occurrence of these shifts be recognized and considered by the forest vegetation manager.

REFERENCES

Barnes, V. G., Jr., Martin, P. and Tietjen, H. P. (1970). J. For. 68(7):433-435.

Black, H. C., and Hooven, E. F. (1977). Proc. 29th Annual California Weed Conf. pp. 119-127.

Borrecco, J. E., Blake, H. C., and Hooven, E. F. (1979). Northwest Sci. 53:97-106.

Brown, C. L. (1974). In "Trees Structure and Function," (M. H. Zimmerman and C. L. Brown, eds.), Springer-Verlag, New York. 336 p.

Cleary, B. D. (1970). In "Regeneration of Ponderosa Pine: Symposium held 9/11-12/69," (R. K. Hermann, ed.), School of Forestry, Oregon State University, Corvallis.

Conard, S. G. (1980). Species interactions, succession, and environment in montane chaparral – white fir vegetation of the northern Sierra Nevada, California. Ph.D. Thesis, University of California, Davis. 124 p.

Conard, S. G., and Radosevich, S. R. (1981a). For. Sci, (in press).

Conard, S. G., and Radosevich, S. R. (1981b). For. Sci., (in press).

Crouch, G. L. (1971). Northwest Sci. 45(4):252-256.

Crouch, G. L. and Hafenstein, E. (1977). USDA Forest Service Research Note PNW-309.

Fowells, H. A. (1965). USDA Agric. Handb. 271. 762 p.

Fowells, H. A. (1965). USDA Agric. Handb. 271. 762 p.
Fowells, H. A., and G. H. Schubert. (1951). J. For. 49:192-196.
Gratkowski, G., Hopkins, D., and Lauterbach, P. (1973). J. For. 71:138-143.
Harper, J. L. (1977). In "Population Biology of Plants," Academic Press, London, New York. 892 p.
Helms, J. A. (1970). Photosynthetica 4:242-253.
Helms, J. A. (1972). Ecol. 53:92-101.
Hermann, R. K., and Chilcote, W. W. (1965). Res. Paper (For. Magmt. Res.), Oregon For. Res. Lab. 4:1-28.
Howard, W. E. and Childs, H. E., Jr., (1959). Hilgardia 29(7):277-358.
Hull, A. C., Jr. (1971). J. Range Manag. 24(3)230-232.
Keith, J. O., Hansen, R. M., and Ward, A. L. (1959). J. Wildl. Manage. 23(2):137-145.
Kozlowski, T. T. (1962). In "Tree Growth," Ronald Press Co., New York. 442 p.
Kramer, P. J. (1962). In "Tree Growth," (T. T. Kozlowski, ed.), pp. 171-183. Ronald Press Co., New York.
Lindquist, J. L. (1977). USDA Serv. Res. Note PSW-325. 5 p.
Maguire, W. P. (1955). For. Sci. 1:277-284.
Newton, M. (1973). J. For. 71:159-162.
Pharis, R. P. (1966). Ecol. 47:211-22.
Quick, C. R. (1944). J. For. 42:827-832.
Roberts, C. (1980). In "Second Year Report Cooperative Brush Control Study," Catherine Roberts, Reforestation Consultant, Corvallis, Oregon. 32 p.
Ronco, F. (1975). J. For. 31-35.
Scholander, P. F., Hammel, H. T. Bradstreet, E. D., and Hemmingsen, E. A. (1965). Science 148:339-348.
Schubert, G. H., and Adams, R. S. (1971). In "Reforestation practices for conifers in California." Calif. Resour. Agency Div. of For., Sacramento. 359 p.
Shimshi, D. (1969). J. Exp. Bot. 20:381-401.
Slatyer, R. O. (1967). In "Plant-Water Relationships," Academic Press, New York. 366 p.
Skau, C. M., Meuwig, R. O., and Townsend, T. W. (1970). Max. C. Fleishmann Coll. Agric. R 71, University of Nevada. 14 p.
Spurr, S. H. (1964). In "Forest Ecology," The Ronald Press Company, New York.
Stewart, R. E. (1980). In "Effects of Weed Trees and Shrubs on Conifers - A Bibliography with Abstracts." USDA Forest Management Research. Washington, D.C. 166 p.
Stewart, R. E., and Beebe, T. (1974). Proc. Ann. Meet. West. Soc. Weed Sci. 27:55-58.
Tappeiner, J. C., II, and Helms, J. A. (1972). Am. Midl. Nat. 86:358-370.
Tieszon, L. L., Johnson, D. A., and Caldwell, M. M. (1974). Photosynthetica 8:151-160.
Waring, R. H., and Cleary, B. D. (1967). Science. 155:1248-1254.

Subject Index

Scientific Name Index